Magnetic Materials, Thin Films and Nanostructures

Magnetic Materials, Thin Films and Nanostructures

Editors

Catalin Constantinescu
Lucian Petrescu

MDPI • Basel • Beijing • Wuhan • Barcelona • Belgrade • Manchester • Tokyo • Cluj • Tianjin

Editors
Catalin Constantinescu
LP3 - UMR CNRS 7341
Aix-Marseille Université
Marseille
France

Lucian Petrescu
"Technical and Applied Magnetism" Laboratory
UPB–University "Politehnica" of Bucharest
Bucharest
Romania

Editorial Office
MDPI
St. Alban-Anlage 66
4052 Basel, Switzerland

This is a reprint of articles from the Special Issue published online in the open access journal *Magnetochemistry* (ISSN 2312-7481) (available at: www.mdpi.com/journal/magnetochemistry/special_issues/magnetic_materials_thin_film_nanostructures).

For citation purposes, cite each article independently as indicated on the article page online and as indicated below:

LastName, A.A.; LastName, B.B.; LastName, C.C. Article Title. *Journal Name* **Year**, *Volume Number*, Page Range.

ISBN 978-3-0365-8367-9 (Hbk)
ISBN 978-3-0365-8366-2 (PDF)

Contents

About the Editors

Catalin Constantinescu

Dr. Catalin-Daniel Constantinescu is a senior scientific researcher affiliated with the CNRS (Centre National de la Recherche Scientifique) in Marseille, France. His research interests cover aspects related to laser and plasma processing of materials, particularly for developing thin films and nanostructures, with applications in microelectronics, photonics, and electrical engineering. Three main processing techniques are worth mentioning, including pulsed laser deposition (PLD), matrix-assisted pulsed laser evaporation (MAPLE), and laser-induced forward transfer (LIFT). However, several other physical/chemical deposition and processing methods, as well as lithographic methods, should be noted as well, such as colloidal lithography using an adapted Langmuir–Blodgett technique and laser dewetting.

With 77 published papers, 1,037 citations, and an h-index of 21 (Scopus, July 2023), the research of Dr. Catalin-Daniel Constantinescu covers various aspects of materials sciences, due to his background in chemistry, physics, and engineering. He has also co-organized several conferences and symposia, including the EMRS 2021 Spring Meeting (symposium H—Laser material processing: from fundamental interactions to innovative applications), has delivered over 100 lectures and speeches at national/international meetings in over 15 countries, and is a member and chairperson of different international conferences, a member of different expertise and jury committees, and an editor. He has coordinated several regional, national, and international research projects, including two European infrastructure project proposals, and has actively participated and/or organized several events to promote science to the public, including "Fête de la Science" and "La Nuit Européenne des Chercheur·e·s".

Lucian Petrescu

Dr. Lucian-Gabriel Petrescu is a tenured Professor at the Politehnica University of Bucharest, Romania. With over two decades of teaching activities and research projects, his main research interests focus on magnetic materials' characteristics using vibrating-sample magnetometers (VSMs) and single sheet testers (SSTs) in order to connect magnetic properties with their engineering utility. Another important aspect of his research is related to the modeling of magnetic materials using different models for predetermining power losses at a high frequency.

Dr. Lucian-Gabriel Petrescu is an author of 79 research papers published in prestigious journals or presented at international conferences. His work has over 206 citations, with an h-index of 9 (Scopus, July 2023). He has led 4 research projects and been a part of 17 other national research projects and 1 international research project. He is the leader of the "Technical Magnetism Laboratory" at the Politehnica University of Bucharest.

Preface to "Magnetic Materials, Thin Films and Nanostructures"

Ever since antiquity, there have been legends circulating about the phenomenon that we generally call "magnetism". In modern societies, we frequently come into contact with real magnets in the course of our daily lives. Thus, it goes without saying that magnets play a very important role in our lives. The modern world has found many applications for magnets compared to what they have been known for more than 2500 years.

There are two types of magnets depending on their magnetic properties and composition. Permanent magnets emit a magnetic field without the need for any external source of magnetism or electrical power. Temporary magnets are magnets made of soft iron, which is usually easy to magnetize; however, temporary magnets lose most of their magnetic properties when the magnetizing cause is discontinued.

Modern permanent magnets are composed of special alloys that have been discovered through research to create increasingly better magnets. The most common families of permanent magnet materials are composed of strontium–iron and/or barium–iron (ferrites, also known as "ceramic magnets"), aluminium–nickel–cobalt ("alnicos", or metallic magnets), neodymium–iron–boron (neodymium magnets, or "super magnets"), and samarium–cobalt (samarium magnets, also known as "super magnets").

Soft iron and certain iron alloys, such as *permalloy* (a mixture of iron and nickel), can be very easily magnetized, even in a weak field. However, as soon as the field is removed, their magnetism is lost. These materials make excellent temporary magnets, like those used in telephones and electric motors.

The strongest magnets are located at the Los Alamos National Laboratory in New Mexico and at the Florida State University (FSU). The two US research laboratories have magnets that were designed for conducting experiments and whose magnetic induction can reach 100 T (tesla) and 45 T, respectively. By comparison, the induction of magnets used in the construction of magnetic lifting devices is approximately 2 T.

Almost all the multimedia equipment that we currently use may have magnetic sensors and/or devices, while bank cards and many other types of cards use magnetic strips that include coded personal information about users and allow various types of transactions to be carried out. Nevertheless, how do we make decisions on what magnetic materials to choose?

Magnetic materials can vary significantly in cost from one to another. On a cost-per-kg basis, neodymium magnets may seem very costly. However, on a cost-per-BHmax basis, they are not so costly at all. The reason is that using a more powerful magnet enables the miniaturization of an entire device that the magnet goes into. Thus, this approach yields cost savings, which favors the choice of more powerful magnet materials. Furthermore, standards have been established by the "*Magnetic Materials Producers Association*" (MMPA) and the "*Magnet Distributors and Fabrications Association*" (MDFA). Both associations are now a part of the "*International Magnetics Association*".

In this first volume, we cover relevant aspects of the chemical and physical processes involved in the production and characterization of magnetic materials in bulk, thin films, nanostructures, and/or nanocomposites, as well as the modeling aspects involving such structures. We present eleven original research studies and review works on current challenges and trends, covering fundamental and experimental work, with a special focus on the design, synthesis, and characterization of various types of magnetic materials, and the study of their structure–property relationships.

State-of-the-art results on the development of new experimental concepts that can lead to the transfer, chemical transformation, and high-resolution patterning of advanced thin films and nanomaterials, as well as to the design and fabrication of devices, are also presented and discussed.

Catalin Constantinescu and Lucian Petrescu
Editors

magnetochemistry

Editorial

Magnetic Materials, Thin Films and Nanostructures

Catalin-Daniel Constantinescu [1,*] and **Lucian-Gabriel Petrescu** [2,*]

1 CNRS, University Aix-Marseille, LP3/UMR 7341 (Laboratoire "Lasers, Plasmas et Procédés Photoniques), Campus de Luminy, F-13009 Marseille, France
2 UPB—University "Politehnica" of Bucharest, Faculty of Electrical Engineering, "Technical and Applied Magnetism" Laboratory, RO-060042 Bucharest, Romania
* Correspondence: catalin.constantinescu@cnrs.fr (C.-D.C.); lucian.petrescu@upb.ro (L.-G.P.)

Abstract: In this first volume, we cover relevant aspects of chemical and physical processes of the production and characterization of magnetic materials in bulk, thin films, nanostructures, and/or nanocomposites, as well as modeling aspects involving such structures. Accordingly, this volume presents eleven original research and review works on the challenges and trends covering fundamental and experimental work, with a special focus on the design, synthesis, and characterization of various types of magnetic materials, and the study of their structure–property relationships. State-of-the-art results on the development of new experimental concepts, leading to the transfer, chemical transformation, and high-resolution patterning of advanced thin films and nanomaterials, and to the design and fabrication of devices, are also presented and discussed.

Keywords: magnetism; nanomaterials; nanostructures; nanoparticles; thin films; electromagnetic shielding; magnetotransport; single-molecule-magnets; magnetic shape memory polymers; magnetic soft materials

Citation: Constantinescu, C.-D.; Petrescu, L.-G. Magnetic Materials, Thin Films and Nanostructures. *Magnetochemistry* **2023**, *9*, 133. https://doi.org/10.3390/magnetochemistry9050133

Received: 5 May 2023
Revised: 12 May 2023
Accepted: 16 May 2023
Published: 19 May 2023

Ever since antiquity, there have been legends circulating about the phenomenon that we generally call today "magnetism". The best known is of a young shepherd from the island of Crete, named Magnes [1]. At one point in his life, while tending the sheep on Mount Ida, he noticed that the metal part of the sole and the nails of his boots were attracted to the ground on which he was treading. To find out what was going on, he began to dig, revealing a strange earth-black stone: a piece of *lodestone*, or *"magnetite"*. Magnetite is a natural ferrite exhibiting magnetic properties. Today, we know that is a stoichiometric mixed oxide consisting of FeO and Fe_2O_3, with the chemical formula Fe_3O_4, and its structure is of the "spinel" type. It is a *"ferrimagnetic"* material, and its Curie temperature is 858 K. Just to exemplify for the reader the importance of this mineral, the strongest natural magnet: it is critical in the understanding and evolution of terrestrial paleomagnetism, in the formation and evolution of rocks, and other areas [2].

Of course, observations of such phenomena have been made long before the appearance of the first forms of writing [3], so we cannot rely on such legends. This story might be apocryphal, but in this case, there are many similarities with historic reality, the most important being that the empirical discovery of magnetism and its mention was made by the Greeks, Indians, and Chinese through observing the properties of the rocks that contained the magnetite mineral [4]. Pliny the Elder (23–79 CE) himself wrote about a hill, somewhere on the banks of the Indus River, which was apparently made entirely of something mysterious that attracted iron. The inexplicable nature of magnetic attraction has been exploited over time by storytellers, meaning it has become difficult to separate truth from fiction. It was believed for a long time that there were islands that were completely made up of magnetite and that, by virtue of the "special" properties of this mineral that is part of some rocks, could attract ships to the shore (they contained nails and iron beams in their structure). One of the explanations for the disappearances of the ships was that they had been destroyed on impact with such magnetic islands—a kind of ancient

Greek equivalent of today's "*Bermuda Triangle*". About Archimedes (287–212 BCE), one of the legends says that he also used stones containing magnetite, very strong, to attract and remove nails from enemy ships during wars, and to sink them (very likely that this is just a simple legend). Over time, the term magnetism remained when one explained the phenomenon of attraction between magnetite and objects containing iron. Today, by the concept of "*magnetism*", we understand the nature of this type of interaction and we can explain the phenomenon of attraction and repulsion. Moreover, today, we know of life forms that are guided by terrestrial magnetism: magnetic bacteria, for example, such as those from the Geobacter family (the first discovered was called "*Geobacter Metalloreducens*"), or even certain birds and insects [5].

Since the very first evidence of "scientific" investigations of magnetic phenomena, dating from ancient Greece, when Aristotle (384–322 BCE) recalls the fascinating teachings of Thales of Miletus (625–545 BCE) on magnetic phenomena and his intuitive way (for that time) to explain the cause of their appearance [6], to Peter Peregrinus de Maricourt's "*Epistola de magnete*" (1269 CE)—the first extensive treatise describing the properties of magnets [7], to William Gilbert's published work "*De Magnete, Magneticisqve Corporibvs et de Magno Magnete Tellure*" (1600 CE)—on magnetism, magnetic bodies, and on the great magnet Earth, using his model "terela" [8], to Hans Christian Ørsted, André-Marie Ampère, Carl Friedrich Gauss, Wilhelm Eduard Weber, and Michael Faraday, to James Clerk Maxwell's equations on the "*Theory of electromagnetism*" [9] and to Albert Einstein's paper "*On the Electrodynamics of Moving Bodies*" [10], electromagnetism continued to develop in the 20th century, being incorporated into the much more fundamental theories of the time, e.g., calibration theory, quantum electrodynamics, electroweak theory, and the Standard Model. Our understanding of Earth's magnetism, and magnetism in general, has come a long way in the last 423 years since "*De Magnete*" was first printed. Today, we live in the future: we ride on MAGLEV trains at more than 500 km/h, do magic with the Meissner effect [11] by using strong NdFeB magnets and (close to) room temperature superconductors [12], and we observe the strongest magnets known, pulsars, while taking glimpses at the Universe. All we have left is "*. . . to boldly go where no one has gone before*"!

Chemistry and magnetism go hand in hand, and magnetochemistry is concerned with the magnetic properties of chemical compounds. Magnetic properties arise from the spin and orbital angular momentum of the electrons contained in a compound. Compounds are diamagnetic when they contain no unpaired electrons [13]. The fundamentals of materials science and magnetism, both experimental and theoretical, encouraged us to edit this first volume as a Special Issue of MDPI's journal *Magnetochemistry*, emphasizing the relevant aspects of chemical and physical processes of the production and characterization of magnetic materials in bulk, thin films, nanostructures and/or nanocomposites, as well as modeling aspects involving such structures. Eleven original research and review works on the challenges and trends are included herein, while covering fundamental and experimental work, with special focus on the design, synthesis, and characterization of various types of magnetic materials, and the study of their structure–property relationships.

An important starting discussion is presented by Robert A. Lawrence et al. in their work entitled "*Magnetic Transition State Searching: Beyond the Static Ion Approximation*" [14]. The effect of structural relaxations on the magnetocrystalline anisotropy energy (MAE) is investigated by using density functional theory (DFT). The theory of the impact of magnetostructural coupling on the MAE is discussed, including the effects on attempt frequency. Two materials are chosen—FePt and PtMn. The MAE for ferromagnetic FePt (3.45 meV/formula unit) and antiferromagnetic PtMn (0.41 meV/formula unit) are calculated within the local density approximation (LDA). The effects of the structural relaxation are calculated and found to give a $<0.5\%$ reduction to the MAE for the ferromagnet and $\sim 20\%$ for the antiferromagnet. This divergence is attributed to the difference in coupling between the magnon responsible for the transition and either an acoustic or optical phonon (for FePt and PtMn, respectively), revealing the importance of considering magnetostructural effects when evaluating the MAE of systems in which the magnetic structure cannot be adequately expressed in the same primitive unit cell as the geometric structure, such as an antiferromagnet.

Nanocomposite FePt-based magnets have been under scrutiny as a new class of permanent magnets due to their high corrosion resistance and their high working temperature. An FePt-based nanocomposite is presented and discussed by Alina D. Crisan et al., in their work entitled "*Role of Disordered Precursor in $L1_0$ Phase Formation in FePt-Based Nanocomposite Magnet*" [15]. In order to prove the usefulness of having a structurally disordered precursor to the formation of the FePt $L1_0$ phase and to facilitate the co-existence of exchange-coupled hard and soft magnetic phases with optimized magnetic properties in various conditions of annealing, an Fe-Pt-Zr-B melt-spun alloy has been synthesized and detailed structural and magnetic investigations have been undertaken to probe its phase evolution during annealing. The dynamics of the formation of the hard magnetic $L1_0$ phase during the gradual disorder–order phase transformation has been monitored by using a complex combination of X-ray diffraction methods and ^{57}Fe Mössbauer spectroscopy methods over a wide range of annealing temperatures. It is shown that the formation of the hard magnetic phase, emerging from the chemically disordered precursor, is gradual and occurs via complex mechanisms, involving the presence of a disordered Fe-Zr-B-rich intergranular region which contributes to an increase in the abundance of the $L1_0$ phase for higher annealing temperatures. Magnetic measurements confirm the good performances of these alloys in terms of coercivity and remanence, contributing to the development of these alloys as the next generation of "rare-earth-free" permanent magnets.

In their work entitled "*Electric Field Control of Magnetic Properties by Means of Li+ Migration in FeRh Thin Film*" [16], Gengfei Li et al. investigated the control of the magnetic properties of FeRh films by means of Li^+ migration in FeRh/MgO heterostructures, and found that the migration of Li^+ could reduce the phase transition temperature by 2 K with an applied voltage of 1 V. Meanwhile, the voltage-dependent saturated magnetization exhibited a repetitive switching behavior from high to low magnetization values while the voltage was switched from 4 to −4 V, indicating that the migration of Li^+ in the FeRh film can be reversible, providing a means to control the magnetic properties of FeRh films.

Mohammad Yaghoub Abdollahzadeh Jamalabadi offers a glimpse into their "*Feasibility Study of Cooling a Bulk Acoustic Wave Resonator by Nanoparticle Enhanced Phase Change Material*" [17]. The coupling of a cooling problem with the electromagnetic resonance of a bulk acoustic wave (BAW) material is investigated. The BAW is a device usually composed of a piezoelectric material (i.e., zinc oxide or aluminium nitride) between two electrodes, while the use of nanoparticles facilitates heat removal and decreases the delay time and thermal inertia of the cooling system.

The magnetic, optical, and phonon properties of ion-doped $CuAlO_2$ (CAO) nanoparticles on the Cu or Al site are theoretically investigated by Iliana Naumova Apostolova et al. in their work, entitled "*Size and Ion-Doping Effects on Magnetic, Optical, and Phonon Properties of $CuAlO_2$ Nanoparticles*" [18], and discuss diluted magnetic semiconductors (DMS) that play an important role in interdisciplinary materials science and future spintronics. Ferromagnetic DMS has been studied from first principles within mean-field approximation, revealing that the origin of ferromagnetism in DMS is not yet clear, even if various methods have been proposed. Here, the size and ion-doping effects on the magnetic, optical, and phonon properties of transition metal (TM = Fe, Co, Mn, and Ni) ion-doped CAO nanoparticles on the Cu and Al site are discussed by using a microscopic model and Green's function theory; the results are in good agreement with existing experimental data.

SiO_2(Co) granular films exhibit unusual magnetic and magnetotransport properties which are strongly dependent on the composition of the film and the material of a substrate. For example, the injection magnetoresistance (IMR) coefficient reaches a giant (GIMR) value of 10^5% at room temperature in SiO_2(Co) films on an n-GaAs substrate. However, the IMR effect is negligible in the case of a similar granular film deposited on the n-Si substrate. Natalia A. Grigoryeva et al. discuss the structural and magnetic properties of granular-film SiO_2(Co) on a Si substrate are studied with the aim to understand the cause of the difference in IMR coefficients for SiO_2(Co) thin films deposited on n-GaAs and on n-Si substrates in their work entitled "*Mesostructure and Magnetic Properties of SiO_2-Co Granular Film on Silicon Substrate*" [19]. Investigations are carried out using complementary methods of polarized neutron reflectometry, grazing incidence small-angle X-ray scattering, X-ray reflectometry, scanning electron microscopy, and SQUID magnetometry. It is shown that the interface layer between the granular film and the Si substrate exhibits metallic rather than magnetic properties and eliminates the GIMR effect.

Yuliana de Jesús Acosta-Silva et al. present their "*Study of the Effects of Er Doping on the Physical Properties of CdSe Thin Films*" [20], revealing that the Er incorporation in CdSe nanocrystals provokes drastic changes in the $Cd_{1-x}Er_xSe$ lattice structure, depending on the level of Er doping. Er^{3+} in Cd^{2+} sites induces a crystalline quality of CdSe; thus, Er^{3+} reduces the structural disorder but increases the chemical one. When the lattice expands, the 1LO phonon softens and, as expected, hardens if the lattice contracts.

Antoine Hoëz et al. present a "New Manufacturing Process for Granular Texture Management in Polycrystalline BaM Hexaferrites through the Goethite Crystallite Laths Aspect Ratio, and a Specialized Law of Approach to the Magnetic Saturation for Partly Polarized Uniaxial Materials" [21]. Their aim is focused on the manufacture and the magnetic properties of polycrystalline M-type hexaferrite $BaFe_{12}O_{19}$ (barium ferrite, or BaM) materials of different magnetic texturing grades, going from a random distribution of the BaM crystallites to their almost complete stacking. The Rietveld refinements of powder diffractograms clearly revealed a particle-stacking enhancement, which is dependent not only on the hematite:goethite ratio but also on the optimal aspect ratio of goethite crystallites. This optimization resulted in a significant improvement of the remanent magnetization value, increasing it to 0.82 compared with the most recent literature. Based on this study, BaM materials are further manufactured with a controlled magnetic texture and, therefore, are partly self-polarized.

Andrea Amaro et al. discuss the "Shielding Effectiveness Measurement Method for Planar Nanomaterial Samples Based on Carbon Nanotubes (CNT) Materials up to 18 GHz" [22]. The proposed measurement methodology demonstrates significant advantages, including the simplicity of the sample machining, which means that very specific geometries or tiny dimensions are not required. Some of the samples analyzed have demonstrated to provide a significant attenuation in the 700 MHz–18 GHz frequency range. For those composites based on a polymer matrix with different concentrations of CNT, a value of −81.30 dB has been obtained for the frequency of 7.125 GHz for the 15 w%CNT composite, which is very significant considering the nature of the materials. They highlighted that these types of characterizations are very relevant from a technological and industrial point of view, specifically for those sectors related to 5G technology, since the use of EMI shielding based on plastic materials has numerous advantages, i.e., manufacturing cost-efficiency.

Next, single-molecule magnets (SMMs)—which are low dimensional molecules consisting of coupled paramagnetic metal ions, possessing unique magnetic properties (magnetic hysteresis, slow magnetic relaxations, quantum tunneling effects) available at the molecular level—are presented and discussed by Oleksandr Pastukh et al. in their work entitled "*AC Susceptibility Studies of Magnetic Relaxation in Mn_{12}-Stearate SMMs on the Spherical Silica Surface*" [23].

Finally, in their "*Review of Magnetic Shape Memory Polymers and Magnetic Soft Materials*", Sanne J. M. van Vilsteren et al. emphasize the development of magnetic soft materials (MSMs) and magnetic shape memory polymers (MSMPs), with a specific focus on the role of the magnetic particles which affect the shape memory recovery and programming behavior of these materials [24]. In addition, the synthesis and application of these materials are addressed, since they may exceed conventional shape memory materials such as shape memory alloys or shape memory polymers. The parameters that have been studied are their material composition, shape recovery, and shape recovery simulation, with a specific emphasis on manufacturing methods and applications. The magnetic shape memory effect activation makes it able to activate shape change remotely and with (almost) no surface heat. This makes shape memory materials useful for more applications than heat-activated shape memory polymers activated by conduction or convection. MSMPs often have only one remembered shape and no reversible shape change activated by the magnetic field. It can also take some time to activate the shape change. However, these drawbacks can be avoided by MSMs which offer fast and reversible shape changes. In addition, the effect of the magnetic properties of the magnetic components (Curie temperature, coercivity, remanence) on the shape memory function of these materials is as yet largely unknown and needs to be the subject of future research. Although several successful applications have already been reported (predominantly in the biomedical field), it can be concluded that magnetic shape memory alloys (both MSMs and MSMPs) are merely at the beginning of their development and still need to find their way to large-scale market applications.

To conclude, magnetic materials have the potential for use in information storage and processing, spintronics, drug delivery, cooling technology, etc. [25,26]. Various magnetic transitions are induced by temperature variation, crystal volume change, substitution of other elements, and applications of high magnetic fields and high pressure [27]. Thus, this first volume presents various materials and techniques which are highly relevant to modern electrical engineering and industry. Applications of the magnetization reversal phenomenon in magnetoelectronic and magnetocaloric devices, such as magnetic memory and magnetic cooling/heating-based constant-temperature baths, will be presented and discussed in a second volume.

As a final word of this Editorial for the first volume of "*Magnetic Materials, Thin Films and Nanostructures*", we would like to present the reader with a simple, yet profound thought from one of the greatest minds of the last century: "*The whole of science is nothing more than a refinement of everyday thinking* ... " (Albert Einstein).

Author Contributions: All authors have read and agreed to the published version of the manuscript.

Conflicts of Interest: The authors declare no conflict of interest.

References

1. Frith, H.; Stepney Rawson, W. *Coil and Current; or the Triumphs of Electricity*; Ward, Lock and Co.: London, UK, 1986; p. 4. Available online: https://books.google.fr/books/about/Coil_and_Current_Or_the_Triumphs_of_Elec.html (accessed on 5 May 2023).
2. Selley, R.C.; Cocks, L.R.M.; Plimer, I.R. (Eds.) *Encyclopedia of Geology*; Academic Press: Cambridge, MA, USA, 2005. Available online: https://www.sciencedirect.com/referencework/9780123693969/encyclopedia-of-geology (accessed on 5 May 2023).
3. Bacon, B.; Khatiri, A.; Palmer, J.; Freeth, T.; Pettitt, P.; Kentridgee, R. An Upper Palaeolithic Proto-writing System and Phenological Calendar. *Camb. Archaeol. J.* **2023**, 1–19. [CrossRef]
4. Shu-Hua, L. Origine de la Boussole II. Aimant et Boussole. *Isis* **1954**, *45*, 175–196. (In French) [CrossRef]
5. Qin, S.; Yin, H.; Yang, C.; Dou, Y.; Liu, Z.; Zhang, P.; Yu, H.; Huang, Y.; Feng, J.; Hao, J.; et al. A magnetic protein biocompass. *Nat. Mater.* **2016**, *15*, 217–226. [CrossRef]
6. Fowler, M. Historical Beginnings of Theories of Electricity and Magnetism. 1997. Available online: http://galileoandeinstein.physics.virginia.edu/more_stuff/E&M_Hist.html (accessed on 5 May 2023).
7. Minecan, A.M.C. Analysis and bilingual translation of epistle of Peter Peregrinus of Maricourt to Sygerus of Foncaucourt: Concerning the magnet (De magnete). *Disput. Philos. Res. Bull.* **2017**, *6*, 277–307. (In Latin) [CrossRef]
8. William Gilbert's Work on "De Magnete, Magneticisqve Corporibvs et de Magno Magnete Tellure" (1600). Bilingual Translation. Available online: https://gallica.bnf.fr/ark:/12148/bpt6k33585/f2.item.r=de+magnete.langFR# (accessed on 5 May 2023). (In Latin).
9. Maxwell, J.C. A Dynamical Theory of the Electromagnetic Field. *Philos. Trans. R. Soc. Lond.* **1865**, *155*, 459–512. [CrossRef]
10. Einstein, A. Zur Elektrodynamik bewegter Körper. *Ann. Der Phys.* **1905**, *4*, 891–921. (In German) [CrossRef]
11. Meissner, W.; Ochsenfeld, R. Ein neuer Effekt bei Eintritt der Supraleitfähigkeit. *Naturwissenschaften* **1933**, *21*, 787–788. (In German) [CrossRef]
12. Saunders, P.J.; Ford, G.A. *The Rise of the Superconductors*; CRC Press: Boca Raton, FL, USA, 2005. Available online: https://www.routledge.com/The-Rise-of-the-Superconductors/Ford-Saunders/p/book/9780748407729 (accessed on 5 May 2023).
13. Orchard, A.F. *Magnetochemistry*; Oxford Chemistry Primers; Oxford University Press: Oxford, UK, 2003. Available online: https://global.oup.com/academic/product/magnetochemistry-9780198792789 (accessed on 5 May 2023).
14. Lawrence, R.A.; Donaldson, S.J.; Probert, M.I.J. Magnetic Transition State Searching: Beyond the Static Ion Approximation. *Magnetochemistry* **2023**, *9*, 42. [CrossRef]

15. Crisan, A.D.; Dan, I.; Crisan, O. Role of Disordered Precursor in L10 Phase Formation in FePt-Based Nanocomposite Magnet. *Magnetochemistry* **2021**, *7*, 149. [CrossRef]
16. Li, G.; Xie, Y.; Wang, B.; Yang, H.; Li, R.-W. Electric Field Control of Magnetic Properties by Means of Li+ Migration in FeRh Thin Film. *Magnetochemistry* **2021**, *7*, 45. [CrossRef]
17. Abdollahzadeh Jamalabadi, M.Y. Feasibility Study of Cooling a Bulk Acoustic Wave Resonator by Nanoparticle Enhanced Phase Change Material. *Magnetochemistry* **2021**, *7*, 144. [CrossRef]
18. Apostolova, I.N.; Apostolov, A.T.; Wesselinowa, J.M. Size and Ion-Doping Effects on Magnetic, Optical, and Phonon Properties of CuAlO2 Nanoparticles. *Magnetochemistry* **2022**, *8*, 169. [CrossRef]
19. Grigoryeva, N.A.; Ukleev, V.; Vorobiev, A.A.; Stognij, A.I.; Novitskii, N.N.; Lutsev, L.V.; Grigoriev, S.V. Mesostructure and Magnetic Properties of SiO_2-Co Granular Film on Silicon Substrate. *Magnetochemistry* **2022**, *8*, 167. [CrossRef]
20. Acosta-Silva, Y.d.J.; Godínez, L.A.; Toledano-Ayala, M.; Lozada-Morales, R.; Zelaya-Angel, O.; Méndez-López, A. Study of the Effects of Er Doping on the Physical Properties of CdSe Thin Films. *Magnetochemistry* **2023**, *9*, 107. [CrossRef]
21. Hoëz, A.; Mattei, J.-L.; Chevalier, A. New Manufacturing Process for Granular Texture Management in Polycrystalline BaM Hexaferrites through the Goethite Crystallite Laths Aspect Ratio, and a Specialized Law of Approach to the Magnetic Saturation for Partly Polarized Uniaxial Materials. *Magnetochemistry* **2023**, *9*, 30. [CrossRef]
22. Amaro, A.; Suarez, A.; Torres, J.; Martinez, P.A.; Herraiz, R.; Alcarria, A.; Benedito, A.; Ruiz, R.; Galvez, P.; Penades, A. Shielding Effectiveness Measurement Method for Planar Nanomaterial Samples Based on CNT Materials up to 18 GHz. *Magnetochemistry* **2023**, *9*, 114. [CrossRef]
23. Pastukh, O.; Konieczny, P.; Laskowska, M.; Laskowski, Ł. AC Susceptibility Studies of Magnetic Relaxation in Mn12-Stearate SMMs on the Spherical Silica Surface. *Magnetochemistry* **2021**, *7*, 122. [CrossRef]
24. Van Vilsteren, S.J.M.; Yarmand, H.; Ghodrat, S. Review of Magnetic Shape Memory Polymers and Magnetic Soft Materials. *Magnetochemistry* **2021**, *7*, 123. [CrossRef]
25. Abdelbasir, S.A.; Shalan, A.E. Intriguing Properties and Applications of Functional Magnetic Materials. In *Functional Materials*; Sahu, D., Ed.; IntechOpen: London, UK, 2019. [CrossRef]
26. Krishnan, K.M. *Fundamentals and Applications of Magnetic Materials*; Oxford Academic Press: Oxford, UK, 2016. [CrossRef]
27. Paes, V.Z.C.; Varalda, J.; Mosca, D.H. Strain-induced magnetization changes and magneto-volume effects in ferromagnets with cubic symmetry. *J. Magn. Magn. Mater.* **2019**, *475*, 539. [CrossRef]

magnetochemistry

MDPI

Article

Magnetic Transition State Searching: Beyond the Static Ion Approximation

Robert A. Lawrence, **Scott J. Donaldson** and **Matt I. J. Probert** *

School of Physics, Engineering and Technology, University of York, Heslington, North Yorkshire YO10 5DD, UK
* Correspondence: matt.probert@york.ac.uk

Abstract: The effect of structural relaxations on the magnetocrystalline anisotropy energy (MAE) was investigated by using density functional theory (DFT). The theory of the impact of magnetostructural coupling on the MAE was discussed, including the effects on attempt frequency. The MAE for ferromagnetic FePt (3.45 meV/formula unit) and antiferromagnetic PtMn (0.41 meV/formula unit) were calculated within the local density approximation (LDA). The effects of the structural relaxation were calculated and found to give a <0.5% reduction to the MAE for the ferromagnet and ∼20% for the antiferromagnet.

Keywords: magnetocrystalline anisotropy; FePt; PtMn; transition state search; magnetism; density functional theory

Citation: Lawrence, R.A.; Donaldson, S.J.; Probert, M.I.J. Magnetic Transition State Searching: Beyond the Static Ion Approximation. *Magnetochemistry* **2023**, *9*, 42. https://doi.org/10.3390/magnetochemistry9020042

Academic Editors: Cătălin-Daniel Constantinescu and Lucian Petrescu

Received: 30 November 2022
Revised: 16 January 2023
Accepted: 20 January 2023
Published: 27 January 2023

1. Introduction

The modern world has been transformed since the advent of the computer due to ever-improving data storage, enabling more complex devices and programmes to be manufactured and written. Huge advances have been made possible due to the development of and improvements in magnetic data storage from the primitive core-rope memory (∼4 Kb) of the Apollo era [1], to modern multi-Tb storage drives [2]. This progress has depended upon the construction of an ever-greater areal bit density (number of data storage "sites" per unit area) [2] and higher performance bits.

For high performance (high clock speed) memory, such as is found in modern random access memory (RAM), the storage must typically be electric rather than magnetic, because conventional magnetic storage techniques take too long to write a bit. This causes the memory to be volatile [3], and has a sizable effect on the power consumption of that memory, because constant refreshing of the bits is required. One potential way around this is antiferromagnetic magnetoelectric RAM (AFM-MERAM) [4], which makes use of the high-switching speeds of antiferromagnetic (AFM) materials to create memory that offers equivalent performance to standard charge-based RAM while still having the low energy requirements and nonvolatility of magnetic storage.

Now, considering the areal bit density, the greatest current limit is the requirement that the memory itself is thermally stable [5]; it is not much use if a warm day causes the data to be erased due to random bits flipping. Therefore, it is necessary to have a high barrier against bit-flipping, such that the probability of a random flip caused by thermal fluctuations is very small, even over long timescales (there must be a high activation energy for the bit flipping at $k_B T$ for typical operational temperatures).

In modern memory-storage devices that rely upon the magnetisation orientation of a magnetic material, this barrier to bit flipping is controlled by the energy required to reorient all of the spin moments within the bit. This barrier is therefore extrinsic, and increases with the size of the device. Any increase of the areal bit density, however, requires shrinking the size of a bit, and therefore to maintain a constant barrier to flipping needs a higher intrinsic barrier to flipping.

This intrinsic barrier to the flipping of individual bits is the magnetocrystalline anisotropy. This property, typically evaluated as a magnetocrystalline anisotropy energy (MAE), is a relativistic effect induced by spin-orbit coupling linking together orbital and spin angular momenta. One way of thinking about this effect is that, in the absence of relativity, spins exist in a spherical orbital. This may then be convolved with an electronic orbital to produce the standard "spin-orbitals" of, for example, Hartree–Fock theory, without affecting the spins. In the relativistic limit, however, l and s are no longer acceptable quantum numbers, and therefore we cannot neglect the effect of the convolution on the spin-like degrees of freedom. Because the original electronic orbitals are both directionally dependent and oriented with respect to the crystal lattice (due to the crystal field effect), this means that the effect of spin-orbit coupling is to make the spins experience an environment which is directionally dependent — magnetocrystalline anisotropy.

Much effort has gone into evaluating MAE over the years, both in ferromagnetic (FM) and antiferromagnetic (AFM) materials through both experimental [6,7] and theoretical [8–10] approaches based on *ab initio* calculation. For antiferromagnetic materials in particular, this is made especially difficult by their lack of apparent response to an applied field, and therefore the MAE of AFM materials tends to be inferred rather than directly measured [7].

Reliable simulation techniques have huge potential within the field of materials discovery because they enable a cheap initial search of likely candidate materials and novel structures [11,12], which may then be synthesised to confirm the findings of the simulations. For AFM materials in particular, in which the MAE is not directly accessible, this ability to directly access a magnetic property through a simulation is a potentially important avenue for materials discovery. Current simulation methods, however, typically significantly overestimate the anisotropy value compared with experiment [8,9].

One reason for this discrepancy between theory and experiment may be that the theoretical simulations do not calculate precisely the same observable that is measured experimentally, such as different timescales of experiment and theory. This sort of error cannot be corrected by using more sophisticated treatments of the electronic structure — such as additional treatments of strong correlations, or better exchange-correlation functionals — because these better methods for electronic structure would still be applied toward more accurately calculating the wrong number. Additionally, by finding a more experimentally relevant description it is possible to gain greater insight into the underlying physical processes, thereby both enhancing understanding and potentially opening up new routes for device optimisation and material parameters tuning.

A potential cause of the overestimation of MAE may be that these techniques typically rely on optimising the ground-state structure and then either sampling different constrained spin orientations [13] or calculating the spin-torque and using that to evaluate the anisotropy [14]. These methods, however, do not typically allow for a relaxation of the crystal structure in response to the changing spin orientation. In this paper, we propose a new strategy to evaluate the magnetic anisotropy which also includes the magnetostructural coupling and apply these to ferromagnetic FePt and antiferromagnetic PtMn.

2. Theory

2.1. Magnetic Transition State Searching

Rather than the narrow definition of MAE as the difference in energy between orientation of the spins along the magnetic easy and hard axes, it is possible to extend the definition to a higher dimensional case (i.e., also considering structural relaxations) by instead defining it as the maximum energy of the minimum energy pathway — the energy of the transition state — between the magnetic ground state and its symmetrically identical reversed state (for example turning all "up" spins into "down" spins and vice versa). When we only consider magnetic relaxations, this definition yields the same MAE as before. However, when simultaneous structural and magnetic relaxations are considered (as must be the case in a real experiment) the extra degrees of freedom make further, potentially lower, energy routes valid.

Transition state searching within chemistry, which considers only the movement of atoms, is a well-established field with many techniques for evaluating the energy barrier for transition between two local minima within a wider energy landscape (in chemical contexts, these are the reactant and product of a reaction step). Much work has been done on developing procedures for these, such as the nudged elastic band method [15,16]. However, an older, and more computationally efficient, method — known as linear synchronous transit/quadratic synchronous transit [17] (LST/QST) — may be sufficient for accurate magnetic transition state searching, and so it is used here.

In LST/QST, a linear pathway between the two minima is constructed by interpolation, and the maximum along this pathway is found through a bisection search as a first approximation to the transition state (this is the LST component). Because this state has no obligation to be a saddle point — which all transition states must be [18] — a further optimisation in directions orthogonal to the original search direction is performed until a minimum is found (the QST component). By identifying a maximum in the direction between minima which is a minimum in all other directions, the saddle point associated with the transition between local minima will have been identified.

For magnetic systems, we may consider the spin orientation of the system to be an orthogonal search direction to the structural configuration of the system (see Figure 1), because any spin configuration and any structural configuration may coexist (even if with a significant energy penalty). This suggests that performing an LST search in spin space, followed by a structural optimisation under the constraint of fixed spin orientations (in the manner of the QST component of the search) would lead to an improved estimation of the transition state. This QST style search also has benefits over standard structural-only LST/QST in that the QST relaxation is guaranteed as the constrained spin orientation cannot be altered by the change of atomic positions or lattice vectors. It is this QST-like optimisation of the crystal structure that we propose as a necessary minimum correction to evaluate the true transition state between two magnetic configurations (a more sophisticated evaluation of the transition state would require a more sophisticated technique).

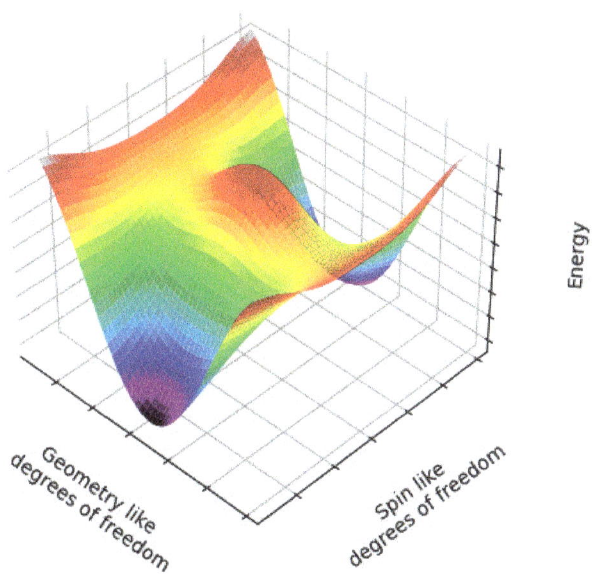

Figure 1. A cartoon schematic of the energy landscape for inverting the magnetic ordering in a domain. Note that the lowest energy route passes through a saddle-point which has a different geometry to the ground state of the magnetic easy axis. Note that all magnetic degrees of freedom are compressed into a single axis, and all geometric degrees of freedom are compressed into another.

Furthermore, considering only the spin-like directions, we know that our initial and final states are identical by symmetry (a naïve rotation of the crystal by 180° maps one to the other — and this does not affect the internal energy of the system). Through Hammond's postulate [19], we find the result that for degenerate minima the transition state lies halfway along the reaction coordinate (which for our case ranges from $\theta = 0$ to $\theta = 180°$) — thereby explaining why the easy and hard axes are perpendicular for a simple collinear magnet. This relationship is unaffected by structural relaxations due to the orthogonality of the spin and structural dimensions.

Finally, we note that the highest quality transition state search estimations of the barrier height also include a correction for the zero-point energy of the phonon mode [20,21] associated with the transition (see Figure 2). These zero-point energy corrections are typically on the order of meV, and may be safely neglected for chemical barriers (which are on the order of eV) but must be considered for these magnetic barriers, which are on the order of meV themselves.

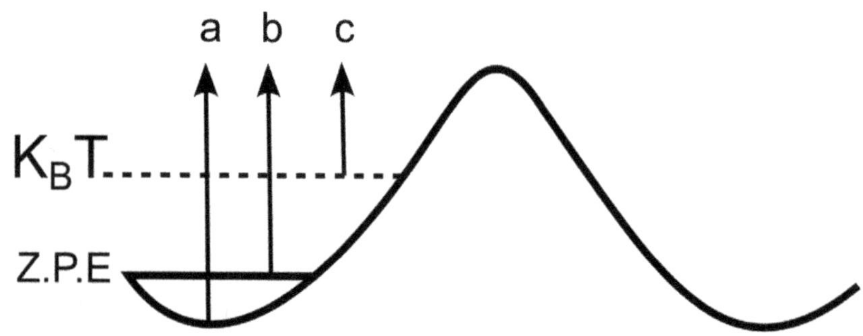

Figure 2. The effect of zero-point energy and temperature on the MAE of an arbitrary barrier. The energy of the electronic state of the system is increased such that the bottom of the potential well is no longer accessible, leading to a decrease in the effective barrier, which change the MAE from being represented by arrow a to arrow b. Increasing the temperature will lead to arrow c.

2.2. *Quasiparticle Picture of Magnetic Anisotropy*

The standard picture of magnetic anisotropy — that of an external field "pushing" spins from one configuration to another — is somewhat troubling when one considers the linear magnetic field alignment in experimental measures of anisotropy [7], as this cannot produce an angular rotation of spins. This is because there is never a component of the applied field perpendicular to the spins. It may then be preferable to consider a quasiparticle picture of the anisotropy, in an analogous way to considering soft phonons to model phase transitions [22]. This does not change the underlying physics but simply uses different language to describe it, which makes the physical processes more apparent.

In this picture, any arbitrary magnetic structure can be written as a sum of the ground-state magnetic structure and a collection of symmetry-appropriate magnons. For the rotation of the ground state to its symmetric partner (exchanging up and down labels), this corresponds to a long-wavelength ($q \to 0$) acoustic magnon.

This gives rise to the current strategy for evaluating MAE; calculate (or measure) the creation energy of this particular excitation to the spin structure with a static structure. However, if the minimum energy (transition) pathway (MEP) follows a route that requires a structural relaxation, then a quasiparticle responsible for this (a phonon) must be generated simultaneously to the magnon. This scenario is shown in Figure 3.

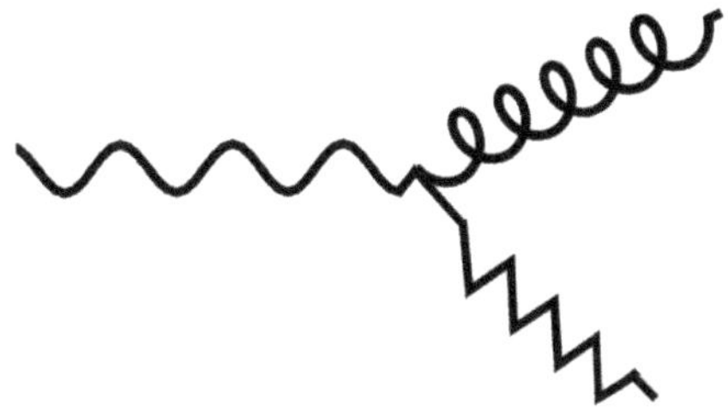

Figure 3. A schematic Feynmann diagram for the calculation of MAE. An incoming photon enables the creation of a magnon (and a phonon if magnetostuctural coupling provides the MEP), thereby driving the magnetic transition.

In terms of quasiparticles, the experimentally measured MAE is the energy loss due to the generation of either the magnon directly or the magnon–phonon pair (for the bound pair where they act as a single quasiparticle rather than as two coexisting quasiparticles, we use the term "magnetophonon"). Simulating the MAE then requires one to evaluate the creation energy of all of these pathways. The authors note that further expansions in terms of additional interactions are possible, such as multiple phonon and multiple magnon excitations; however the contribution of these higher order terms to the expansion are smaller, and therefore are neglected here.

One final subtlety of note is that magnons and phonons are Majorana particles — that is, they are their own antiparticles [23] — and accordingly, one can read the Feynman diagram in Figure 3 in multiple ways. The most obvious are the two solutions that come from considering the phonon as an emitted (see Equation (1)) or absorbed particle (Equation (3)). Additionally, one may consider the case where no phonon interactions are involved (Equation (2)). We have

$$E^{+} = E_{magnon} + E_{phonon} = E_{magnetophonon} \tag{1}$$

$$E^{0} = E_{magnon} \tag{2}$$

$$E^{-} = E_{magnon} - E_{phonon}. \tag{3}$$

This makes the experimentally measured MAE an expectation value of the multiple pathways with appropriate weighting with the final experimentally measured barrier given as

$$\langle MAE \rangle = w^{+}E^{+} + w^{-}E^{-} + w^{0}E^{0}, \tag{4}$$

where $w^{\pm}$ are the Arrhenius factors for each route and may be expressed as

$$w^{i} = A_{i}e^{\frac{E^{i}}{K_{b}T}}, \tag{5}$$

where A_i is the Arrhenius prefactor for route i. This gives an explicit temperature dependence for the MAE and means that when $k_B T < E^i$, the lowest energy route will dominate the kinetics. In order to evaluate the MAE at any arbitrary temperature (that is, below the critical temperature for the relevant magnetic phase), all that is required is to evaluate the energy of the individual routes and the Arrhenius prefactor. This prefactor may be thought of as the rate at which a system may attempt to overcome the transition barrier.

We also note that although a continuum of pathways for the spin flip exists, because the energy landscape is continuous, we only consider the routes that are lowest in energy for any given conditions. An experiment will average over all possible pathways, with those that are lower in energy dominating due to the exponential factor within Equation (5) (the Arrhenius equation).

2.3. Attempt Frequency

It is well known that a further important property when interpreting experimental data relating to the anisotropy is the attempt frequency [24,25]. This is typically stated to be a constant characteristic frequency for spin-flipping events (and therefore corresponds to the Arrhenius prefactor). It is notable that the attempt frequency for antiferromagnetic materials (10^{12} Hz) [25] is three orders of magnitude higher than for ferromagnetic materials (10^9 Hz) [26] and also that these are of the same order of magnitude as the reciprocal phonon–phonon lifetimes τ_{ph-ph} for acoustic and optical phonons, respectively [27].

A possible explanation for this is outlined in schematic form in Figure 4. In this scenario, the specific magnon mode commensurate with a spin-transition and its commensurate phonon (which must have the appropriate symmetry to couple strongly with this magnon mode) are abstracted from two general baths of magnons and phonons respectively, and specific transition rates between all four (and internal conversions for the bath) are considered. In the strong magnon–phonon coupling limit ($k_{e,ph} \to \infty$), these two modes form a single magnetophonon and may be considered as a single quasiparticle, which may be scattered by either magnons or phonons. By considering rate equations and requiring the system to be in a dynamic equilibrium, we may determine rates of conversion between and the populations of all levels.

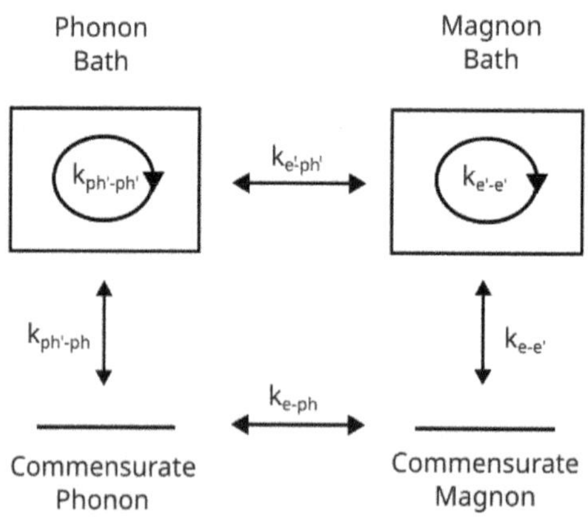

Figure 4. Schematic of the possible interactions between the magnon associated with the selected spin transition, its commensurate phonon, and the baths of available phonon and magnon states.

First, we define an attempt as the creation of a magnon-like quasiparticle (either a magnon or magnetophonon) that is commensurate with changing the state of the electronic system to the barrier in spin-space between up and down configurations. Because these quasiparticles are bosonic, there is no limit to the amount of population of a single mode, so accordingly there are no density-of-states considerations in the transition rates, only requirements for conservation of energy and momentum which then yield selection rules.

By the principle of detailed balance, for a system in equilibrium (or near to equilibrium in the adiabatic limit) the rates of creation and annihilation of the magnon/magnetophonon mode must be equal (otherwise the population of the mode will change and the system is not in equilibrium). This suggests that the attempt frequency is the lifetime of the magnon/magnetophonon, which for the case of the magnetophonon may be calculated from the magnon and phonon scattering rates by using Matthiessen's rule,

$$\frac{1}{\tau_{magnetophonon}} = \frac{1}{\tau_{magnon}} + \frac{1}{\tau_{phonon}}, \quad (6)$$

where $\tau_{quasiparticle}$ is the lifetime of the respective quasiparticle. The lifetime of the magnon seems unlikely to change significantly for different spin structures, as the magnon will remain acoustic (an optical mode corresponds to an FM-AFM transition). The magnon and phonon lifetimes will be dominated by scattering into their respective baths, and therefore the lifetime of the magnetophonon will be lower than the shorter of the phonon and magnon lifetimes. This magnetophonon picture (rather than the bare-magnon) explains the finding that AFM materials have a higher attempt frequency; they couple to optical phonons with a shorter lifetime, thereby increasing the attempt frequency, which may be represented as $\tau^{-1}_{magnetophonon}$.

2.3.1. Symmetry Considerations

There are symmetry considerations with regard to the selection of which magnon and phonon modes couple together to form the magnetophonon associated with the MAE. For the magnon modes, the magnon drives the reorientation of the spins with respect to the crystal axes, and the lowest energy magnon mode is an acoustic mode in the long wavelength limit ($q = 0$). This mode rotates all spins across all space uniformly.

Any phonon mode coupled to this must affect all of space in the same way as the magnon ($q = 0$), but whether it is an acoustic or an optical mode depends on whether the system is ferromagnetic (in which case it will be acoustic) or antiferromagnetic (in which case the opposite responses of the different sublattices will lead to an optical phonon).

One result of this is that applying a magnetic field to a ferromagnetic system will also generate an acoustic phonon, leading to a volume change (magnetostriction). Conversely, applying a magnetic field to an antiferromagnetic system will generate an optical phonon (which conserves the volume) and so there will not be any magnetostriction. This behaviour is in agreement with previous studies on other magnetic systems [28].

It is also worth noting that the energy of a $q \to 0$ acoustic phonon is vanishingly small, which suggests that the overall effect on the MAE of the magnon–phonon coupling will also be very small. However, this is not true for an optical phonon, and therefore the effect should be significantly larger in AFM materials than in FM materials.

2.3.2. Conservation of Energy

The energy for the creation of the magnon (or magnetophonon) that drives the transition cannot come from the applied magnetic field. The total energy of a 1 T magnetic field, as typically used in experiments, in a volume of free space equal to the size of a crystallographic unit cell is ~0.3 meV — and the energy per photon in this region will be even smaller. Hence, the external field only serves to couple different electronic states together, thereby enabling the stimulated rather than spontaneous generation of our commensurate magnon (magnetophonon). The energy to drive the transition, therefore, comes not from the external applied field but rather from internal sources of energy (and therefore, indirectly, the environment), namely the magnon heat bath and the phonon heat bath. Hence, this is a case of photon-assisted magnon–phonon conversion (in contrast to [29], which use phonons under illumination to generate magnons). The applied field enables the transition between different electronic states of the system to occur, whereas the energy comes from internal conversion due to scattering from phonon and magnon modes that have been thermally populated.

Considering that these two heat baths are connected [30,31] and the material is at a constant temperature (in thermal equilibrium), the magnon and phonon bath may be expected to have an equal distribution of energy, under the principle of equipartition. This suggests that there are two types of routes leading to the creation of the $q \to 0$ acoustic magnon that corresponds to the MAE. The first of these routes is the "standard" E^0 route, which is purely magnonic, and involves no structural relaxations (except as a subsequent potential decay path for the commensurate magnon). The second type of route is the phononic routes identified in Section 2.2 (E^+ or E^-). These routes involve scattering within the phonon bath to populate a mode that obeys the same symmetries as the commensurate magnon (or in the magnetophonon case, is the same quasiparticle).

These two phononic routes differ in that the E^+ route treats the magnon and phonon as a single quasiparticle, whereas the E^- route treats them as two independent quasiparticles. Which of these regimes dominates (and therefore controls the "fastest switching pathway" MAE for an AFM material) depends on the coupling strength between the magnon associated with the MAE and its associated phonon [31]. For $q \to 0$ magnons and phonons, the low momentum leads to a large uncertainty in the position of both quasiparticles and therefore, within the relaxation time approximation, the coupling should be extremely large ($\to \infty$), leading to E^+ being the dominant route. In this strong coupling limit (E^+), it does not make sense to consider magnons and phonons separately, but rather as a single magnetophonon [32].

2.4. Recap

1. The MAE is the minimum energy barrier to transition between two magnetic states which correspond to easy axis alignment.
2. This corresponds to the creation of a $q \to 0$ magnon mode, which may or may not be simultaneously coupled to a phonon mode (a magnetophonon).
3. For FM materials, this phonon mode is a long-wavelength acoustic mode and consequently should only give a small correction to the MAE, but will generate magnetostriction.
4. For AFM materials, this phonon mode is a long-wavelength optical mode, and consequently will provide a more significant change to the MAE due to the higher energy of the optical mode, and no magnetostriction.
5. When the magnon–phonon coupling between these modes is strong, the MAE is represented by a magnetophonon that has a different energy to either the magnon or phonon when they are considered separately.
6. Coupling to an optical rather than an acoustic mode has a three-orders of magnitude effect on the attempt frequency due to the difference in phonon lifetimes, leading to faster switching speeds for AFM-based devices.

3. Methodology

3.1. Density Functional Theory Parameters

The plane-wave DFT [33,34] code CASTEP [35] was used to simulate the MAE by evaluating the total energy of a conventional unit cell as the orientation of the spin moments on the non-Pt (Fe or Mn) ions — where the majority of the spin in the system is located — were rotated by applying spin constraints. Fully relativistic pseudopotentials from the CASTEP SOC19 pseudopotential library were used with vector spin-modelling and spin-orbit coupling in order to capture this effect. Because both FePt and PtMn are metals, the Perdew–Zunger formulation [36] of the local density approximation (LDA) was used and may be expected to perform acceptably well.

Although more sophisticated treatments of exchange-correlation are available for the collinear limit, many do not yet have explicit noncollinear reformulations [37]. However, the methodological improvements suggested in this paper are independent of the choice of exchange-correlation functional, and therefore it is sufficient to demonstrate these benefits for the LDA, despite the fact that other choices might improve other aspects of the physical model.

Numerical convergence of the plane wave basis set was performed to ensure total energy convergence of 0.01 meV/atom (2100 eV plane wave cutoff), and the same tolerance was used to converge the k-space sampling (a $25 \times 25 \times 25$ Monkhorst-Pack grid [38] for PtMn and $20 \times 20 \times 20$ for FePt).

Applied Constraints

In order to sample the potential energy surface away from the global minimum, spin constraints were applied to ensure that the desired spin configuration was achieved through a method of Lagrangian minimization [39]. To this end, a constraint that vanishes at the desired spin-configuration but was positive (destabilising) elsewhere was applied. This ensured that the correct value of the potential energy landscape was sampled for the constraint configuration, but also that this was an artificial minimum in the energy landscape such that the minimisation of the total energy could identify this as the appropriate ground state of the system. This technique is conceptually similar to metadynamics [40], and ensures that both the true potential energy surface as well as non-ground-state spin configurations can be sampled.

3.2. Geometry Optimisation Parameters

The structures were relaxed by using BFGS minimisation until stresses were lower than 1×10^{-4} GPa and forces were less than 1×10^{-5} eV/Å, and a finite basis correction was used to ensure constant basis quality.

During the geometry optimisation, spatial symmetry operations were not enforced. This is because the magnetic ordering lowers the symmetry of the crystal structure (always for the AFM, and for the FM whenever the spins are not aligned along the c-axis), and it was necessary to remove any potential "false symmetries" that may be present which could lead to force cancellation by symmetry.

Additionally, a collinear LDA+U relaxation of the structures was performed, with an effective U correction of 4.5 eV applied to the Mn 3d orbitals and 3.5 eV to the Fe 3d orbitals. The MAE for this LDA+U relaxed structure was then evaluated by using the previous method with no Hubbard correction applied.

4. Results and Discussion

4.1. Materials

FePt is an exemplar high-anisotropy ferromagnetic material which has an L_10 structure (see Figure 5a). PtMn is the AFM equivalent of FePt, with the only differences emerging from substituting Fe for Mn (see Figure 5b). This substitution yields significant differences to the magnetic structure (which changes from collinear FM to collinear C-type AFM) but not the crystal structure or the magnitude of the spin-orbit coupling introduced by the "heavy metal ion" Pt, which is constant between both.

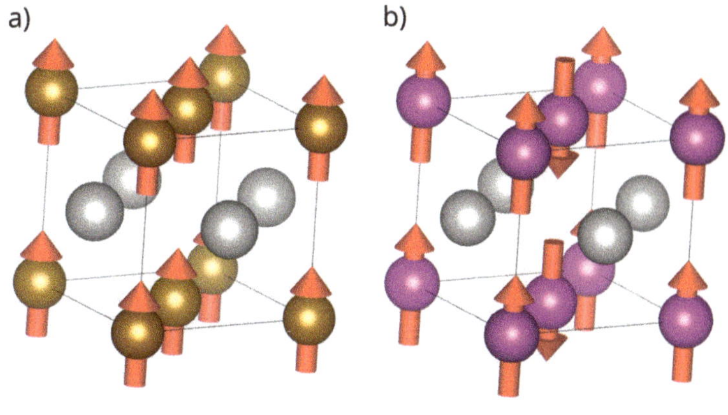

Figure 5. (**a**) The ground-state magnetic and geometric structure of FePt. (**b**) The ground-state magnetic and geometric structure of PtMn. Brown atoms are Fe, purple atoms are Mn and silver are Pt. Note that the major difference between these structures is in the magnetic coupling between the Fe atoms and the Mn atoms.

4.2. *Magnetic-Only Transition State*

In order to perform the magnetic-only (LST-like) transition search — corresponding to the standard method with clamped ions — a series of "sweeps" were performed by varying the spin orientation from $\theta = 0$ ($+\vec{c}$) to $\theta = 180°$ ($-\vec{c}$). These sweeps varied by their ϕ orientation (angle to $\vec{a}$) in order to ensure a good coverage of the magnetic phase space. Due to the symmetry of the crystals, the ϕ variation is periodic with a period of 90 degrees. It is also a minor variation compared with the variation in θ, and accordingly in the interest of clarity only the ϕ sweeps at 0 and 45 degrees have been reported.

4.2.1. FePt

For FePt, the sweeps are shown in Figure 6. We observe the expected $K\sin^2(\theta)$ behaviour (including higher order terms, as expected from an *ab initio* model). Additionally, the ϕ dependence is observed to be very weak, although with a very slightly softer $\phi = 45°$ axis, which is the minimum energy pathway through magnetic space only for the LDA unit cell. Nevertheless, this ϕ-variation is so weak that FePt may be said to exhibit uniaxial anisotropy with a hard plane.

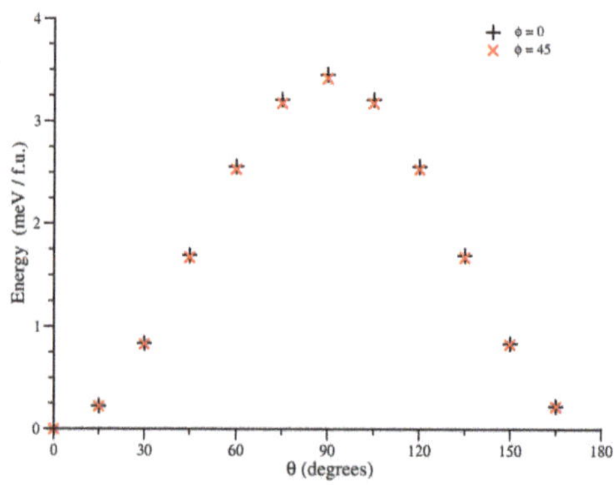

Figure 6. Variation of total energy (in meV / f.u.) with angle from the easy axis, θ, for two values of the azimuthal angle, $\phi = 0$ and $\phi = 45°$ for FePt. θ is expected to be the dominant term under the uniaxial approximation.

The MAE from these results (3.45 meV/f.u., see Figure 6) were found to agree with certain literature results [41] but to be in relatively poorer agreement with others [8]. This was driven by the choice of lattice parameters — in this work we have used the relaxed LDA geometry with no Hubbard-U correction, but other works [8] have used values nearer the bulk experimental lattice parameters (the relaxed LSDA+U geometry). Using the same lattice parameters as these other works ($|\vec{a}| = |\vec{b}| = 3.863$ Å, $|\vec{c}| = 3.783$ Å) yielded the same values as they reported (2.68 meV/f.u. using our method compared with 2.68 meV/f.u. for the calculation on the same cell [8]). Thus, the choice of geometry clearly has an effect on the evaluated MAE; the overbinding of LDA leads to shorter "bonds" and therefore increased effective spin-orbit interactions for the Fe orbitals (which have hybridised with Pt orbitals), and also a higher crystal field splitting of the orbitals — thereby increasing the calculated anisotropy. Nevertheless, in order to avoid mixing different approximations to the exchange-correlation functional, we have chosen to evaluate all of our MAEs at the bulk LDA lattice parameters with no Hubbard-U corrections.

In addition, we calculated the variation of the MAE with applied biaxial strain (keeping the $\vec{c}$ lattice vector fixed and measuring the value of the MAE at several $|\vec{a}| = |\vec{b}|$ values) as shown in Figure 7. This clearly demonstrates that the MAE is dependent upon the strain of the system. Although not perfectly linear, there is a clear linear trend, which may be indicative of decreasing effective spin-orbit coupling for electrons predominantly centred on Fe atoms as the overlap between Fe and Pt *d*-orbitals decreases.

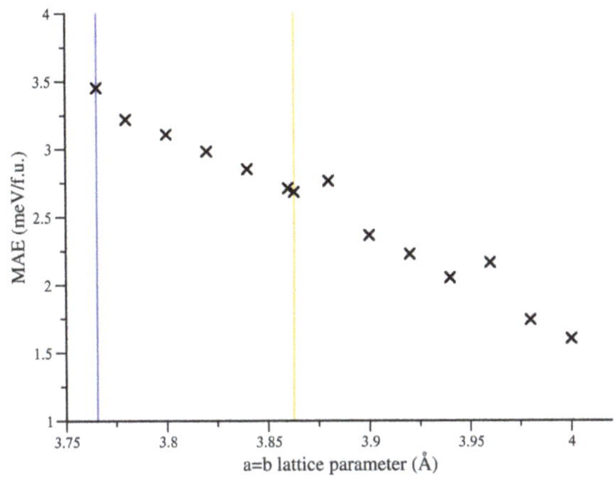

Figure 7. Change of energy difference between $\vec{c}$- and $\vec{a}$-aligned spin structures with strain for FePt. This figure includes both the LDA (blue line) and LDA+U (yellow line) structures at their exact in-plane lattice parameter.

4.2.2. PtMn

The angular sweeps for PtMn are shown in Figure 8. The predicted anisotropy per formula unit is lower than for FePt (as expected), and shows the same $\sin^2(\theta)$ behaviour with strongly uniaxial behaviour. In this case, the $\phi = 0$ orientation provides a fractionally lower energy pathway; however, this is not resolvable in Figure 8. We also note that the energy per formula unit is not the same as the energy for the unit cell, although a conventional unit cell is required to model the magnetic structure the formula unit is explicitly 1 Pt and 1 Mn.

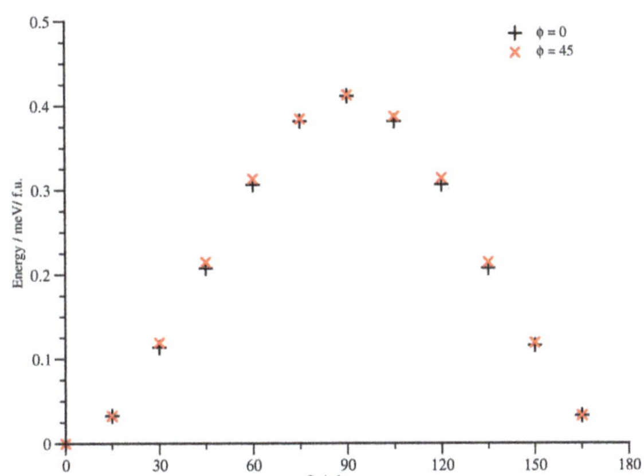

Figure 8. Variation of total energy (in meV) with angle from the easy axis, θ, for two values of the azimuthal angle, $\phi = 0$ and $\phi = 45°$.

In addition to the full angular sweep shown in Figure 8, the MAE was evaluated for various systems with biaxial strain (along $\vec{a}$ and $\vec{b}$ axes) applied in order to mimic the effects of growth seed layer choice upon the MAE. These results are shown in Figure 9. The total range of strains evaluated, of the order of $\pm 3\%$, is large, but not unreasonable in the thin film limit typically used to grow samples. As for FePt, the MAE was evaluated for all systems (including the LDA+U geometry) without applied Hubbard-U. This does not, however, capture the direct effects of the applied-U to the MAE beyond the effects induced by the change in lattice parameters.

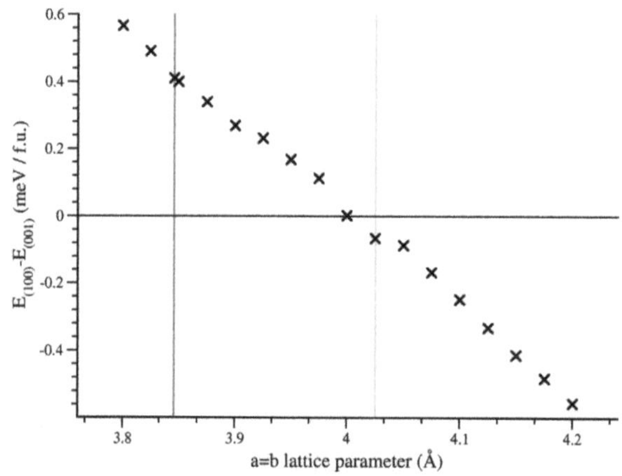

Figure 9. Change of energy difference between $\vec{c}$- and $\vec{a}$-aligned spin structures with strain. Note that the MAE is the absolute value of this number, and the sign change indicates a strain-induced alteration of the easy and hard magnetic axes. The blue line represents the LDA lattice parameter, and the yellow line indicates the LDA+U lattice parameter.

The change of MAE with strain shown in Figure 9 indicates a strong dependence of the MAE on the underlying geometry. These are large changes — and so the underlying assumptions of the phonon-based model of small amplitude oscillations are not applicable — and accordingly a large change in anisotropy is seen. In addition to the obvious ability this gives to engineer thin-film anisotropy based upon a careful choice of seed layer, it should also be able to create electrical control of the MAE by coupling a PtMn layer with a piezoelectric layer.

We also note that these results are in qualitative agreement with other recent theoretical efforts which considered varying both the $\vec{a}$ and $\vec{b}$ lattice parameters [42]. On the other hand, it seems likely that this change is driven by the changing crystal field splitting at the Mn sites driven by the change of interatomic distances rather than the $|\vec{c}|/|\vec{a}|$ ratio *per se*.

4.3. Magnetostructural Transition State

Following the identification of the peak of the magnetic anisotropy (the hard axis), the structure was relaxed in order to perform the QST-like search throughout the orthogonal structure space (as opposed to the spin space). For a maximum, such as the magnetic-only transition state, there are formally no forces, because it is a stationary point. Accordingly, several small initial perturbations were applied to this structure before the relaxation was performed. This is to ensure that the symmetry of the system is broken to enable the unit cell to relax (as the forces are zero at a maximum).

4.3.1. FePt

The relaxed geometry of L_10 FePt was found for both the easy and hard magnetic configurations through geometry optimisation of the lattice parameters with the magnetic configuration constrained to the easy and hard axes respectively. The ions remained at high symmetry points. The lattice parameters given by these relaxation are given in Table 1.

Table 1. Lattice parameters for relaxed PtMn dependant upon the different spin configurations. LDA+U and experimental lattice parameters are also provided for comparison.

	$\|\vec{a}\|/\text{Å}$	$\|\vec{b}\|/\text{Å}$	$\|\vec{c}\|/\text{Å}$
Easy Geometry	3.7656	3.7656	3.6841
Hard Geometry	3.7688	3.7644	3.6835
LDA+U, Easy	3.863	3.863	3.783
Experimental [43]	3.87	3.87	3.77

These relaxations show a small (0.02 Å^3) increase in total cell volume and a breaking of the symmetry of the unit cell for the hard magnetic configuration. This is in agreement with the previous arguments about magnetostriction in FM materials.

Following relaxation, the total energy of the system for all combinations of these spin and geometry configurations were evaluated (results shown in Table 2).

Table 2. Relative energies (in meV/formula unit) for the various configurations of spin and geometry in FePt. The easy geometry is the ground-state geometry of the system, and the hard geometry is the relaxed geometry of the system under the constraint that the system is aligned along the magnetic hard axis.

	Easy Geometry	**Hard Geometry**
Easy Axis	0	0.014
Hard Axis	3.453	3.439

It is clear that the structural relaxation is a small effect ($\sim$0.5%) — in line with the $\sim$0.1% $\vec{a}$ lattice parameter change. However, this is per formula unit; for a macroscopic layer of FePt, this effect is multiplied up until these apparently small energy barriers become significant — with even the small difference between easy and hard geometries for the easy spin configuration becoming a barrier larger than k_BT when 200 conventional unit cells are present.

Finally, it is worth noting that in addition to the thermodynamic arguments here, there are also dynamical considerations. As discussed in Sections 2.2 and 2.3, there are three possible routes from the magnetic easy to the magnetic hard axis: E^0, corresponding to a fixed easy axis geometry; E^+ corresponding to a direct change from the easy geometry to the hard geometry; and E^- corresponding to a fixed hard axis geometry. The attempt frequency for these different routes varies significantly and therefore, in a short time period such as in a data read or write process, the route with the highest attempt rate that is thermally accessible is likely to dominate. In a data-storage scenario, however, the long time period means that the lowest energy route is likely to be favoured.

For FePt these considerations are small. The difference in the barrier between E^0 and E^- — the routes with the largest difference in energy — is less than 1%, making the choice of protocol a smaller effect than other errors associated with the choice of functional or geometry used. This is in accord with the very low energy of the $q = 0$ acoustic phonon mode. For AFM materials, however, the relatively large energy of the optical phonon mode must be considered.

4.3.2. PtMn

The same procedure was carried out for antiferromagnetic $L1_0$ PtMn, and the structural changes are shown in Table 3. As anticipated, the observed magnetostriction was small (<2 mÅ) and may be attributed to remnant FM couplings between the component of the spins, which are canted by the Dzyaloshinskii–Moriya interaction (DMI). The magnitude of DMI is expected to be small because it leads to the presence of a spin-spiral ground state when it is strong [44]. The largest effect on the structure is a very small movement of the Pt ion in $\pm\vec{c}$, which corresponds to the 22 meV transverse optical (TO) mode of the system. At first it may seem surprising that the Pt ions should move in response to the reorientation of the spins at the Mn sites, but this may be explained by considering that the movement of the Pt ions affects both the strength and symmetry of the crystal field experienced at the Mn sites. This perturbation changes the energy of individual bands, thereby leading to a reduction in energy of the system when the spins are aligned along the hard axis.

Table 3. Lattice parameters and atomic positions for relaxed PtMn dependant upon the different spin configurations.

	$\|\vec{a}\|/Å$	$\|\vec{b}\|/Å$	$\|\vec{c}\|/Å$
Easy Geometry	3.8464	3.8464	3.6593
Hard Geometry	3.8464	3.8464	3.6593
LDA+U, Easy	4.025	4.025	3.665
Experimental [45]	3.96	3.96	3.73
Easy Geometry		Fractional co-ordinates	
	$\vec{a}$	$\vec{b}$	$\vec{c}$
Pt 1	0.5	0.0	0.5
Pt 2	0.0	0.5	0.5
Mn 1	0.0	0.0	0.0
Mn 2	0.5	0.5	0.0
Hard Geometry		Fractional co-ordinates	
	$\vec{a}$	$\vec{b}$	$\vec{c}$
Pt 1	0.5	0.0	0.5008
Pt 2	0.0	0.5	0.4992
Mn 1	0.0	0.0	0.0
Mn 2	0.5	0.5	0.0

The key change in the structure is the perturbation of the Pt atoms (unlike the FePt case in which they remained stationary at the high symmetry positions). This change, corresponding to a TO phonon mode of 22 meV, is in agreement with previously reported values of 18 meV (4.3 THz) [42]. DFT LDA calculations (as used here) overbind and hence give higher phonon frequencies, whereas PBE (as used in ref [42]) underbinds and hence gives slightly lower phonon frequencies.

With PtMn, unlike FePt, the energy of the system with spin aligned along the hard axis does not significantly change when the geometry is relaxed (in agreement with the fact that antiferromagnetic systems do not respond strongly to applied fields). However, when the hard-axis relaxed geometry is measured in the easy-axis configuration, a significant increase in the energy is observed, as shown in Table 4.

Now, the displacement that leads to this increase in energy is very small (≈2.9 mÅ). This is actually smaller than the zero-point amplitude of the phonon mode associated with (≈18.7 mÅ). The fact that this displacement is smaller than the zero-point motion—which is an effective permanent population of the relevant phonon mode—means that this lower energy route, E^-, is always achievable. Also unlike FePt, the effect of this small displacement is to significantly reduce the MAE from 0.41 meV/f.u. to 0.32 meV/f.u., which is an ∼20% reduction in the size of the theoretical barrier.

Table 4. Relative energies (in meV/f.u.) for the various configurations of spin and geometry in PtMn. The easy geometry is the ground-state geometry of the system, and the hard geometry is the geometry of the system having been relaxed under the constraint that the system is aligned along the magnetic hard axis.

	Easy Geometry	Hard Geometry
Easy Axis	0	0.0945
Hard Axis	0.4108	0.4108

Although the absolute magnitude of these effects will vary with the level of theory used in DFT for the system under investigation — for example, the use of a GGA or Hubbard-U correction might be expected to remove the overbinding of LDA and make the model of the system more accurate—the effect of the magnetostructural coupling on the MAE will remain. Although it is a negligible effect in ferromagnetic systems, this work shows that it is a significant effect for antiferromagnetic systems. Hence, it should be included in all calculations of MAE in AFMs, even with higher levels of theory, if the MAE is to be accurately modelled and not systematically overestimated.

5. Conclusions

The MAE was evaluated for ferromagnetic FePt and antiferromagnetic PtMn and found to be 3.45 meV/f.u. and 0.41 meV/f.u., respectively, for the LDA approximation in the static-ion limit. The choice of lattice parameters (for example LDA vs LDA+U geometries) was found to provide a significant change in the calculated MAE, up to an exchange of the easy and hard axes of the system. Relaxing the systems to find the magnetostructural rather than just the magnetic transition states yielded a $\sim$1% reduction in the barrier height of FePt (to 3.44 meV/f.u.), and a $\sim$20% reduction for PtMn (to 0.32 meV/f.u.) when the lowest energy pathway was considered. This difference is attributed to the difference in coupling between the magnon responsible for the transition and either an acoustic or optical phonon (for FePt and PtMn respectively). This shows the importance of considering magnetostructural effects when evaluating the MAE of systems in which the magnetic structure cannot be adequately expressed in the same primitive unit cell as the geometric structure, such as an antiferromagnet.

Author Contributions: Conceptualization and investigation, R.A.L.; writing—original draft preparation, R.A.L.; writing—review and editing, S.J.D.; visualization, R.A.L. and S.J.D.; supervision, resources, project administration and funding acquisition, M.I.J.P. All authors have read and agreed to the published version of the manuscript.

Funding: Computational support was provided by the UK national high performance computing service, ARCHER2, for which access was obtained via the UKCP consortium and funded by EPSRC grant ref EP/P022561/1. The authors also acknowledge EPSRC grant EP/V047779/1 for financial support.

Institutional Review Board Statement: Not applicable.

Informed Consent Statement: Not applicable.

Data Availability Statement: The data presented in this study are openly available in the York Research Database at doi:10.15124/b1ec96de-4796-45a6-89f9-dc0c34d0f431.

Acknowledgments: The authors would like to thank Jerome Jackson, Kevin O'Grady and Gonzalo Vallejo Fernandez for their kind comments and valuable conversations in preparing this work.

Conflicts of Interest: The authors declare no conflict of interest.

References

1. Mattioli, M. The Apollo Guidance Computer. *IEEE Micro* **2021**, *41*, 179–182. [CrossRef]
2. Nordrum, A. The fight for the future of the disk drive. *IEEE Spectr.* **2019**, *56*, 44–47. [CrossRef]
3. Wang, K.L.; Alzate, J.G.; Amiri, P.K. Low-power non-volatile spintronic memory: STT-RAM and beyond. *J. Phys. D: Appl. Phys.* **2013**, *46*, 074003. [CrossRef]

4. Kosub, T.; Kopte, M.; Hühne, R.; Appel, P.; Shields, B.; Maletinsky, P.; Hübner, R.; Liedke, M.O.; Fassbender, J.; Schmidt, O.G.; et al. Purely antiferromagnetic magnetoelectric random access memory. *Nat. Commun.* **2017**, *8*, 13985. [CrossRef]
5. Charap, S.; Lu, P.L.; He, Y. Thermal stability of recorded information at high densities. *IEEE Trans. Magn.* **1997**, *33*, 978–983. [CrossRef]
6. Martins, A.; Trippe, S.; Santos, A.; Pelegrini, F. Spin-wave resonance and magnetic anisotropy in FePt thin films. *J. Magn. Magn. Mater.* **2007**, *308*, 120–125. [CrossRef]
7. O'Grady, K.; Fernandez-Outon, L.; Vallejo-Fernandez, G. A new paradigm for exchange bias in polycrystalline thin films. *J. Magn. Magn. Mater.* **2010**, *322*, 883–899. [CrossRef]
8. Shick, A.B.; Mryasov, O.N. Coulomb correlations and magnetic anisotropy in ordered $L1_0$ CoPt and FePt alloys. *Phys. Rev. B* **2003**, *67*, 172407. [CrossRef]
9. Khan, S.A.; Blaha, P.; Ebert, H.; Minár, J.; Šipr, O. Magnetocrystalline anisotropy of FePt: A detailed view. *Phys. Rev. B* **2016**, *94*, 144436. [CrossRef]
10. Nauman, M.; Kiem, D.H.; Lee, S.; Son, S.; Park, J.G.; Kang, W.; Han, M.J.; Jo, Y. Complete mapping of magnetic anisotropy for prototype Ising van der Waals $FePS_3$. *2D Mater.* **2021**, *8*, 035011. [CrossRef]
11. Higgins, E.J.; Hasnip, P.J.; Probert, M.I. Simultaneous Prediction of the Magnetic and Crystal Structure of Materials Using a Genetic Algorithm. *Crystals* **2019**, *9*, 439. [CrossRef]
12. Zhu, B.; Lu, Z.; Pickard, C.J.; Scanlon, D.O. Accelerating cathode material discovery through *ab initio* random structure searching. *APL Mater.* **2021**, *9*, 121111. [CrossRef]
13. Burkert, T.; Eriksson, O.; Simak, S.I.; Ruban, A.V.; Sanyal, B.; Nordström, L.; Wills, J.M. Magnetic anisotropy of $L1_0$ FePt and $Fe_{1-x}Mn_xPt$. *Phys. Rev. B* **2005**, *71*, 134411. [CrossRef]
14. Shick, A.B.; Máca, F.; Lichtenstein, A.I. Magnetic anisotropy of single 3d spins on a CuN surface. *Phys. Rev. B* **2009**, *79*, 172409. [CrossRef]
15. Jónsson, H.; Mills, G.; Jacobsen, K.W. Nudged elastic band method for finding minimum energy paths of transitions. In *Classical and Quantum Dynamics in Condensed Phase Simulations: Proceedings of the International School of Physics, Lerici, Villa Marigola, 7–18 July 1997*; Berne, B., Ciccoti, G., Coker, D.F., Eds.; World Scientific: Singapore, 1998; ._0016. [CrossRef]
16. Henkelman, G.; Uberuaga, B.P.; Jónsson, H. A climbing image nudged elastic band method for finding saddle points and minimum energy paths. *J. Chem. Phys.* **2000**, *113*, 9901–9904. [CrossRef]
17. Halgren, T.A.; Lipscomb, W.N. The synchronous-transit method for determining reaction pathways and locating molecular transition states. *Chem. Phys. Lett.* **1977**, *49*, 225–232. [CrossRef]
18. Peterson, A.A. Acceleration of saddle-point searches with machine learning. *J. Chem. Phys.* **2016**, *145*, 074106. [CrossRef] [PubMed]
19. Hammond, G.S. A Correlation of Reaction Rates. *J. Am. Chem. Soc.* **1955**, *77*, 334–338. [CrossRef]
20. Zhu, Y.A.; Dai, Y.C.; Chen, D.; Yuan, W.K. First-principles calculations of CH_4 dissociation on Ni(100) surface along different reaction pathways. *J. Mol. Catal. A: Chem.* **2007**, *264*, 299–308. [CrossRef]
21. Lechner, B.A.J.; Hedgeland, H.; Ellis, J.; Allison, W.; Sacchi, M.; Jenkins, S.J.; Hinch, B.J. Quantum Influences in the Diffusive Motion of Pyrrole on Cu(111). *Angew. Chem. Int. Ed.* **2013**, *52*, 5085–5088. [CrossRef]
22. Rudin, S.P. Generalization of soft phonon modes. *Phys. Rev. B* **2018**, *97*, 134114. [CrossRef]
23. Wilczek, F. Majorana returns. *Nat. Phys.* **2009**, *5*, 614–618. [CrossRef]
24. Vallejo-Fernandez, G.; Aley, N.P.; Chapman, J.N.; O'Grady, K. Measurement of the attempt frequency in antiferromagnets. *Appl. Phys. Lett.* **2010**, *97*, 222505. [CrossRef]
25. Roca, A.G.; Vallejo-Fernández, G.; O'Grady, K. An Analysis of Minor Hysteresis Loops of Nanoparticles for Hyperthermia. *IEEE Trans. Magn.* **2011**, *47*, 2878–2881. [CrossRef]
26. Sharma, S.; Shallcross, S.; Elliott, P.; Dewhurst, J.K. Making a case for femto-phono-magnetism with FePt. *Sci. Adv.* **2022**, *8*. [CrossRef] [PubMed]
27. Maldonado, P.; Carva, K.; Flammer, M.; Oppeneer, P.M. Theory of out-of-equilibrium ultrafast relaxation dynamics in metals. *Phys. Rev. B* **2017**, *96*, 174439. [CrossRef]
28. Liu, S.; del Águila, A.G.; Bhowmick, D.; Gan, C.K.; Do, T.T.H.; Prosnikov, M.; Sedmidubský, D.; Sofer, Z.; Christianen, P.C.; Sengupta, P.; et al. Direct Observation of Magnon-Phonon Strong Coupling in Two-Dimensional Antiferromagnet at High Magnetic Fields. *Phys. Rev. Lett.* **2021**, *127*, 097401. [CrossRef]
29. fan Qi, S.; Jing, J. Magnon-assisted photon-phonon conversion in the presence of structured environments. *Phys. Rev. A* **2021**, *103*, 043704. [CrossRef]
30. Aguilar, R.V.; Sushkov, A.B.; Zhang, C.L.; Choi, Y.J.; Cheong, S.W.; Drew, H.D. Colossal magnon-phonon coupling in multiferroic $Eu_{0.75}Y_{0.25}MnO_3$. *Phys. Rev. B* **2007**, *76*, 060404. [CrossRef]
31. Man, H.; Shi, Z.; Xu, G.; Xu, Y.; Chen, X.; Sullivan, S.; Zhou, J.; Xia, K.; Shi, J.; Dai, P. Direct observation of magnon-phonon coupling in yttrium iron garnet. *Phys. Rev. B* **2017**, *96*, 100406. [CrossRef]
32. Shen, P.; Kim, S.K. Magnetic field control of topological magnon-polaron bands in two-dimensional ferromagnets. *Phys. Rev. B* **2020**, *101*, 125111. [CrossRef]
33. Hohenberg, P.; Kohn, W. Inhomogeneous electron gas. *Phys. Rev.* **1964**, *136*, B864–B871. [CrossRef]

34. Kohn, W.; Sham, L.J. Self-consistent equations including exchange and correlation effects. *Phys. Rev.* **1965**, *140*, A1133–A1138. [CrossRef]
35. Clark, S.J.; Segall, M.D.; Pickard, C.J.; Hasnip, P.J.; Probert, M.J.; Refson, K.; Payne, M. First principles methods using CASTEP. *Z. Kristall.* **2005**, *220*, 567–570. [CrossRef]
36. Perdew, J.P.; Zunger, A. Self-interaction correction to density-functional approximations for many-electron systems. *Phys. Rev. B* **1981**, *23*, 5048–5079. [CrossRef]
37. Scalmani, G.; Frisch, M.J. A New Approach to Noncollinear Spin Density Functional Theory beyond the Local Density Approximation. *J. Chem. Theory Comput.* **2012**, *8*, 2193–2196. [CrossRef]
38. Monkhorst, H.J.; Pack, J.D. Special points for Brillouin-zone integrations. *Phys. Rev. B* **1976**, *13*, 5188–5192. [CrossRef]
39. Cuadrado, R.; Pruneda, M.; García, A.; Ordejón, P. Implementation of non-collinear spin-constrained DFT calculations in SIESTA with a fully relativistic Hamiltonian. *J. Phys. Mater.* **2018**, *1*, 015010. [CrossRef]
40. Laio, A.; Parrinello, M. Escaping free-energy minima. *Proc. Natl. Acad. Sci. USA* **2002**, *99*, 12562–12566. . pnas.202427399. [CrossRef]
41. Barreteau, C.; Spanjaard, D. Magnetic and electronic properties of bulk and clusters of FePt $L1_0$. *J. Phys. Condens. Matter* **2012**, *24*, 406004. [CrossRef]
42. Aissat, D.; Baadji, N.; Mazouz, H.; Boussendel, A. Connection between lattice parameters and magnetocrystalline anisotropy in the case of $L1_0$ ordered antiferromagnetic MnPt. *J. Magn. Magn. Mater.* **2022**, *563*, 170013. [CrossRef]
43. Klemmer, T.J.; Shukla, N.; Liu, C.; Wu, X.W.; Svedberg, E.B.; Mryasov, O.; Chantrell, R.W.; Weller, D.; Tanase, M.; Laughlin, D.E. Structural studies of L10 FePt nanoparticles. *Appl. Phys. Lett.* **2002**, *81*, 2220–2222. [CrossRef]
44. Hervé, M.; Dupé, B.; Lopes, R.; Böttcher, M.; Martins, M.D.; Balashov, T.; Gerhard, L.; Sinova, J.; Wulfhekel, W. Stabilizing spin spirals and isolated skyrmions at low magnetic field exploiting vanishing magnetic anisotropy. *Nat. Commun.* **2018**, *9*, 1015. [CrossRef]
45. Mao, S.; Gao, Z. Characterization of magnetic and thermal stability of PtMn spin valves. *IEEE Trans. Magn.* **2000**, *36*, 2860–2862. [CrossRef]

magnetochemistry

Article

Role of Disordered Precursor in L1$_0$ Phase Formation in FePt-Based Nanocomposite Magnet

Alina Daniela Crisan [1], Ioan Dan [2] and Ovidiu Crisan [1,*]

1 National Institute for Materials Physics, 077125 Magurele, Romania; ad_crisan@yahoo.com
2 R&D Consulting and Services S.R.L., 023761 Bucharest, Romania; ioan_dan@rd-consultanta.ro
* Correspondence: ocrisan@infim.ro

Abstract: In order to prove the usefulness of having a structurally disordered precursor to the formation of FePt L1$_0$ phase and to facilitate the co-existence of exchange coupled hard and soft magnetic phases with optimized magnetic properties in various conditions of annealing, a Fe-Pt-Zr-B melt spun alloy has been synthesized and detailed structural and magnetic investigations have been undertaken to probe its phase evolution during annealing. The dynamics of formation of the hard magnetic L1$_0$ phase during the gradual disorder–order phase transformation has been monitored by using a complex combination of X-ray diffraction methods and ^{57}Fe Mössbauer spectroscopy methods, over a wide range of annealing temperatures. Multiple phases co-existing in the annealed sample microstructures, observed in XRD, have been reconfirmed by the Mössbauer spectra analysis and, moreover, accurate quantitative data have been acquired in what concerns the relative abundance of each of the observed crystalline phases in every stage of annealing. It is shown that the formation of the hard magnetic phase, emerging from the chemically disordered precursor, is gradual and occurs via complex mechanisms, involving the presence of a disordered Fe-Zr-B-rich intergranular region which contributes to an increase in the abundance of the L1$_0$ phase for higher annealing temperatures. Magnetic measurements have confirmed the good performances of these alloys in terms of coercivity and remanence. These results contribute to the development of these alloys as the next generation of rare earth, free permanent magnets.

Keywords: nanocomposite magnets; L1$_0$ phase; Mössbauer spectroscopy; magnetic properties

Citation: Crisan, A.D.; Dan, I.; Crisan, O. Role of Disordered Precursor in L1$_0$ Phase Formation in FePt-Based Nanocomposite Magnet. *Magnetochemistry* **2021**, 7, 149. https://doi.org/10.3390/magnetochemistry7110149

Academic Editor: Cătălin-Daniel Constantinescu

Received: 1 October 2021
Accepted: 9 November 2021
Published: 14 November 2021

1. Introduction

Nanocomposite FePt-based magnets have been under scrutiny as a new class of permanent magnets, due to their high corrosion resistance and their high working temperature. The interest in using FePt alloys is based on the disorder–order structural phase transition [1] that occurs in the alloy. FePt usually has a disordered face-centered-cubic fcc A1 structure. Upon annealing, for quasi-equiatomic stoichiometry, the structure undergoes a structural phase transformation, towards the formation of an L1$_0$ ordered face-centered-tetragonal fct phase that has a high coercivity and a very large magnetocrystalline anisotropy. Non-equilibrium synthesis methods are appropriate for achieving the formation of an L1$_0$ phase in melt-spun FePt-based ribbons without the need for post-synthesis annealing [2]. Another way to create magnetic alloys with good magnetic properties is to achieve a hard–soft magnetic structure, where the hard magnetic FePt phase is exchange-coupled to soft magnetic ones, and to take full advantage from the high magnetization of the soft phase and the high coercivity of the hard magnetic phase, in order to maximize the energy product $(BH)_{max}$. To achieve such a phase structure, the initial composition [3] must be modulated by addition of elements, which can provide the necessary additional soft magnetic phase such as boron. This a well-known glass-forming element which introduced in FePt-based metallic alloys allows formation of a disordered intermetallic precursor. Following appropriate annealing, this precursor may transform in both hard and soft magnetic phases. The advantage of emerging form the same precursor is that the obtained nanocrystalline phases are crystallographically coherent and their exchange coupling is favored. In this way, suitable hard–soft arrangements of phases with well-dispersed nanometric-size grains upon annealing can be obtained. Arrangements of magnetically soft and hard phases for the improvement of hard magnetic properties has been already employed by other researchers. For instance, Chrobak et al. [4] obtained ultra-high coercivity of $(Fe_{86-x}Nb_xB_{14})_{0.88}Tb_{0.12}$ bulk nanocrystalline magnets where such arrangements of different magnetic phases were achieved.

Recently, there have been reports on the preparation and characterization of such ternary Fe-Pt-B alloys, using a plethora of non-equilibrium synthesis methods, to obtain both layered structures, such as thin films and multilayers, and bulk alloys, such as melt spun ribbons [5–7]. Magnetic composites of the multilayer type, made of FeB pre-alloy co-doped within FePt deposited films, were studied by Tsai et al. [8] Such exchange-coupled hard–soft system have shown a perpendicular anisotropy with a single switching field. In sandwiched systems made of soft (CoFeB) and hard (FePtB) layers, it has been shown [9] that oxidation during annealing induces stresses through the capping layer, which facilitates greatly the formation of the $L1_0$ phase in the FePtB hard layer. FePtB melt-spun alloys of various compositions have also been studied in [10]. In this case, the precursor alloy is synthesized in an as-cast state and is then annealed in order to obtain the crystalline state. In this case, hard, tetragonally ordered $L1_0$ FePt and soft Fe_2B phases were formed, but the composition modulation and Pt deficit strongly alters the magnetic performances. Co-sputtered FePt-B multilayers have also been studied [11] and it was found that rapid annealing induces ordering of $L1_0$ FePt via intermixing with B at the interfaces as well as by the introduction of 0.1 at% Ag. Other synthesis procedures, such as pulsed laser deposition, have also been used to construct exchange-coupled FePtB-based composites [12] and it was found that best magnetic performance has been found for very thick (1.7 microns) FePtB where both hard FePt and soft iron boride phases co-exist. Even lower ordering temperatures with higher thicknesses (3 microns) of co-sputtered FePtB are shown to also produce good magnetic results [13]. Optimizing the microstructure is a factor that strongly influences the overall magnetic behavior. For this, the performed annealing is of great importance. The effect of annealing time on the obtained magnetic properties in FePtB melt-spun alloys was investigated in [14]. The compositional effect of introducing boron on the exchange coupling effect in FePtB has also been investigated [15]. It was shown [16] that from application point of view, such exchange-coupled FePt-FePtB composites show good potential as magnetic recording media.

In our previous work [17], we have shown that in Fe-rich Fe-Pt-Nb-B alloys the microstructure is highly sensitive to the stoichiometry of the as-cast alloy [18]. Boron content of about 8–9% is proven to be not enough to ensure an amorphous-like as-cast state and the samples are obtained in their nanocrystalline A1 structure, while for B content of 18–20%, a disordered structure is obtained in the as-cast state, with large Bragg lines of the A1 structure, as seen in the X-ray diffraction. For samples with a lower Fe content, $(Fe_{0.65}Pt_{0.35})_{78+x}Nb_2B_{20-x}$, we have proven that, upon appropriate annealing, very good exchange spring properties are obtained with an energy product $(BH)_{max}$ of 70 kJ/m^3 [19].

In the case of systems exhibiting an $L1_0$ phase, the high ordering temperature hinders their potential as future nanocomposite RE-free exchange-coupled magnets. It is thought that the reduction of this value can be achieved by adding other elements to the composition, such as Au, Ag, Nb or Zr, with the aim of promoting earlier ordering by segregation to the FePt grain boundaries [20–24] or a glass-forming element (B) to allow the formation of a chemically disordered phase as a precursor for both hard and soft magnetic phases, exchange-coupled to optimize in this way the magnetic properties [25–35]. We have previously shown [2] that the Fe-Pt-Ag-B melt-spun alloy shows a direct formation of the $L1_0$ phase from the as-cast state without subsequent annealing. It seems quite clear that the different magnetic features in $L1_0$-based nanocomposite magnets are strongly influenced by their multiphase nanostructure features.

The present work is dedicated to the detailed structural and morphological study of the microstructure modifications and phase evolution during annealing, as well as the magnetic properties of a Fe-Pt-Zr-B melt-spun alloys, co-existence of exchange coupled hard and soft magnetic phases and the magnetic properties in various conditions of annealing.

2. Experimental

An alloy with nominal composition $Fe_{65}Pt_{15}Zr_3B_{17}$ has been synthesized by rapid solidification of the melt. The alloy was synthesized starting with elemental powders and flakes of high purity. They were melted together in an induction furnace with a controlled melting temperature (set at 1300 °C). The primary alloy re-melted 3 times to prevent element segregation and to improve its chemical homogeneity. A total amount of 5 g has been used for each sample. The rapid solidification of the melt is performed on a Buhler Melt Spinner SC with protective Ar atmosphere. The obtained melt is purged onto the surface of a Cu wheel. The wheel has 40 cm in diameter and rotates with 2000 rot/min. The melt is flown away from the quartz tube through a circular nozzle of 0.5 mm using Ar pressure of 40 kPa. The size of the nozzle dictates the width of the obtained ribbons. Since the total mass of the experiment was only 5 g, it was necessary to use such a small nozzle. The melt solidifies with a cooling rate of about 10^6 K/min. Away from the wheel, continuous and homogeneous ribbons are obtained. They are about 30 microns thick, 2–3 mm wide and several decimeters long.

In order to obtain the full crystallization and formation of the hard magnetic phase, the as-obtained ribbons were subjected to isothermal annealing. The procedure has been performed at various temperatures, and were chosen at 100 °C intervals. The chosen temperatures of annealing were: 500 °C, 600 °C, 700 °C and 800 °C. The heating rate was established at 5 K/min and annealing time (time spent at maximum temperature) was set to 30 min. The annealing was performed under a high vacuum (10^{-4} mbar) in order to avoid oxidation.

The detailed structural and morphological study of the microstructure modifications, phase evolution during annealing, and magnetic properties of the annealed alloys, was performed by X-ray diffraction (XRD), Mössbauer spectrometry (MS) and magnetic characterizations. For XRD, a Bruker D8 Advance (Bruker AXS GmbH, Karlsruhe, Germany) with Cu K_α radiation wavelength of 1.54 Å was used. The ^{57}Fe Mössbauer spectrometry is performed at 300 K and 77 K using a conventional set-up in transmission geometry with a ^{57}Co source in a Rh matrix. Magnetic characterization has been performed with a SQUID (Superconducting QUantum Interference Device) unit of an MPMS (Magnetic Properties Measurement System) from Quantum Design (Quantum Design Europe GmbH, Darmstadt, Germany), under an applied field of up to 5.5 Tesla, in parallel and perpendicular geometry and temperatures ranging from 4.2 K to 300 K.

3. Results and Discussion

3.1. As-Cast State

The ribbon as-cast state has been investigated using XRD and Mössbauer spectrometry. The X-ray diffractogram for the as-cast $Fe_{65}Pt_{15}Zr_3B_{17}$ sample exhibits very broad Bragg lines. They have been indexed as belonging to the fcc A1 phase. As proven by the lines wide broadening, there is chemical disorder within the A1 fcc structure. The full-profile analysis of the XRD patterns has been analyzed using MAUD (Materials Analysis Using Diffraction) software. The structure of the as-cast sample has disordered fcc A1 symmetry, as resulted from the fittings. Figure 1 shows the X-ray diffractogram fitted with MAUD. It is probable that this structure could be a precursor for the formation of FePt and iron boride phases after appropriate crystallization treatment. We have calculated, based on the integral breadth approach, the crystallite size for all the Bragg peaks indexed in the fcc A1 system. The mean grain size is found to be 3.8 ± 1.4 nm.

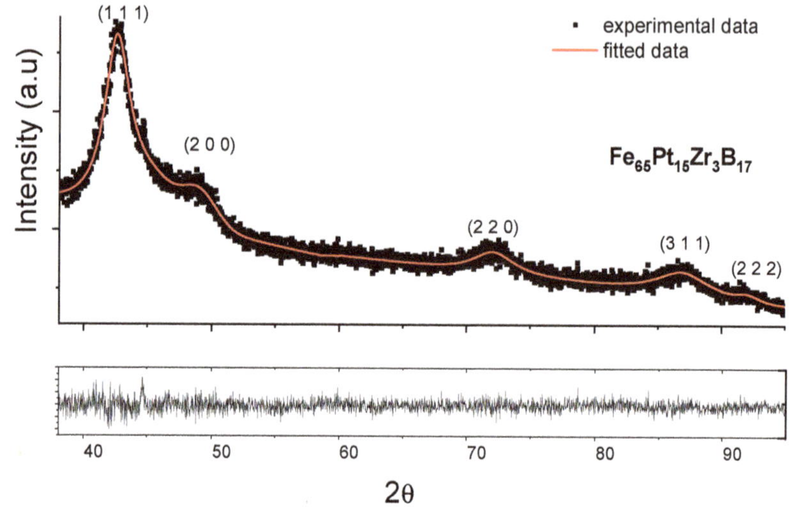

Figure 1. X-ray diffractogram of as-cast $Fe_{65}Pt_{15}Zr_3B_{17}$ sample fitted with the MAUD software. The indexed lines belong to fcc A1 FePt. Bottom graph is the difference between fitted and experimental data.

The 300 K and 77 K Mössbauer spectra of the as-cast sample (Figure 2) exhibit broad magnetic sextets, typical of distributed Fe environments encountered in Fe-rich amorphous ribbons. The shape of the spectra confirms the XRD results where a chemically disordered A1 phase has been identified. The hyperfine field (HF) distributions derived from the fitting of the Mössbauer spectra (Figure 3) show a bimodal-type large Gaussian profile distribution, which is characteristic of a disordered Fe environment, with two main environments for Fe. From the numerical fitting of the size distribution (Figure 3) with two Gaussian profiles, we have obtained for the low field mode an average HF of 16 T and 18 T for 300 K and 77 K respectively and for the high field mode average HF of 26 T and 29 T for 300 K and 77 K respectively. The relative proportion of the high-field to low-field relative contributions to the HF distributions is about 3:1. From the HF values we can presume that the low field mode corresponds to a disordered precursor that would give rise upon annealing to a boride (possibly Fe_2B) phase, while the high field mode would give rise to an fcc FePt phase.

3.2. *Isothermal Annealing*

3.2.1. XRD Analysis

In order to obtain the full crystallization and formation of the hard magnetic phase, isothermal annealing has been performed at various temperatures, chosen at 100 °C intervals, for observing better the degree of formation of the crystalline phases as well as to monitor the evolution of the disorder–order phase transformation. The chosen temperatures of annealing were: 500 °C, 600 °C, 700 °C and 800 °C. The heating rate was established at 5 K/min and annealing time (time spent at maximum temperature) was set to 30 min. The annealing was performed under a high vacuum (10^{-4} mbar) in order to avoid oxidation.

The structure of the resulting samples was characterized by XRD, Mössbauer spectrometry and their magnetic properties by vibrating sample magnetometry and SQUID. The X-ray diagrams of the annealed samples are plotted in Figure 4. For a better comparison, the as-cast X-ray diffractogram is also added. In the as-cast state, the broad Bragg peaks of the disordered A1 FePt-rich solid solution have been identified, as previously discussed. The broad line centered at around 25° is due to the sample holder and it can be seen that this broad feature is preserved at all annealing temperatures, until 800 °C, where the intensity of the main Bragg peak is very high and the broad feature appears strongly diminished due to scaling. At 500 °C, the A1 fcc pattern is better formed, with a smaller linewidth than in the as-cast state. Besides these lines, the main superlattice peaks of the tetragonal $L1_0$ phase appear already formed. At 600 °C the main superlattice peaks of the tetragonal $L1_0$ phase are more pronounced and the process of peak narrowing is furthermore continued, as expected. Starting with 600 °C a small Bragg peak, attributed to the main Bragg reflection of the Fe_2B tetragonal phase, is observed.

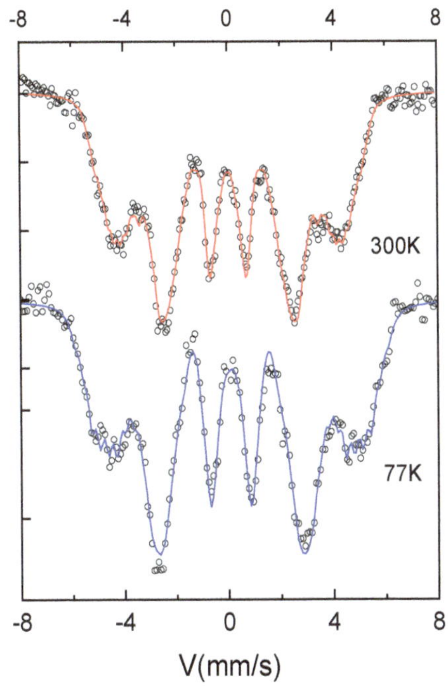

Figure 2. 300K and 77K Mössbauer spectra of as-cast $Fe_{65}Pt_{15}Zr_3B_{17}$ sample.

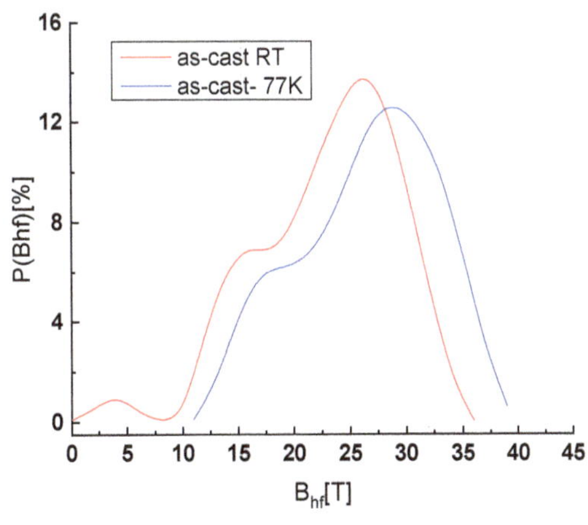

Figure 3. 300K and 77K Mössbauer HF distribution of as-cast $Fe_{65}Pt_{15}Zr_3B_{17}$ sample.

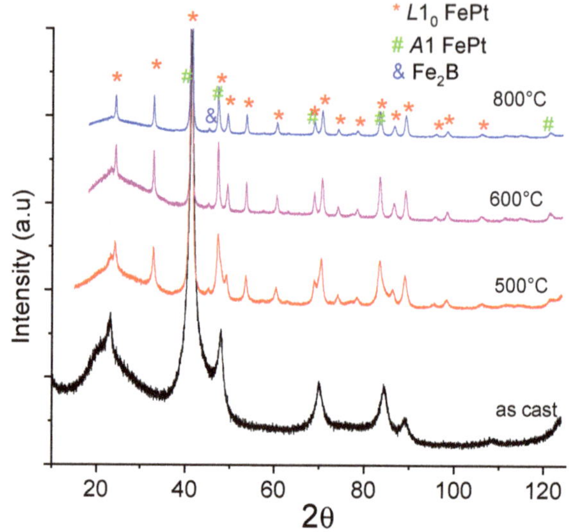

Figure 4. X-ray diffractograms of the $Fe_{65}Pt_{15}Zr_3B_{17}$ sample, in its as-cast state as well as annealed at various temperatures (500 °C, 600 °C and 800 °C).

Then at 800 °C the lines of the crystalline phases are even narrower, and the tetragonal $L1_0$ phase is the most predominant in the sample. The XRD diagram at 800 °C shows a completely crystallized pattern and besides the $L1_0$ phase, the Fe_2B is also visible in the diagram.

It has to be mentioned that the two main superlattice peaks of tetragonal $L1_0$, the (0 0 1) and (1 1 0) reflections occurs for the Cu Kα radiation we used, at about 24° and 32° respectively. From the observation of the Bragg line linewidths, it can be seen that the main Bragg lines observed in the fcc FePt solid solution in the as-cast state undergo a refinement process as the annealing temperature is increased, from bottom to top. The main 5 Bragg lines of the fcc A1 FePt solid solution in the as-cast state, from lower to higher angles, are, in order, attributed to: (1 1 1), (2 0 0), (2 2 0), (3 1 1) and (2 2 2). Upon annealing, these main reflections become overlapped with the main Bragg reflections of the $L1_0$ FePt tetragonal phase. It is very interesting to observe that already, from the first annealed sample at 500 °C, the (2 0 0), (2 2 0) and (3 1 1) peaks split into two. It seems that the peaks from the fcc phase are now spectrally separated from those of the tetragonal phase.

For the other two annealing treatments, at 700 °C (Figure 5) and 800 °C, the separation continues and the peaks become more and more narrow and better separated between one and another (see for instance the fitting from Figure 5 for the sample annealed at 700 °C). The specific signatures of the tetragonal $L1_0$ phase, the superlattice peaks (0 0 1) and (1 1 0), are clearly observed for all the annealed samples at around 24° and 33° (in 2-theta), respectively.

The good separation of the spectral lines allowed us an accurate determination by numerical fitting of the peak positions and linewidths, a calculation of the lattice parameters, the tetragonality or ordering parameter c/a and of the mean grain sizes for each of the observed crystalline phases. The obtained results are schematized in Table 1.

It can be seen that the lattice parameters, as revealed from the fit of the experimental spectra, do not change drastically with the increase of the annealing temperature; however, a steady increase of the ordering parameter c/a, associated with the tetragonality or the degree of ordering of the $L1_0$ phase, is observed upon increasing the annealing temperature.

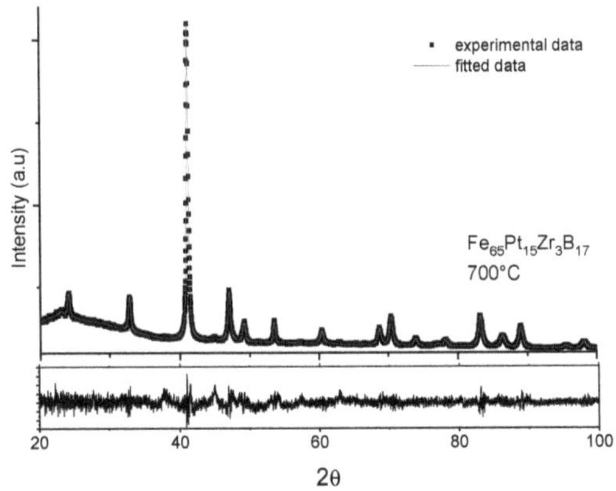

Figure 5. MAUD fitting of the XRD diagram of the sample annealed at 700 °C for 30 min.

Table 1. Lattice parameters, ordering parameter and grain sizes for the as-cast and annealed $Fe_{65}Pt_{15}Zr_3B_{17}$ samples.

Sample	Anneal.	$L1_0$ FePt			A1 FePt	Grain Size $L1_0$ FePt (nm)	Grain Size A1 FePt (nm)
		a (Å)	c (Å)	c/a	a (Å)		
$Fe_{65}Pt_{15}Zr_3B_{17}$	As-cast	-	-	-	3.8411 ± 0.02416	-	3.8 ± 1.4
	500 °C	3.8542 ± 0.0043	3.7127 ± 0.0067	0.9632	3.8315 ± 0.0308	15 ± 2	17 ± 3
	600 °C	3.8537 ± 0.0027	3.7134 ± 0.0042	0.9636	3.8268 ± 0.0284	30 ± 3	36 ± 4
	700 °C	3.8524 ± 0.0049	3.7147 ± 0.0071	0.9642	3.8379 ± 0.0247	46 ± 5	49 ± 5
	800 °C	3.8516 ± 0.0052	3.7166 ± 0.0051	0.9649	3.8292 ± 0.0303	57 ± 3	54 ± 3

The grain size for both $L1_0$ and A1 FePt phases in the annealed samples increases continuously from the as-cast values 15 nm (17 nm respectively) as the annealing temperature increases. At 800 °C annealing, the grain size in the $L1_0$ FePt phase is about 54 nm, while for the same annealing in the A1 FePt phase the average grain size is estimated to be around 57 nm. In conclusion, in the annealed samples, using structural data, we have unambiguously proven the coexistence of the hard and soft magnetic phases with well-refined grain sizes. These observations may be made quantitative with the help of the Mössbauer spectroscopy, where the relative fraction of each of the phases present in the samples can be obtained.

3.2.2. Mössbauer Analysis

^{57}Fe Mössbauer analysis on all the samples has been performed using a conventional setup in transmission geometry. The Mössbauer source we used was a ^{57}Co source embedded in a Rh matrix. Measurements were carried out at an ambient temperature. The annealed $Fe_{65}Pt_{15}Zr_3B_{17}$ samples have been investigated and their Mössbauer spectra, recorded at −10 +10 mm/s velocity range, have been fitted, with a procedure involving several subspectra, magnetic sextets. All the hyperfine parameters, i.e, the hyperfine field (HF), quadrupole splitting (QS) and the isomer shift (IS), were kept as free parameters during the fitting. In Figure 6 the Mössbauer spectra of three of the annealed $Fe_{65}Pt_{15}Zr_3B_{17}$ samples are presented together with the subspectra, as identified from the fitting. All the hyperfine field parameters obtained from the fit are depicted in Table 2. The components used for the fitting are of the same color as their hyperfine parameters, for easier identification.

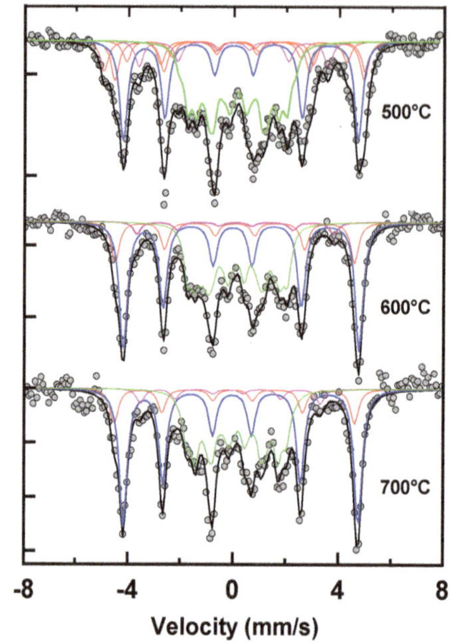

Figure 6. Mössbauer spectra of the $Fe_{65}Pt_{15}Zr_3B_{17}$ samples, annealed at 500 °C, 600 °C and 700 °C together with the fitted subspectra.

Table 2. The obtained R-factors, as resulted from the full-profile refinement XRD patterns of annealed samples. Profile Rp, weighted profile Rwp and goodness-of-fit χ^2 are shown.

Annealing T	Profile R-Factor R_p	Weighted Profile R_{wp}	Goodness-of-Fit χ^2
500 °C	4.89	2.11	1.37
600 °C	4.31	2.24	1.23
700 °C	3.78	2.19	1.31
800 °C	3.54	2.05	1.29

Mössbauer spectra of the annealed samples show different and more complex hyperfine features than the as-cast ones. Indeed, the Mössbauer patterns consist of sextets with much narrower lines, which indicate that the samples are crystallized. The fitting model chosen to adjust the spectra of annealed samples takes into account the XRD results where a microstructure consisting of multiple phases was proven. The spectra have been fitted with a number of subspectra that were indexed and assigned to the phases identified in the XRD.

The Mössbauer spectra of the sample annealed at 500 °C has been fitted with 5 subspectra, a fitting model that corresponds to the situation observed in the XRD data. The two subspectra designed by the red line in Figure 6 have been assigned to the A1 fcc FePt phase. The hyperfine parameters are depicted in Table 3. The quadrupole shift value suggests a cubic symmetry, while the hyperfine field of 30 T is typical for A1 FePt phase which is consistent with the results on the as-cast sample. The relative proportion of the fcc phase, as resulted from the fit, is 18%. The HF parameters obtained are consistent with the results reported in the literature [36] and with those obtained from the investigation on the as-cast sample. The subspectrum designed by the blue line in Figure 6 has been assigned to the $L1_0$ FePt phase. The quadrupole splitting of 0.27 mm/s suggests a tetragonal symmetry and the hyperfine field of 27.6 T corresponds to the value found for the $L1_0$ phase in the literature. For the annealing at 500 °C, this phase is the most abundant in the sample. Its relative proportion, as revealed from the fit, is 28%. This indicates that the disorder–order phase transition has started already at 500 °C and an important fraction of the initial fcc FePt solid solution is already transformed into tetragonal $L1_0$ FePt. The pink lines are the components of the boride phases. Compositionally, there is an Fe excess which does not crystallize in the FePt phases; therefore, in principle, during annealing there are conditions that Fe and B form borides, as there is a large affinity for these two elements to alloy. Fe_2B has been already observed in the sample annealed at 500 °C and Fe_3B is a metastable phase that may be formed in intermediate annealing temperatures. Their cumulate relative intensity amounts to 14%. The central part of the spectrum was fitted with an HF distribution between 3.5 and 18.5 T. These contributions are represented to be convoluted as one contribution (green) in Figure 6.

Table 3. Hyperfine parameters as resulted from fitting of the Mössbauer spectra (Figure 6).

Samples	IS (mm/s)	Γ/2 (mm/s)	QS (mm/s)	B_{hf} (T)	%	Phase
500 °C	0.21	0.18	−0.1	30.5	8	A1(FePt)
	0.30	0.18	0.1	29.2	10	A1(FePt)
	0.27	0.18	0.27	27.6	28	$L1_0$(FePt)
	0.23	0.18	0.1	25.8	7	Fe_3B
	0.12	0.18	0	22.1	7	Fe_2B
	0.26	0.18	0.03	9.9	40	Disordered Fe-Zr-B
600 °C	0.20	0.18	0	28	13	A1(FePt)
	0.28	0.18	0.33	27.6	43	$L1_0$(FePt)
	0.12	0.18	0	22.8	5	Fe_2B
	0.28	0.18	0.06	9.5	39	Disordered Fe-Zr-B
700 °C	0.17	0.18	0.06	28.1	12	A1(FePt)
	0.28	0.18	0.33	27.5	44	$L1_0$(FePt)
	0.11	0.18	0	22	6	Fe_2B
	0.29	0.18	0.1	9.5	38	Disordered Fe-Zr-B

The estimated errors are: ±0.02 mm/s for IS and QS/2ε, ±0.1 T for B_{hf} and ±1 for the relative proportion.

The average hyperfine field of these low-field contributions is 9.9 T and its relative intensity amounts to 40%. From the HF value we can assume that this low-field contribution to the Mössbauer spectrum may be attributed to the disordered Fe-Zr-B-rich phase. The quadrupole shift close to 0 sustains the argument of a cubic symmetry of this phase. The low hyperfine field of this contribution suggests that it is a Fe-poor phase. From all of the arguments we consider that the green contribution may be a disordered Fe-Zr-B phase with few Pt atoms interstitially inserted or randomly occupying the Fe sites. Most of this phase is probably located in interfacial regions between the nanocrystalline grains and the in-grain boundaries. We have to emphasize that the disordered phase, as fitted in the Mössbauer spectra, accounts for all Fe-distributed ionic chemical environments. For the sake of fitting consistency, every contribution that does not belong to the magnetic subspectra (sextets) is included in one disordered sublattice. In Mössbauer terms this is a hyperfine field distribution. Usually, a hyperfine field distribution is used to fit the Mössbauer spectra of Fe-disordered components or Fe-amorphous alloys. Therefore, the disordered part, illustrated by the green component, represents in fact the contribution of all of the chemically distributed Fe sites, each of them with different probabilities, as shown, for instance, in Figure 3. We have named it disordered Fe-Zr-B, to account in fact for all of the disordered Fe chemical environments. This is in agreement with other interpretations of Mössbauer spectra in Fe-Zr-B alloys [37].The Mössbauer spectra of the sample annealed at 600 °C and 700 °C have the same complex pattern as the previous one, but with better resolved lines. The spectra were fitted using the same fitting model. The parameters were consistent throughout the series of measurements. From Table 3 one can observe the phase evolution with the annealing conditions, as well as the relative abundance of the A1 and $L1_0$ FePt phases. The tetragonal phase increases slightly up to 44%, while Fe_3B decomposes. It can be seen that Fe_2B relative abundance does not increase, therefore we may assume that from the decomposition of Fe_3B, the obtained Fe helps the formation of $L1_0$ FePt and B is incorporated into the Fe-Zr-B-rich residual phase. With an increasing annealing temperature, all the phases that decompose help the formation of the $L1_0$ phase, which is the most ordered one, in the ribbon microstructure.

3.2.3. Magnetic Properties

The hysteresis loops for the $Fe_{65}Pt_{15}Zr_3B_{17}$ samples, as-cast and annealed at 600 °C and 700 °C, have been recorded at 300 K with a magnetic field applied parallel to the ribbon plane, and are plotted in Figure 7. The hysteresis loop for the as-cast ribbons is typical for soft ferromagnets. The magnetization saturates almost immediately after applying a small magnetic field. The saturation reaches 1.3 T, while the loop shows a very low hysteresis (around 30 mT). The annealed samples show, however, a completely different behavior. The magnetization has a slower approach to saturation than in the as-cast state. This may be explained by the presence of a significant fraction of hard $L1_0$ grains. The value of the maximum magnetization is about 15% lower than that of the as-cast sample, reaching about 0.9 T. Upon decreasing the field, a very large value of remanent magnetization is noticed, as is the case for hard magnetic materials. The sample annealed at 600 °C shows a large openness of the loop.

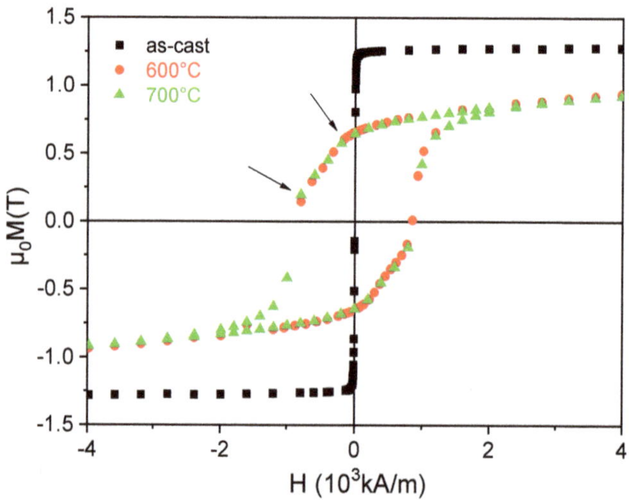

Figure 7. RT hysteresis loops for the as-cast and annealed samples.

The remanent magnetization is in this case 0.63 T, thus giving a remanence-to-saturation ratio of about 0.7. The shape of the hysteresis loop presents two vaguely observed inflection points (at around −190 and −840 kA/m applied field), typical for systems with both hard and soft magnetic phases in co-existence; however, here the hard magnetic phase is predominant and the exchange coupling between the hard and the soft magnetic phases is not complete. As a consequence, a quite strong coercive field is observed. The coercivity reaches in this case 847 kA/m, which is more than 3 times increased, compared with the annealing at 500 °C. The switching field of the soft magnetic phase is smaller than the coercive field, which indicates a better exchange coupling between the grains. The value of the switching field is 155 kA/m for the soft magnetic grains, increasing substantially compared with the annealing at 500 °C. The sample annealed at 700 °C presents an almost similar hysteresis loop as the sample annealed at 600 °C. While the coercivity is slightly higher than before (877 kA/m), the remanent and the maximum magnetization are almost identical to the previous annealing, with a remanent-to-saturation ratio of about 0.7. The energy product for all the samples was calculated using the B-H loops. The highest $(BH)_{max}$ values of 64 kJ/m^3 are obtained for the sample annealed at 700 °C. All the magnetic parameters obtained for these samples at 300 K in the parallel field are depicted in Table 4.

Table 4. Magnetic parameters obtained for annealed samples at 300 K in parallel applied field.

Annealing Conditions	Hc (kA/m)	$\mu_0 M_S$ (T)	$\mu_0 M_r$ (T)	M_r/M_s	$(BH)_{max}$ (kJ/m^3)
As-cast	2.94	1.26	1.2	0.95	-
500 °C	194	1.1	0.85	0.77	61
600 °C	847	0.9	0.63	0.70	64
700 °C	877	0.9	0.63	0.70	64

4. Conclusions

We have synthesized a nanocomposite magnet based on the FePt-Zr-B alloy, by using the non-equilibrium melt spinning method. The stoichiometry was chosen in such a way as to obtain a chemically disordered as-cast state. From this state, by appropriate annealing, we have been able to obtain a mixture of hard $L1_0$-FePt and soft A1 FePt and boride magnetic phases. The dynamics of formation of the hard magnetic $L1_0$ phase during the gradual disorder–order phase transformation has been monitored by using a complex combination of X-ray diffraction methods and ^{57}Fe Mössbauer spectroscopy methods over a large range of annealing temperatures. Multiple phases co-existing in the annealed sample microstructures, observed in XRD, have been reconfirmed by the Mössbauer spectra analysis and moreover, accurate quantitative data have been acquired in what concerns the relative abundance of each of the observed crystalline phases in every stage of annealing. It has been shown that the formation of the hard magnetic phase, emerging from the chemically disordered as-cast precursor, is gradual and occurs via complex mechanisms involving the presence of a disordered Fe-Zr-B-rich intergranular region, as well as the decomposition of iron borides, providing excess Fe, which contributes furthermore to the increase of the hard magnetic phase abundance, for higher annealing temperatures. Magnetic measurements have confirmed the good performances of these alloys in terms of coercivity and remanence, a maximum energy product of 64 kJ/m^3 being obtained for annealing at 700 °C. These findings open good perspectives for a future class of permanent magnets made from these alloys.

Author Contributions: Conceptualization, O.C.; methodology, I.D.; writing—original draft preparation, A.D.C.; writing—review and editing, O.C. All authors have read and agreed to the published version of the manuscript.

Funding: This research was funded through POC Project: "Materiale multifuncționale inteligente pentru aplicații de înaltă tehnologie (MATI2IT)" (code MySMIS 105726) funded by the Romanian Ministry of European Funds.

Institutional Review Board Statement: Not applicable.

Informed Consent Statement: Not applicable.

Data Availability Statement: The data are not publicly available due to IPR protection measures.

Conflicts of Interest: The authors declare no conflict of interest. The funders had no role in the design of the study; in the collection, analyses, or interpretation of data; in the writing of the manuscript, or in the decision to publish the results.

References

1. Lyubina, J.; Khlopkov, K.; Gutfleisch, O.; Müller, K.-H.; Schultz, L. Intergrain interactions in nanocomposite Fe–Pt powders. *J. Appl. Phys.* **2006**, *99*, 08E903. [CrossRef]
2. Crisan, A.D.; Crisan, O. Direct formation of $L1_0$ FePt in as-cast FePt-based magnetic nanocomposite ribbons without post-synthesis annealing. *J. Phys. D Appl. Phys.* **2011**, *44*, 365002. [CrossRef]
3. Von Haeften, K.; Binns, C.; Brewer, A.; Crisan, O.; Howes, P.B.; Lowe, M.P.; Sibbley-Allen, C.; Thornton, S.C. A novel approach towards the production of luminescent silicon nanoparticles: Sputtering, gas aggregation and co-deposition with H_2O. *Eur. Phys. J. D* **2009**, *52*, 11–14. [CrossRef]
4. Chrobak, A.; Ziółkowski, G.; Randrianantoandro, N.; Klimontko, J.; Chrobak, D.; Prusik, K.; Rak, J. Ultra-high coercivity of $(Fe_{86-x}Nb_xB_{14})_{0.88}Tb_{0.12}$ bulk nanocrystalline magnets. *Acta Mater.* **2015**, *98*, 318. [CrossRef]
5. Pietrusiewicz, P.; Nabialek, M.; Jez, B. Evolution of the structural and magnetic properties of bulk $Fe_{61}Co_{10}B_{20}W_1Y_{8-X}Pt_x$ alloys through the partial substitution of Pt for Y. *Materials* **2020**, *13*, 4962. [CrossRef]
6. Ma, D.G.; Wang, Y.M.; Li, Y.H.; Umetsu, R.Y.; Ou, S.L.; Yubuta, K.; Zhang, W. Structure and properties of nanoporous FePt fabricated by dealloying a melt-spun $Fe_{60}Pt_{20}B_{20}$ alloy and subsequent annealing. *J. Mater. Sci. Technol.* **2020**, *36*, 128–133. [CrossRef]
7. Grabias, A.; Kopcewicz, M.; Latuch, J.; Oleszak, D.; Pekala, M.; Kowalczyk, M. Influence of cobalt content on the structure and hard magnetic properties of nanocomposite (Fe,Co)-Pt-B. alloys. *J. Magn. Magn. Mater.* **2017**, *434*, 126–134. [CrossRef]
8. Tsai, J.L.; Huang, J.; Chen, L.H.; Lin, C.S. Magnetic properties and microstructure of exchange coupled FePt-FeB films. *Surf. Coat. Technol.* **2013**, *231*, 456–459. [CrossRef]
9. Choi, G.M.; Min, B.C.; Shin, K.H. $L1_0$ ordering of FePtB layers by oxidation-induced stress of capping layer. *Appl. Phys. A* **2013**, *111*, 389–394. [CrossRef]
10. Grabias, A.; Kopcewicz, M.; Oleszak, D.; Latuch, J.; Kowalczyk, M.; Pekala, M. Structural transformations and magnetic properties of $Fe_{60}Pt_{15}B_{25}$ and $Fe_{60}Pt_{25}B_{15}$ nanocomposite alloys. *J. Magn. Magn. Mater.* **2010**, *322*, 3137–3141. [CrossRef]
11. Tsai, J.L.; Huang, J.C.; Tai, H.W.; Tsai, W.C.; Lin, Y.C. Magnetic properties and microstructure of FePtB, FePt(B-Ag) granular films. *J. Magn. Magn. Mater.* **2013**, *329*, 6–13. [CrossRef]
12. Kaushik, N.; Sharma, P.; Nagar, S.; Rao, K.V.; Kimura, H.; Makino, A.; Inoue, A. Exchange-coupled FePtB nano-composite hard magnets produced by pulsed laser deposition. *Mater. Sci. Eng. B* **2010**, *171*, 62–68. [CrossRef]
13. Sharma, P.; Waki, J.; Kaushik, N.; Louzguine-Luzgin, D.V.; Kimura, H.; Inoue, A. High coercivity characteristics of FePtB exchange-coupled nanocomposite thick film spring magnets produced by sputtering. *Acta Mater.* **2007**, *55*, 4203–4212. [CrossRef]
14. Chang, C.W.; Chang, H.W.; Chiu, C.H.; Chen, C.H.; Chang, W.C.; Ouyang, H.; Chang, C.W.; Liu, C.C. The effect of annealing time on the magnetic properties and microstructure of $(Fe_{0.675}Pt_{0.325})_{84}B_{16}$ ribbons. *J. Magn. Magn. Mater.* **2006**, *310*, 2593–2595. [CrossRef]
15. Chang, C.W.; Chang, H.W.; Chiu, C.H.; Chang, W.C. Effect of boron on the magnetic properties and exchange-coupling effect of FePtB-type nanocomposite ribbons. *J. Appl. Phys.* **2005**, *97*, 10N117. [CrossRef]
16. Li, N.; Lairson, B.M. Magnetic recording on FePt and FePtB intermetallic compound media. *IEEE Trans. Magn.* **1999**, *35*, 1077–1082.
17. Crisan, A.D.; Nicula, R.; Crisan, O.; Burkel, E. Thermally and pressure activated phase evolution in Fe–Pt–Nb–B melt spun ribbons. *Mater. Sci. Eng. C* **2007**, *27*, 1280–1282. [CrossRef]
18. Crisan, O.; Crisan, A.; Randrianantoandro, D.N.; Nicula, R.; Burkel, E. Crystallization processes and phase evolution in amorphous Fe–Pt–Nb–B alloys. *J. Alloys Compd.* **2007**, *440*, L3–L7. [CrossRef]
19. Crisan, A.D.; Crisan, O.; Randrianantoandro, N.; Valeanu, M.; Morariu, M.; Burkel, E. Crystallization processes in Fe–Pt–Nb–B melt spun ribbons. *Mater. Sci. Eng. C* **2007**, *27*, 1283–1285. [CrossRef]
20. Bruck, E.; Xiao, Q.F.; Thang, P.D.; Toonen, M.J.; de Boer, F.R.; Buschow, K.H.J. Influence of phase transformation on the permanent-magnetic properties of Fe–Pt based alloys. *Physica B* **2001**, *300*, 215–229. [CrossRef]
21. Makino, A.; Bitoh, T. High coercivity of melt-spun $(Fe_{0.55}Pt_{0.45})_{78}Zr_{2-4}B_{18-20}$ nanocrystalline alloys with $L1_0$ structure. *J. Appl. Phys.* **2004**, *95*, 7498–7500. [CrossRef]
22. Makino, A.; Bitoh, T.; Inoue, A.; Hirotu, Y. Magnetic properties and structure of Fe–Pt–M–B (M = Zr, Nb and Ti) alloys produced by quenching technique. *J. Alloys Compd.* **2007**, *434*, 614–617. [CrossRef]
23. Randrianantoandro, N.; Crisan, A.D.; Crisan, O.; Marcin, J.; Kovac, J.; Hanko, J.; Greneche, J.M.; Svec, P.; Chrobak, A.; Skorvanek, I. The influence of microstructure on magnetic properties of nanocrystalline Fe–Pt–Nb–B permanent magnet ribbons. *J. Appl. Phys.* **2010**, *108*, 093910. [CrossRef]
24. Wang, S.; Kang, S.S.; Nikles, D.E.; Harrell, J.W.; Wu, X.W. Magnetic properties of self-organized $L1_0$ FePtAg nanoparticle arrays. *J. Magn. Magn. Mater.* **2003**, *266*, 49–56. [CrossRef]
25. Seqqat, M.; Nogues, M.; Crisan, O.; Kuncser, V.; Cristea, L.; Jianu, A.; Filoti, G.; Dormann, J.L.; Sayah, D.; Godinho, M. Magnetic properties of $Fe_{100-x}Sm_x$ thin films and $Fe_{80-x}Sm_xB_{20}$ thin films and ribbons. *J. Magn. Magn. Mater.* **1996**, *157*, 225. [CrossRef]
26. Crisan, O.; Angelakeris, M.; Flevaris, N.K.; Filoti, G. Magnetism and Anisotropy in Core-Shell Nanoparticles. *J. Optoelectron. Adv. Mater.* **2003**, *5*, 959.

27. Crisan, O.; Labaye, Y.; Berger, L.; Greneche, J.M.; Coey, M.D. Exchange coupling effects in nanocrystalline alloys studied by Monte Carlo simulation. *J. Appl. Phys.* **2002**, *91*, 8727. [CrossRef]
28. Gonzalez, J.A.; Andres, J.P.; De Toro, J.A.; Muniz, P.; Munoz, T.; Crisan, O.; Binns, C.; Riveiro, J.M. Co–CoO nanoparticles prepared by reactive gas-phase aggregation. *J. Nanopart. Res.* **2009**, *11*, 2105–2111. [CrossRef]
29. Reddy, V.R.; Crisan, O.; Gupta, A.; Kuncser, V. Tuning exchange spring effects in FePt/Fe(Co) magnetic bilayers. *Thin Solid Film.* **2012**, *520*, 2184. [CrossRef]
30. Crisan, A.D. Compositional studies and thermal analysis in amorphous and nanocrystalline FePtNbB melt spun ribbons. *J. Optoelectron. Adv. Mater.* **2010**, *12*, 250–256.
31. Rosenberg, M.; Kuncser, V.; Crisan, O.; Hernando, A.; Navarro, E.; Filoti, G. A Mössbauer spectroscopy and magnetic study of FeRh. *J. Magn. Magn. Mater.* **1998**, *177*, 135–136. [CrossRef]
32. Iwata, S.; Yamashita, S.; Tsunashima, S. Perpendicular magnetic anisotropy and magneto-optical Kerr spectra of MBE-grown PtCo alloy films. *IEEE Trans. Magn.* **1997**, *33*, 3670. [CrossRef]
33. Chang, C.W.; Chang, H.W.; Chiu, C.H.; Hsieh, C.C.; Fang, Y.K.; Chang, W.C. Magnetic property improvement of Pt-lean FePt/Fe–B-type nanocomposites by Co substitution. *J. Appl. Phys.* **2008**, *103*, 07E133. [CrossRef]
34. Makino, A.; Bitoh, T.; Nakagawa, M. Direct synthesis of $L1_0$-(Fe,Co)Pt nanocrystallites from (Fe,Co)–Pt–Zr–B liquid phase by melt-spinning. *J. Non-Cryst. Sol.* **2007**, *353*, 3655–3660. [CrossRef]
35. Inoue, A.; Zhang, W. Nanocrystalline Fe-Pt-B base hard magnets with high coercive force obtained from amorphous precursor. *J. Appl. Phys.* **2005**, *97*, 10H308. [CrossRef]
36. Goto, T.; Utsugi, H.; Watanabe, K. Mössbauer study of permanent magnets Fe-Pt. *Hyperfine Interact.* **1990**, *54*, 539. [CrossRef]
37. BBrzozka, K.; Slawska-Waniewska, A.; Jezuita, K. Mössbauer studies of FeZrB(Cu) amorphous alloys. *J. Magn. Magn. Mater.* **1996**, *160*, 255–256. [CrossRef]

magnetochemistry

MDPI

Article

Electric Field Control of Magnetic Properties by Means of Li^+ Migration in FeRh Thin Film

Gengfei Li [1,2], Yali Xie [1,2,*], Baomin Wang [1,2], Huali Yang [1,2] and Run-Wei Li [1,2,3]

1 CAS Key Laboratory of Magnetic Materials and Devices, Ningbo Institute of Materials Technology and Engineering, Chinese Academy of Sciences, Ningbo 315201, China; ligengfei@nimte.ac.cn (G.L.); wangbaomin@nimte.ac.cn (B.W.); yanghl@nimte.ac.cn (H.Y.); runweili@nimte.ac.cn (R.-W.L.)
2 Zhejiang Province Key Laboratory of Magnetic Materials and Application Technology, Ningbo Institute of Materials Technology and Engineering, Chinese Academy of Sciences, Ningbo 315201, China
3 Center of Materials Science and Optoelectronics Engineering, University of Chinese Academy of Sciences, Beijing 100049, China
* Correspondence: xieyl@nimte.ac.cn

Abstract: Recently, the electric control of magnetism by means of ion migration has been proven to be effective with nonvolatility and low energy consumption. In this work, we investigated the control of the magnetic properties of FeRh films by means of Li^+ migration in FeRh/MgO heterostructures. We found that the migration of Li^+ could reduce the phase transition temperature by 2 K with an applied voltage of 1 V. Meanwhile, the voltage-dependent saturated magnetization exhibited a repetitive switching behavior from high to low magnetization values while the voltage was switched from 4 to −4 V, indicating that the migration of Li^+ in the FeRh film can be reversible. This provides a means to control the magnetic properties of FeRh films.

Keywords: electric control of magnetic phase transition; FeRh; Li^+ migration

Citation: Li, G.; Xie, Y.; Wang, B.; Yang, H.; Li, R.-W. Electric Field Control of Magnetic Properties by Means of Li^+ Migration in FeRh Thin Film. *Magnetochemistry* **2021**, *7*, 45. https://doi.org/10.3390/magnetochemistry7040045

Academic Editors: Cătălin-Daniel Constantinescu and Lucian Petrescu

Received: 1 March 2021
Accepted: 25 March 2021
Published: 26 March 2021

1. Introduction

Fallot et al. discovered that the CsCl-type FeRh alloy undergoes a first-order phase transition from an antiferromagnetic (AF) to ferromagnetic (FM) phase at around room temperature accompanied by an approximately 1% volume expansion of the crystal lattice [1]. Due to this magnetic transition, FeRh has attracted extensive attention for its potential applications in heat-assisted magnetic recording, AF memory, and magnetic refrigeration [2–7]. In order to realize its applications under different conditions, many studies have been conducted to tune the magnetic properties of the FeRh film by means of the strain, magnetic field, electric field, etc. [8]. However, the manipulation of FeRh properties by means such as the strain and magnetic field is difficult to implement in spintronics devices, limiting its practical applications [9,10]. In recent years, the electric control of magnetism has been proven to be effective with low energy consumption and repeatability [11–16]. For example, Cherifi et al. found that a moderate electric field can produce a large magnetization variation in FeRh/$BaTiO_3$ [14]. Lee et al. reported a large resistivity modulation by applying an electric field to FeRh/PMN-PT heterostructures [16]. However, these methods usually require a high-quality interface or very thin film. Recently, the electric control of magnetism by means of ion migration has been proven to be effective with nonvolatility and low energy consumption [17–19]. For instance, Dhanapal et al. demonstrated reversibly controlled magnetic domains of Co via electric-field driven oxygen migration at the nanoscale [18]. In this work, we investigated the electric control of magnetic properties of FeRh films by means of Li^+ migration in FeRh/MgO heterostructures. We found that this method can not only manipulate the phase transition temperature, but also change the saturated magnetization of FeRh film.

2. Experiment

First, 30 nm thick FeRh films were deposited by sputtering from an FeRh alloy target using a DC power of 100 W and an argon pressure of 5 mTorr. The (100)-oriented MgO substrates were preheated to 530 °C for 1 h in a vacuum chamber and then held at this temperature during deposition. After growth, the samples were annealed at 700 °C for about 1 h at a base pressure below 1.0×10^{-5} Pa in order to obtain the chemically ordered CsCl-type FeRh. The film thicknesses were calibrated by X-ray reflectivity (XRR) and the rate of growth was 5 nm/min. The surface topographies of FeRh films were measured using a Bruker Icon atomic force microscope. The crystal structure was characterized using a Bruker Discover X-ray diffractometer. The temperature-dependent magnetizations of the FeRh films were determined using a superconducting quantum interference device (SQUID) magnetometer. The electric field was applied using a commercial three-electrode experimental setup [20]. In this work, FeRh/MgO, a standard calomel electrode, and a platinum sheet were used as the working electrode, reference electrode, and counter electrode, respectively. In addition, 0.1 M of $LiClO_4$ dissolved in propylene carbonate was used as the electrolytic solution. When voltage was applied, the Li^+ could migrate into or out of the FeRh films. When we performed magnetic measurements, the FeRh/MgO sample was taken out of the experimental setup and the electrode setup was also removed.

3. Results and Discussion

An AFM image of the FeRh films is displayed in Figure 1a, which indicates that the roughness of the film was very small (~0.3 nm). Figure 1b shows the XRD θ–2θ pattern of the FeRh film. As seen in Figure 1b, only the reflections at 29.90° and 62.11° could be observed, which demonstrates that the obtained FeRh film had the required CsCl-type structure. The X-ray φ-scan of FeRh/MgO in Figure 1c indicates that the FeRh films were epitaxially grown with a 45° in-plane lattice rotation with respect to the MgO(001) substrates. The temperature-dependent magnetization of the FeRh film measured with an in-plane magnetic field of 2 kOe is plotted in Figure 1d. When warming up, the magnetization steeply increased from 100 to 1.1×10^6 A/m at around 350 K, exhibiting a typical characteristic of the transition from the AF to the FM phase in FeRh. By contrast, the magnetization of the FeRh film gradually decreased to the original value of the AF state in the cooling process. A temperature hysteresis in the magnetization of about 25 K could be observed, indicating the first-order phase transition of FeRh.

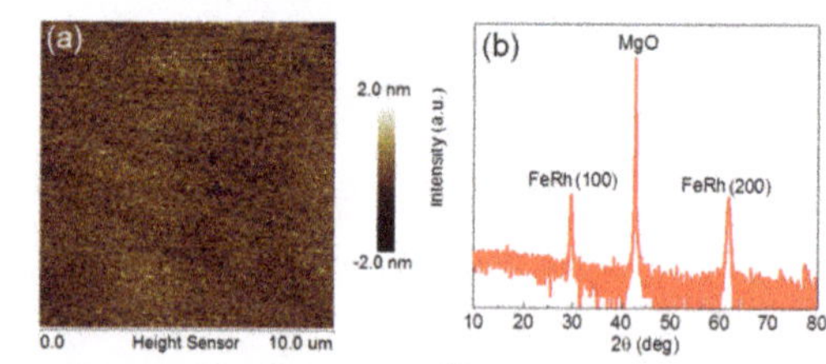

Figure 1. (**a**) The AFM image of the FeRh films. (**b**) θ-2θ scan and (**c**) φ-scans of FeRh/MgO. (**d**) The temperature-dependent magnetization of the FeRh film measured with an in-plane magnetic field of 2 kOe.

Figure 2a shows the temperature-dependent magnetization of the FeRh film at 2 kOe without and with applied voltages of 1 V through the electrochemical method. It can be seen that with an applied voltage of 1 V, the behavior of the temperature-dependent magnetization was almost unchanged compared with that for 0 V. Meanwhile, the phase transition shifted to the lower temperature after the applied voltage of 1 V. The voltage-dependent critical phase transition temperatures of FeRh, which were defined as $T_{AFM\text{-}FM}$ on heating and $T_{FM\text{-}AFM}$ on cooling processes, are plotted in Figure 2b. $T_{AFM\text{-}FM}$ exhibited a decrease of about 2 K with an applied voltage of 1 V, indicating that the Li^+ migrated into the FeRh film. However, no further change was observed after increasing the applied voltages. $T_{FM\text{-}AFM}$ behaved similarly to that of $T_{FM\text{-}AFM}$. This kind of behavior may be ascribed to the small size of Li^+. When voltage was applied, the Li^+ could migrate into the FeRh film and the lattice volume could expand, which have been confirmed in our previous work [21]. The magnetization and phase transition temperature of FeRh are strongly dependent on the lattice parameter [22–24]. Consequently, the $T_{AFM\text{-}FM}$ and $T_{FM\text{-}AFM}$ change due to the migration of Li^+. However, the lattice volume cannot change by very much, due to the small size of Li^+, leading to a modulation of about 2 K in $T_{AFM\text{-}FM}$ and $T_{FM\text{-}AFM}$ when a voltage of 1 V was applied, as well as an easy saturation.

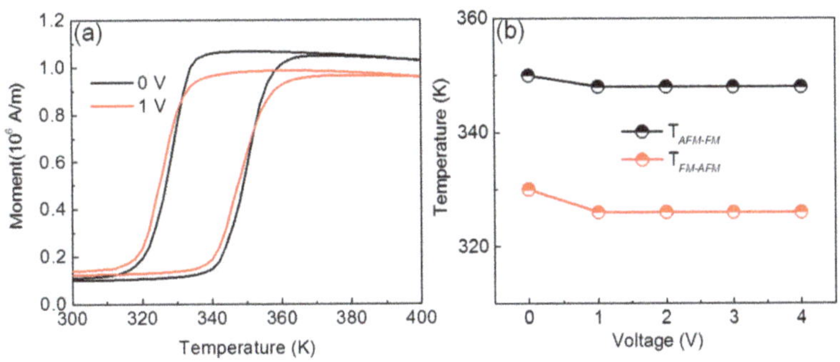

Figure 2. (**a**) Temperature-dependent magnetization without and with an applied voltage of 1 V for the FeRh thin film at 2 kOe. (**b**) Voltage-dependent $T_{AFM\text{-}FM}$ and $T_{FM\text{-}AFM}$ of FeRh on heating and cooling processes.

Figure 3a shows the reversibility of the temperature-dependent magnetization of the FeRh film measured with applied voltages of 4 and −4 V. The applied magnetic field was 2 kOe. It can be seen that when 4 V voltages were applied, the saturated magnetization of the first temperature-dependent magnetization (1) at 400 K could reach about 0.8×10^6 A/m. Meanwhile, when voltages of −4 V were applied, the saturated magnetization of the first temperature-dependent magnetization (2) at 400 K decreased to about 0.7×10^6 A/m, indicating that applying an electric field of −4 V can lead to the decrease in the saturated magnetization compared with that for a applied voltages of 4 V. When applied voltages of 4 and −4 V were applied again, the change in the saturated magnetization (3 and 4) was repeated. This kind of behavior can be seen more clearly in the inset of Figure 3a. It can also be ascribed to the migration of Li^+. When 4 V voltages were applied, the Li^+ can migrate into the FeRh film. However, when voltages of −4 V were applied, the Li^+ migrated out of the FeRh film, leading to a decrease in the saturated magnetization. The saturated magnetizations under different voltages are summarized in Figure 3b. It can be seen that the voltage-dependent saturated magnetization exhibited a repetitive switching behavior from high to low magnetization values when switching the voltage from 4 to −4 V repetitively, indicating that the migration of Li^+ in FeRh film can be reversible.

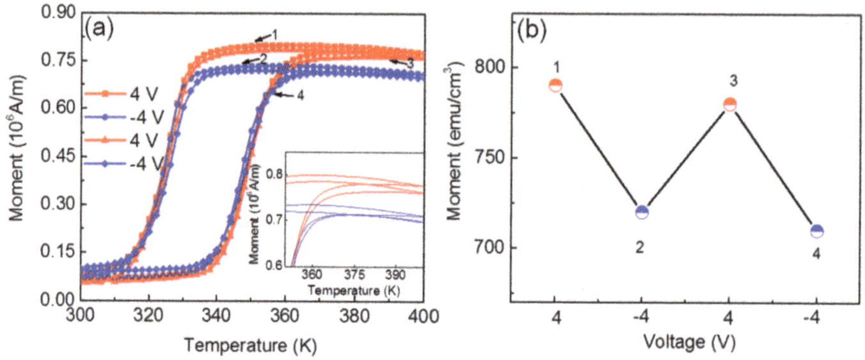

Figure 3. (**a**) Temperature-dependent magnetization with applied voltages of 4 and −4 V for FeRh thin films. The inset shows the enlarged figure of the temperature-dependent magnetization. (**b**) Voltage dependence of saturated magnetization measured with a magnetic field of 2 kOe in FeRh thin films.

4. Conclusions

In summary, the magnetic properties of FeRh films can be controlled by means of electrochemistry in FeRh/MgO heterostructures. Both the phase transition temperature and saturated magnetization of FeRh films can be modulated by varying the voltage. This kind of modulated behavior can be ascribed to the migration of Li^+ ions by applying voltage. This work provides a new method to control the magnetic properties of FeRh films.

Author Contributions: Conceptualization, Y.X. and B.W.; Methodology, Y.X. and B.W.; Investigation, G.L., Y.X., H.Y.; Writing—Original Draft Preparation, G.L.; Writing—Review and Editing, Y.X. and B.W.; Supervision, B.W.; Funding Acquisition, Y.X., B.W. and R.-W.L. All authors have read and agreed to the published version of the manuscript.

Funding: This work was supported by a National Key Technologies R&D Program of China (2016YFA0201102), the National Natural Science Foundation of China (51871233, 51871232, 51931011, and 51525103), the Youth Innovation Promotion Association of the Chinese Academy of Sciences (2019299), the Ningbo Science and Technology Bureau (2018B10060), and the Ningbo Natural Science Foundation (2019A610054, 2019A610051).

Institutional Review Board Statement: Not applicable.

Informed Consent Statement: Not applicable.

Data Availability Statement: All data generated or analyzed during this study are included in this article.

Conflicts of Interest: The authors declare no conflict of interest.

References

1. Fallot, M.; Horcart, R. Sur l'apparition du ferromagnétisme par élévation de température dans des alliages de fer et de rhodium. *Rev. Sci.* **1939**, *77*, 498.
2. Kouvel, J.S.; Hartelius, C.C. Anomalous Magnetic Moment and Transformations in the Ordered Alloy FeRh. *J. Appl. Phys.* **1962**, *33*, 1343. [CrossRef]
3. Zakharov, A.I.; Kadomtseva, A.M.; Levitin, R.Z.; Ponyatovskii, E.G. Magnetic and magnetoelastic properties of a metamagnetic iron–rhodium alloy. *Sov. Phys. JETP* **1964**, *19*, 1348.
4. Thiele, J.-U.; Maat, S.; Fullerton, E.E. FeRh/FePt exchange spring films for thermally assisted magnetic recording media. *Appl. Phys. Lett.* **2003**, *82*, 2859. [CrossRef]
5. Marti, X.; Fina, I.; Frontera, C.; Liu, J.; Wadley, P.; He, Q.; Paull, R.J.; Clarkson, J.D.; Kudrnovský, J.; Turek, I.; et al. Room-temperature antiferromagnetic memory resistor. *Nat. Mater.* **2014**, *13*, 367. [CrossRef]
6. Liu, Y.; Phillips, L.C.; Mattana, R.; Bibes, M.; Barthelemy, A.; Dkhil, B. Large reversible caloric effect in FeRh thin films via a dual-stimulus multicaloric cycle. *Nat. Commun.* **2016**, *7*, 11614. [CrossRef]

7. Maat, S.; Thiele, J.-U.; Fullerton, E.E. Temperature and field hysteresis of the antiferromagnetic-to-ferromagnetic phase transition in epitaxial FeRh films. *Phys. Rev. B* **2005**, *72*, 214432. [CrossRef]
8. Song, C.; Cui, B.; Li, F.; Zhou, X.; Pan, F. Recent progress in voltage control of magnetism: Materials, mechanisms, and performance. *Prog. Mater. Sci.* **2017**, *87*, 33. [CrossRef]
9. Suzuki, I.; Itoh, M.; Taniyama, T. Elastically controlled magnetic phase transition in Ga-FeRh/$BaTiO_3$(001) heterostructure. *Appl. Phys. Lett.* **2014**, *104*, 022401. [CrossRef]
10. Barua, R.; Jimenez-Villacorta, F.; Lewis, L.H. Predicting magnetostructural trends in FeRh based ternary systems. *Appl. Phys. Lett.* **2013**, *103*, 102407. [CrossRef]
11. Xuan, H.C.; Wang, L.Y.; Zheng, Y.X.; Li, Y.L.; Cao, Q.Q.; Chen, S.Y.; Wang, D.H.; Huang, Z.G.; Du, Y.W. Electric field control of magnetism without magnetic bias field in the Ni/ $Pb(Mg_{1/3}Nb_{2/3})O_3$ -$PbTiO_3$ /Ni composite. *Appl. Phys. Lett.* **2011**, *99*, 032509. [CrossRef]
12. Weiler, M.; Brandlmaier, A.; Geprags, S.; Althammer, M.; Opel, M.; Bihler, C.; Huebl, H.; Brandt, M.S.; Gross, R.; Goennenwein, S.T.B. Voltage controlled inversion of magnetic anisotropy in a ferromagnetic thin film at room temperature. *New J. Phys.* **2009**, *11*, 013021. [CrossRef]
13. Skumryev, V.; Laukhin, V.; Fina, I.; Martí, X.; Sánchez, F.; Gospodinov, M.; Fontcuberta, J. Magnetization Reversal by Electric-Field Decoupling of Magnetic and Ferroelectric Domain Walls in Multiferroic-Based Heterostructures. *Phys. Rev. Lett.* **2011**, *106*, 057206. [CrossRef] [PubMed]
14. Cherifi, R.O.; Ivanovskaya, V.; Phillips, L.C.; Zobelli, A.; Infante, I.C.; Jacquet, E.; Garcia, V.; Fusil, S.; Briddon, P.R.; Guiblin, N.; et al. Electric-field control of magnetic order above room temperature. *Nat. Mater.* **2014**, *13*, 345. [CrossRef]
15. Xie, Y.L.; Zhan, Q.F.; Shang, T.; Yang, H.L.; Liu, Y.W.; Wang, B.M.; Li, R.W. Electric field control of magnetic properties in FeRh/PMN-PT heterostructures. *AIP Adv.* **2018**, *8*, 055816. [CrossRef]
16. Lee, Y.; Liu, Z.Q.; Heron, J.T.; Clarkson, J.D.; Hong, J.; Ko, C.; Biegalski, M.D.; Aschauer, U.; Hsu, S.L.; Nowakowski, M.E.; et al. Large resistivity modulation in mixed-phase metallic systems. *Nat. Commun.* **2015**, *6*, 5959. [CrossRef]
17. Jiang, M.; Chen, X.Z.; Zhou, X.J.; Cui, B.; Yan, Y.N.; Wu, H.Q.; Pan, F.; Song, C. Electrochemical control of the phase transition of ultrathin FeRh films. *Appl. Phys. Lett.* **2016**, *108*, 202404. [CrossRef]
18. Pravarthana, D.; Zhang, T.; Wang, B.M.; Yang, H.L.; Xuan, H.C.; Bi, C.; Wang, W.G.; Li, R.W. Reversibly controlled magnetic domains of Co film via electric field driven oxygen migration at nanoscale. *Appl. Phys. Lett.* **2019**, *114*, 232401.
19. Pravarthana, D.; Wang, B.M.; Mustafa, Z.; Agarwal, S.; Pei, K.; Yang, H.L.; Li, R.W. Reversible Control of Magnetic Anisotropy and Magnetization in Amorphous Co40Fe40B20 Thin Films via All-Solid-State Li-ion Redox Capacitor. *Phys. Rev. Appl.* **2019**, *12*, 054065. [CrossRef]
20. Wu, J.; Qiu, D.; Zhang, H.L.; Cao, H.T.; Wang, W.; Liu, Z.P.; Tian, T.; Liang, L.Y.; Gao, J.H.; Zhuge, F. Flexible Electrochromic V2O5 Thin Films with Ultrahigh Coloration Efficiency on Graphene Electrodes. *J. Electrochem. Soc.* **2018**, *165*, D183. [CrossRef]
21. Mustafa, Z.; Pravarthana, D.; Wang, B.M.; Yang, H.L.; Li, R.W. Manipulation of Exchange Bias Effect via All-Solid-State Li-Ion Redox Capacitor with Antiferromagnetic Electrode. *Phys. Rev. Appl.* **2020**, *14*, 014062. [CrossRef]
22. Han, G.C.; Qiu, J.J.; Yap, Q.J.; Luo, P.; Kanbe, T.; Shige, T.; Laughlin, D.E.; Zhu, J.G. Suppression of low-temperature ferromagnetic phase in ultrathin FeRh films. *J. Appl. Phys.* **2013**, *113*, 123909. [CrossRef]
23. Xie, Y.L.; Zhan, Q.F.; Shang, T.; Yang, H.L.; Wang, B.M.; Tang, J.; Li, R.W. Effect of epitaxial strain and lattice mismatch on magnetic and transport behaviors in metamagnetic FeRh thin films. *AIP Adv.* **2017**, *7*, 056314. [CrossRef]
24. Phillips, L.C.; Cherifi, R.O.; Ivanovskaya, V.; Zobelli, A.; Infante, I.C.; Jacquet, E.; Guiblin, N.; Ünal, A.A.; Kronast, F.; Dkhil, B.; et al. Local electrical control of magnetic order and orientation by ferroelastic domain arrangements just above room temperature. *Sci. Rep.* **2015**, *5*, 10026. [CrossRef] [PubMed]

magnetochemistry

Article

Feasibility Study of Cooling a Bulk Acoustic Wave Resonator by Nanoparticle Enhanced Phase Change Material

Mohammad Yaghoub Abdollahzadeh Jamalabadi

Chabahar Maritime University, Chabahar 99717-56499, Iran; my.abdollahzadeh@cmu.ac.ir

Abstract: In the current study, the coupling of a cooling problem with the electromagnetic resonance of a bulk acoustic wave (BAW) material is investigated. As well, a new cooling method by the addition of nanoparticles to a phase change material surrounding the BAW resonator is presented. To solve the governing equations of piezoelectric charge and momentum balance, thermal balance, and fluid flow a code with the method of finite element is introduced. After validation of various features of the code with melting profile, heat generation, charge curve, and dispersion curve with benchmarks, the eigenfrequency analysis of the system is done. The thermal behavior of the system at first mode and various boundary conditions are studied. As well, the effect of nanoparticles in fastening the cooling of the BAW resonator is demonstrated.

Keywords: bulk acoustic waves; filter design; thermal management; MEMS; phase change material; nanofluid; electronic packaging

Citation: Abdollahzadeh Jamalabadi, M.Y. Feasibility Study of Cooling a Bulk Acoustic Wave Resonator by Nanoparticle Enhanced Phase Change Material. *Magnetochemistry* **2021**, *7*, 144. https://doi.org/10.3390/magnetochemistry7110144

Academic Editor: Cătălin-Daniel Constantinescu and Lucian Petrescu

Received: 6 September 2021
Accepted: 15 October 2021
Published: 28 October 2021

1. Introduction

Nanotechnology could be a technique that searches about the behavior of materials on a nanometric scale to improve their abilities [1–5]. Nanomaterials have many industrial and engineering applications. One of the current challenges related to nanotechnology lies in how nanomaterials are produced [2]. In quickly creating a universe of IoT, the RFFE of shrewd gadgets should deal with higher information rates and access the full data transmission of 4G/5G remote innovation. The higher bandwidth and data rate of 5G requires smarter technology. The explanation behind this is the developing requests of pervasive low dormancy information (data should transfer fast) at higher working frequencies needed to oblige improved information transmission abilities and quickly develop quantities of clients [3]. Usual electro-acoustic resonators utilized for RF channel applications depend on two strategies for the excitation and proficient energy catching of an acoustic wave (1200 to 4000 m/s) proliferating in a piezoelectric material. The first strategy and technique depends on SAW, where the surface acoustic wave is created on a piezoelectric substrate by metal IDTs by all accounts. The second strategy and technique depends on BAW, where the mass acoustic wave in the bulk of the material is energized by the use of an electric field through terminals above and under a dainty piezoelectric plate [4]. Key execution markers and quality index for resonator configuration are frequency of resonance, quality factor, coupling coefficient, and impedance for RF channel applications [5].

A bulk acoustic wave (BAW) device is usually composed of a piezoelectric material (i.e., zinc oxide or aluminum nitride) between two electrodes [6–10]. The BAW devices are currently used in Wi-fi systems, wireless positioning systems, cell phones, and satellite navigation. The range of working frequency (10^8 Hz to 2×10^{10} Hz) benefits the 6G technologies in small sizes (10^{-6} m) [7]. For acoustic isolation of the BAW device from the surrounding or substrate, two methods usually are used: air cavity (free-standing) and (solidly mounted). The free-standing method is cheaper than using Bragg reflector [8]. Such acoustic mirrors usually used a series of coupled low and high acoustic impedance [9]. The thickness of the Bragg reflector is equal to the quarter wavelength for maximum acoustic reflectivity [10].

Many numerical methods are used in the layout design of the BAW devices to reduce the production costs and increase the accuracy of the device for requested resonance modes [11]. The network of BAW resonators makes a frequency filter that protects the electronic device from unwanted frequencies. When electrostatic discharge can occur in the system, in case of high power requests, lower insertion loss requests, high-frequency requests, ($f > 1.5\ 2.5 \times 10^9$ Hz), or in case of steep stopband attenuation requests, the SAW device is replaced by a BAW device [12]. The design of a BAW device requires a scattering profile, impedance, stage, and Q-factor [13]. The scattering profile distinguishes all the potential modes, energy-related with every mode, the get over between the primary method of interest and other plate modes when the BAW resonator is energized. The impedance and stage contour measure the non-idealities in the BAW resonator plan and the general effect on the Q-values [14]. BAW resonator-based circuits require energy control in the piezoelectric layer dependent on the ideal choice of Bragg reflector layer, piezoelectric film, and terminal thickness.

Figure 1 shows the BAW device geometry which characterizes the ideal resilience on the piezoelectric film and electrodes. The thickness of cathode and anode could be different [15]. That thicknesses will affect the frequency resonances of the system. Five significant decisions in the plan and advancement of high-quality acoustic wave resonator are numbered in the list as follows:

1. Material;
2. Geometry;
3. Mechanical stability;
4. Pad between electrode and transmission line;
5. Temperature expansion coefficient.

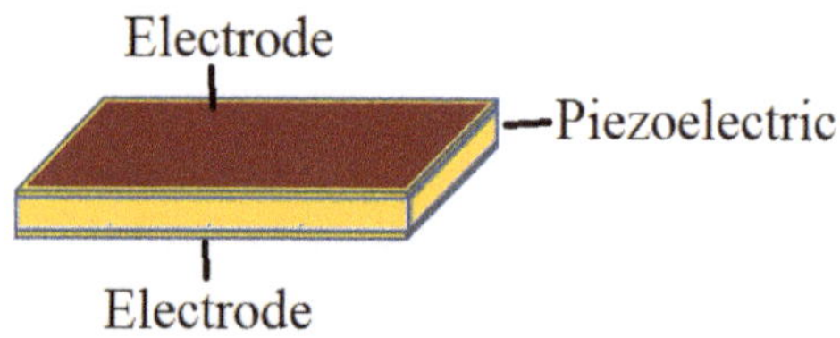

Figure 1. Schematic of piezoelectric and electrodes in a BAW configuration.

The solution of governing equations is not straightforward. This is because of the intricacy of the piezoelectric conditions that depict the gadget and its stacking conditions, delivering scientific arrangements of the conditions for non-trivial 2D and 3D calculations [16]. The first challenge in the plan of thin-film resonators is the concealment of side resonances that can be energized around the recurrence of the ideal mode shapes. The BC applied by IDTs and electrodes can add many extra modes. Some tough layers are added at the end of fingers to control such effects. In any case, the anode measurements directed by electrical coordinating regularly are with the end goal that precise outcomes do not warrant 1D or 2D display. By 1D or 2D analysis, the effect of fingers and spurious modes or fake resonances cannot be detected [17]. Along these lines, one should examine the fake resonances through full 3D FEM calculations. Be that as it may, the solid advantage of FEM is in its natural ability to oblige convoluted calculations, various materials, piezoelectricity, and full precious stone anisotropy. Subsequently, just a few disentangling approximations are required, bringing about an exact and adaptable reproduction strategy [18].

Various nanoparticles such as Ag, TiO_2, CuO, ZnO, Fe_3O_4, Indium, SiO_2, SWCNTs, and MWCNTs, can be mixed by a base fluid throughout sonication to make some improvements in thermo-physical properties [9]. Usual PCMs in the industry are paraffin, wax, and poly-a-olefin [19]. In the process of sonication which takes from 30 min up to 24 h, some surfactants (such as ethanol and sodium dodecyl sulfate) are added for better mixing. In Figure 2, the geometry of the BAW resonator assisted by PCM box is presented.

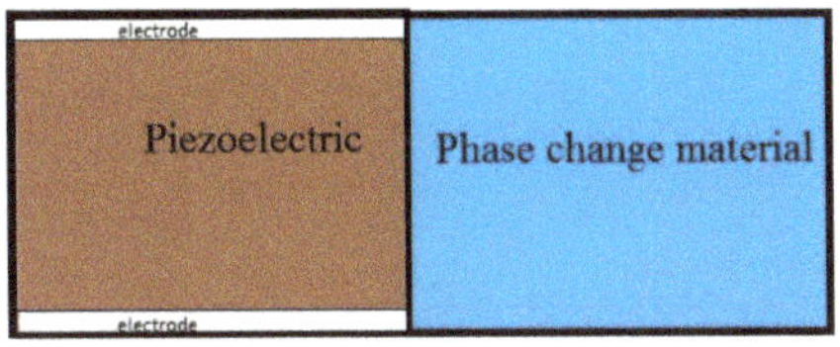

Figure 2. Geometry of the BAW resonator assisted by PCM box in 2D simulation.

As the thermal management and temperature compensation are important, the thermal design of BAW resonators is considered in its design [20]. Since the purpose of the current work is the use of NEPCM for cooling of BAW resonators in its design has significance in BAW development [21]. The main aim of the work is the investigation of nano-added PCM in thermal management of BAW which has the principal conclusions of durability and stability in electronic packaging application [22]. A novel design for PCM is a topic in BAW cooling [23]. Phase change of NEPCM could be used in BAW cooling [24].

Rational design calculations of surface acoustic wave devices were performed in many types of research [25]. However, the use of latent heat of phase change in a thermal energy storage systems especially PCM including nanoparticles for cooling on a small-scale is a new idea. In some batteries for temperature control purposes, the PCM and heat pipes are used throughout the discharge–charge cycle [26]. Property (temperature-dependent) variations in thermal energy storage and Brownian motion can affect freezing and melting problems in PCM. The big issues of thermal stability and compatibility within the context of materials are critical in latent heat storage systems. Before using for practical cases life-cycle assessment of phase change material is required. In such systems which are usually assisted by finned heat pipes (with highly conductive metal foams), high-temperature is a common problem [27]. Nowadays phase change materials are used as thermal energy storage for buildings. Thermal performance enhancement of NePCMs for solar energy heat exchangers and accumulators were tested. Since they can work with multiple phase change materials [28]. In this paper, a finite element method (FEM) code is developed for the mathematical arrangement of the electro–elastic conditions that oversee the directly constrained piezoelectric vibrations. Both methodologies, that of taking care of the field issue (consonant investigation) and that of taking care of the relating eigenvalue issue (modular examination), are portrayed. A FEM programming bundle has been made without any preparation. Significant angles fundamental to the proficient execution of FEM are clarified, for example, memory of the executives and tackling the summed up piezoelectric eigenvalue issue. Calculations for diminishing the necessary PC memory through enhancement of the framework profile, just as matrix calculation for the arrangement of the eigenfrequency issue are connected into the product from outer mathematical estimations. Current FEM programming is applied to to solve the mathematical governing equations of a thin-film mass BAW. Specifically, 3D reproductions are utilized to explore the impact of the top terminal shape. The legitimacy of the displayed strategy is shown by contrasting the recreated and estimated removal profiles at a few frequencies. The outcomes show that valuable data on the exhibition of the flimsy BAW acquired by generally large cross-sections and, subsequently, some assets. The novelty of the current study is that the NEPCM has not been used previously for cooling thin-film bulk acoustic wave resonance. The performance of that method of cooling and its side effects on electrical and mechanical parts are investigated in this paper.

2. Materials and Methods

In this section governing equations and boundary conditions of the system are presented.

2.1. Cooling Part

The governing equations of fluid, the non-slip boundary conditions of fluid on walls, and the boundary condition for thermal energy are presented in Tables 1–3 respectively. Enhancement of solidification rate of latent heat thermal energy storage using corrugated fins (shell-and-tube) was proved but here just a simple PCM cube is used. In some systems the PCM slurries' density changes at phase change temperature, but here a Newtonian model is used. The nanoparticles dispersed in a phase change material can also improve melting characteristics which is neglected here.

Table 1. Governing equations of fluid [2].

Equation	Conservation of
$\frac{\partial u}{\partial x}+\frac{\partial v}{\partial y}=0$, (1)	mass
$\frac{\partial u}{\partial t}+u\frac{\partial u}{\partial x}+v\frac{\partial u}{\partial y}=$	
$\frac{1}{\rho_{\mathrm{nf}}}(-\frac{\partial p}{\partial x}+\mu_{\mathrm{nf}}\nabla^2 u+(\rho\beta)_{\mathrm{nf}}\, f_x(T-T_{\mathrm{ref}}))-\frac{\rho_s}{\rho_{\mathrm{nf}}}\frac{\partial\phi}{\partial x}$, (2)	x momentums
$\frac{\partial v}{\partial t}+u\frac{\partial v}{\partial x}+v\frac{\partial v}{\partial y}$	
$\frac{1}{\rho_{\mathrm{nf}}}(-\frac{\partial p}{\partial y}+\mu_{\mathrm{nf}}\nabla^2 v+(\rho\beta)_{\mathrm{nf}}\, f_y(T-T_{\mathrm{ref}}))-\frac{\rho_s}{\rho_{\mathrm{nf}}}\frac{\partial\phi}{\partial y}$, (3)	y momentums
$\frac{\partial T}{\partial t}+u\frac{\partial T}{\partial x}+v\frac{\partial T}{\partial y}$	
$\frac{\partial}{\partial x}\left[\frac{(k_{\mathrm{nf0}}+k_{\mathrm{d}})}{(\rho c_{\mathrm{p}})_{\mathrm{nf}}}\frac{\partial T}{\partial x}\right]+\frac{\partial}{\partial y}\left[\frac{(k_{\mathrm{nf0}}+k_{\mathrm{d}})}{(\rho c_{\mathrm{p}})_{\mathrm{nf}}}\frac{\partial T}{\partial y}\right]$, (4)	energy

Table 2. Non-slip boundary conditions of fluid on walls [4].

x-Velocity	y-Velocity
$u(x=0,L)=0$, (5)	$v(x=0,L)=0$, (6)
$u(y=0,H)=0$, (7)	$v(y=0,H)=0$, (8)

Table 3. Thermal boundary conditions of fluid [8].

Wall	Equation
left	$q(x=0)=-k_{nf}\frac{\partial T(x=0)}{\partial x}$, (9)
right	$\frac{\partial T(x=L)}{\partial x}=0$, (10)
top	$-k_{nf}\frac{\partial T(y=H)}{\partial y}=h(T(y=H)-T_\infty)$, (11)
bottom	$k_{nf}\frac{\partial T(y=0)}{\partial y}=h(T(y=0)-T_\infty)$, (12)

Figure 1 shows the resonator configuration. The geometry of the BAW cell is equal to one of that circuit lines. Figure 2 illustrate the resonator and cooling part. The effective characteristics are shown in Table 4 while the material thermophysical properties are presented in Table 5. The effect of the addition of nanoparticle on the thermal behavior of PCM is obtainable from Table 4. Many fluids and solids such as organic materials, carb-oxy-methyl cellulose carbon foams, paraffin, and expanded graphite are used as phase change material previously for a thermal storage system. However, here TH29 was selected as [2,29].

Graphs of heat flow (in joule per second) as a function of temperature which is called in literature the differential scanning calorimeter (DSC), usually provided by PCM producers. TH29 (Calcium-chloride hexahydrate) performance in DSC form is presented in [29] for the pure material and its mixing with cellulose. As well, the TH29 shows good performance in aged cases which exposes to ambient air for two weeks. From the mass measurements, the hydrated salt-based TH29 absorbed moisture at about 50% of their weight and lost their ability in eight days.

As well, various nanoparticles were used such as Ag, carbon nanofiber, Al, aluminum, and carbon nanotubes. However, here copper nanoparticle was used. The governing equations and thermal boundary conditions for heat transfer in solid part are presented in Table 6.

Table 4. Nano-fluid property formulas [4].

Nano-Fluid Property	Formula
ρ	$(1-\alpha)\rho_f + \alpha\rho_s$, (13)
ρc_p	$(1-\alpha)(\rho c_p)_f + \alpha(\rho c_p)_s$, (14)
$\rho\beta$	$(1-\alpha)(\rho\beta)_f + \alpha(\rho\beta)_s$, (15)
μ	$\frac{\mu_f}{(1-\alpha)^{2.5}}$, (16)
k	$k_f\frac{k_s+2k_f-2\alpha(k_f-k_s)}{k_s+2k_f+\alpha(k_f-k_s)} + C(\rho c_p)_{nf}\sqrt{u^2+v^2}\alpha d_p$, (17)
ρL	$(1-\alpha)(\rho L)_f$, (18)
σ	$\sigma_f\left(1+\frac{3(\sigma_s-\sigma_f)\phi}{(\sigma_s+2\sigma_f)-(\sigma_{np}-\sigma_f)\phi}\right)$, (19)

Table 5. Material thermo-physical properties [2].

Material	k (W/m·K)	C (kJ/kg·K)	ρ (kg/m^3)	β (K^{-1})	H_f (J/kg)	μ (Pa·s)
fluid TH29	0.53	2.2	1530	2×10^{-4}	187	5.33×10^{-3}
solid TH29	1.09	1.4	1719		187	
Cu	400	0.383	8954	1.67×10^{-5}		

Usually, the thermal delay method is used to determine the effective characteristics of micro-encapsulated phase change material. Table 5 mentions TH29, a commercially available PCM.

Table 6. Governing equations and thermal boundary conditions for heat transfer in solid part [6].

Zone	Equation
thermal balance over volume	$\frac{\partial T}{\partial t} = \frac{\partial}{\partial x}\left[\frac{k_s}{\rho_s c_s}\frac{\partial T}{\partial x}\right] + \frac{\partial}{\partial y}\left[\frac{k_s}{\rho_s c_s}\frac{\partial T}{\partial y}\right] + \frac{q'''}{\rho_s c_s}$, (20)
left wall	$\frac{\partial T(x=-B)}{\partial x} = 0$, (21)
right wall	$q(x=0) = -k_s\frac{\partial T(x=0)}{\partial x}$, (22)
top wall	$-k\frac{\partial T(y=H)}{\partial y} = h(T(y=H)-T_\infty)$, (23)
bottom wall	$k\frac{\partial T(y=0)}{\partial y} = h(T(y=0)-T_\infty)$, (24)

2.2. Piezoelectric Induced Wave

The piezoelectric constitutive conditions that couple the mechanical and electrical amounts in the piezoelectric material are communicated utilizing the coupling equations. The conservation of momentum for the bulk acoustic wave in an electrical part is presented in Table 7.

Table 7. Conservation of momentum for bulk acoustic wave in electrical part [6].

Zone	Equation
momentum conservation	$\rho\frac{\partial^2}{\partial t^2}\begin{bmatrix}u & v & w\end{bmatrix} = \begin{bmatrix}\frac{\partial}{\partial x} & \frac{\partial}{\partial y} & \frac{\partial}{\partial z}\end{bmatrix}\begin{bmatrix}T_{11} & T_{12} & T_{13}\\ T_{12} & T_{22} & T_{23}\\ T_{13} & T_{23} & T_{33}\end{bmatrix}$, (25)
periodic left and right walls	$\mathbf{u}_{destination} = \mathbf{u}_{source}e^{(-i\mathbf{k}(\mathbf{r}_{destination} - \mathbf{r}_{source}))}$, (26)
top and bottom free surfaces	$\mathbf{T}.\mathbf{n} = 0$, (27)

The electrical boundary condition for the piezoelectric wave is depicted in Figure 3. The wave motion ($\rho\frac{\partial^2 u_i}{\partial t^2} = \frac{\partial T_{ij}}{\partial x_j}$) which comes from conservation of momentum and presented in in Table 7. The coupled constitutive equations in strain-charge form [6] are

$$\begin{bmatrix}S_{11}\\ S_{12}\\ S_{13}\\ S_{22}\\ S_{23}\\ S_{33}\end{bmatrix} = \begin{bmatrix}s_{11}^E & s_{12}^E & s_{13}^E & 0 & 0 & 0\\ s_{12}^E & s_{22}^E & s_{23}^E & 0 & 0 & 0\\ s_{13}^E & s_{23}^E & s_{33}^E & 0 & 0 & 0\\ 0 & 0 & 0 & s_{44}^E & 0 & 0\\ 0 & 0 & 0 & 0 & s_{55}^E & 0\\ 0 & 0 & 0 & 0 & 0 & s_{66}^E\end{bmatrix}\begin{bmatrix}T_{11}\\ T_{12}\\ T_{13}\\ T_{22}\\ T_{23}\\ T_{33}\end{bmatrix} - \begin{bmatrix}0 & 0 & d_{31}\\ 0 & 0 & d_{32}\\ 0 & 0 & d_{33}\\ 0 & d_{24} & 0\\ d_{15} & 0 & 0\\ 0 & 0 & 0\end{bmatrix}\begin{bmatrix}\frac{\partial\phi}{\partial x}\\ \frac{\partial\phi}{\partial y}\\ \frac{\partial\phi}{\partial z}\end{bmatrix} \tag{28}$$

and

$$\begin{bmatrix}D_1\\ D_2\\ D_3\end{bmatrix} = \begin{bmatrix}0 & 0 & 0 & 0 & d_{15} & 0\\ 0 & 0 & 0 & d_{25} & 0 & 0\\ d_{31} & d_{32} & d_{33} & 0 & 0 & 0\end{bmatrix}\begin{bmatrix}T_{11}\\ T_{12}\\ T_{13}\\ T_{22}\\ T_{23}\\ T_{33}\end{bmatrix} - \begin{bmatrix}\varepsilon_{11}^T & 0 & 0\\ 0 & \varepsilon_{22}^T & 0\\ 0 & 0 & \varepsilon_{33}^T\end{bmatrix}\begin{bmatrix}\frac{\partial\phi}{\partial x}\\ \frac{\partial\phi}{\partial y}\\ \frac{\partial\phi}{\partial z}\end{bmatrix} \tag{29}$$

Consider next the case where the slab is electrically excited by applying a harmonic voltage, v = V_0e^{jt}, across the electrodes. In this case, it is convenient to consider a slab of length L with ends at x = L/2. The stresses at the ends are zero. The fields of interest are expressed by eigenfrequency analysis ($-\rho\omega^2 u = \frac{\partial T_{11}}{\partial x}$). The dielectric permittivity, elastic and piezoelectric constants of ZnO are considered in this study.

As the velocity of the electromagnetic wave is faster than the acoustic wave, the electrostatic equation of Laplace is applied for electric potential ($\nabla^2\phi = 0$). As well, the electrical displacement is the function of strain and the electric field ($D_i = d_{ikl}S_{kl} + \varepsilon_{ik}^T E_k$).

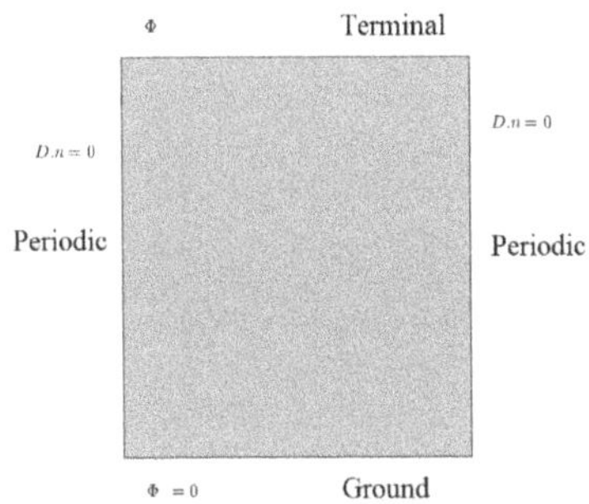

Figure 3. Schematic of BAW piezoelectric material (ZnO) and periodic condition (for real and imaginary k), terminal, and ground boundary condition at aluminum surfaces.

3. Results

Here, predominantly, the instance of voltage stacking is considered, i.e., one terminal is electrically grounded, and a sinusoidal voltage with endorsed consistent plentifulness is associated between the grounded and "hot" anode. Arrangement of the eigenvalue issue at that point brings about arrangement reverberation frequencies (short-circuited terminals) at which the electrical impedance expects low qualities. The equal reverberation frequencies with enormous estimations of impedance can be acquired from the open-circuited case. For the open-circuit limit conditions, the potential is set consistent on the "hot" anode, yet not fixed (drifting potential).

In this section which describes the results, first, the confrontation of the obtained results with other research are presented through validations. Such a suitable comparison significantly raises the meaning of the presented paper. Then the effect of adding PCM to BAW is discussed.

3.1. Validation of Used Code

In this section, various parts of the code are validated against well-known benchmarks and commercial software.

3.1.1. Validation of Fluid Part

For the purpose of the validation of the fluid part, the previous work of Jamalabadi and Park [2] is used. The vertical axis in Figure 4 is solid content that has no unit. As revealed in Figure 4 the solid percent of PCM decreased by the increase of time (melting process). As seen in Figure 4 both methods are in good agreement.

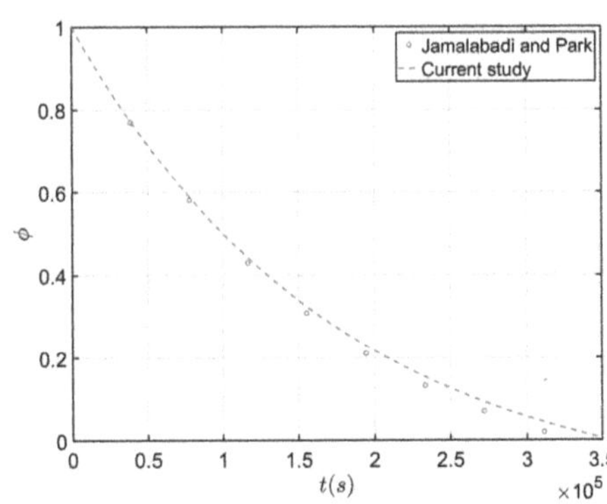

Figure 4. Validation of fluid part by comparing the solid percent of PCM versus time in a melting process.

3.1.2. Validation of Heat Generation

In this section, the validation of heat generation caused by resonance with the analytical solution is performed. The loss mechanisms in the thin film BAW resonators include leakage to the substrate, laterally escaping waves, ohmic losses, viscous losses, dielectric losses, and wave scattering. Additionally mentioned in the literature are eddy current losses and losses due to the formation of conducting channels in the high resistivity substrate interface. The analytical solution of the piezoelectric governing equation for heat generation is given in Appendix A. To calculate the heat generation of the electric source, both mechanical part (elastic energy) and electric part (electric displacement energy) should be considered. The Poynting vector (W/m^2) which is the directional energy flux (the energy transfer per unit area per unit time) usually used as a symbol of conservation of electromagnetic energy ($S = E \times H$). The basic idea is that Poynting energy losses some parts as passed through the piezoelectric material and electrodes. Here the power loss distribution when the slab is electrically excited by an applied voltage is considered. The electrical excitation with the amplitude of 10 V is applied across the terminals. Material and geometrical parameters for the case of heat generation are presented in Figure 5 and Table 8.

Table 8. Material and geometrical parameters for the case of heat generation.

Parameter	Value	Unit
L	100	mm
B	10	mm
h	5	mm
ρ	7500	kg/m^3
d_{31}	$-123 \times 10^{-12}(1 - j \times 0.005)$	C/N
ϵ_{33}^T	$1300 \times 8.854 \times 10^{-12}(1 - j \times 0.004)$	F/m
s_{11}^E	$12.3 \times 10^{-12}(1 - j \times 0.005)$	F/m

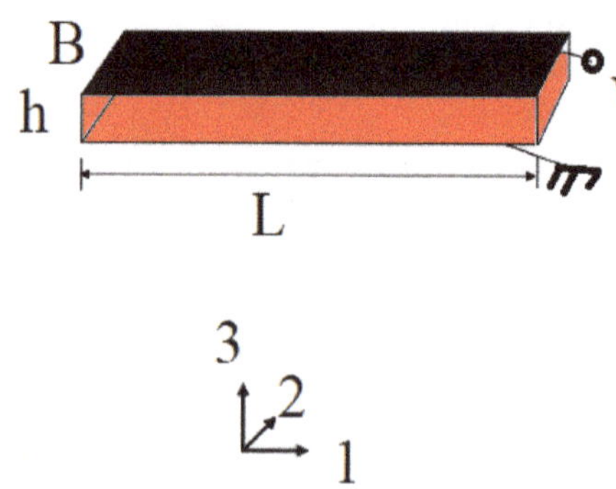

Figure 5. Geometrical parameters for the case of heat generation in a slab.

The comparison of the current code with the analytical solution which is given in Appendix B is plotted in Figure 6. As seen in Figure 6 both numerical and analytical results are in good agreement. The validation results of Figure 6 has the error of simulation near zero as it designed to respond in 1D and the effect of transverse direction is not too much as the electrode extends all over the top and bottom surface. In engineering applications, the electrodes cover the surface partially and some materials are added to help the system to have these ideal mechanical support connections.

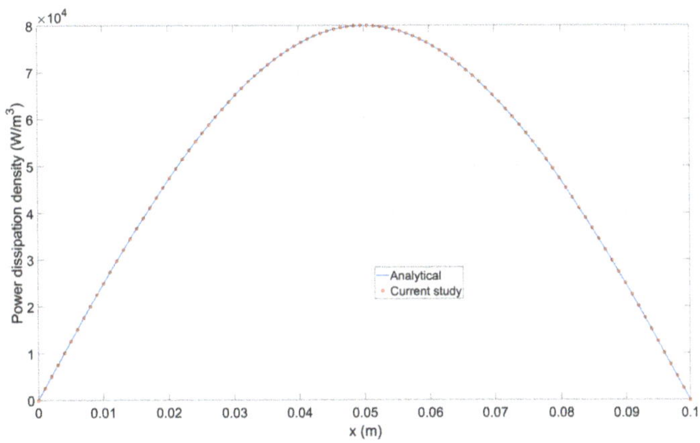

Figure 6. Heat generation in axial direction calculated by FEM and analytical solution.

3.1.3. Validation of Charge Curve

In this section, the 3D problem is considered while in another part of the paper you can consider the 2D model as the third direction has not affected the results. Simulation of a conventional BAW resonator is performed in the appendix code presented at the end of the paper. After a variable definition on substrate, electro, piezo, and resonator parameters. Element types of solid226 and solid95 are used to model the piezoelectric parts. Materials parameters of aluminum nitride for various directions as anisotropic material in the form of tabular data are given. Some damping parameters are also considered for mechanical and electrical losses. Silicon and molybdenum data are entered as well. For building geometry, the bottom substrate, bottom electrode, piezo layers, and top electrode simple block are used. 3D regular mesh is created to apply the mechanical boundary conditions (two symmetry faces, right surface from top view symmetry, and up to surface from top view symmetry, down surface from top view hard boundary as a fixed plate) and electrical boundary conditions (electrode bottom, electrode top). The results of harmonic analysis in a range of 1100 and 1250 Mhz are used to validate the charge collected on the electrode surface. Material parameters definition and geometry of the compared case are presented in Tables 9 and 10, respectively. As seen the piezoelectricity and full anisotropy of the materials are taken into account. The comparison of the current code with the ANSYS-APDL code results which the code is provided in Appendix B is plotted in Figure 7. As seen in the figure, both results are in good agreement. The validation results of Figure 7 have the error of simulation near zero. As both used the same method (FEM) these results were anticipated.

Table 9. Material parameters definition.

Material Name	Parameter	Value
Aluminum Nitride	ρ	3260
	ν_x	8
	ν_y	8
	ν_z	10
Silicon	ρ	2181.5
	E	176.52×10^9
	ν	0.22
	G	72.345×10^9
Molybdenum	ρ	10096
	E	232.36×10^9
	ν	0.362
	G	85.3×10^9

Table 10. Variable definition.

Variable Name	Value
x_{subs}	200×10^{-6}
y_{subs}	200×10^{-6}
z_{subs}	5×10^{-7}
$x_{electro,bot}$	200×10^{-6}
$y_{electro,bot}$	200×10^{-6}
$z_{electro,bot}$	200×10^{-9}
$x_{electro,top}$	150×10^{-6}
$y_{electro,top}$	150×10^{-6}
$z_{electro,top}$	260×10^{-9}
x_{piezo}	200×10^{-6}
y_{piezo}	200×10^{-6}
z_{piezo}	1×10^{-6}

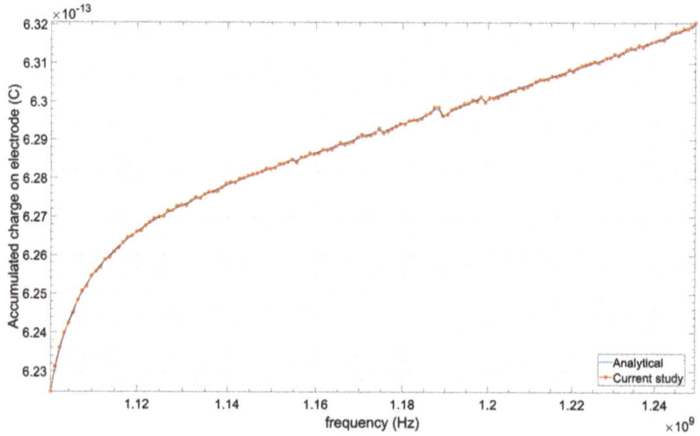

Figure 7. Accumulated charge on electrode as a function of frequency estimated by FEM and ANSYS.

3.1.4. Validation of Dispersion Curve

The behavior of the side resonances in a multilayer structure is a function of Lamb modes curves traveling in a multilayer structure is investigated in this section. The comparison of a dispersion curves presented by Makkonen et al. [3] is given in Figure 8. As depicted in the Figure 8, both methods are in good agreement.

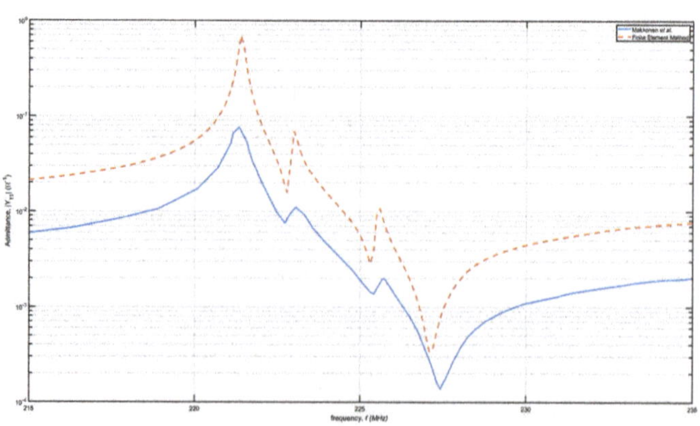

Figure 8. Admittance as a function of frequency estimated by FEM and Makkonen et al. [3].

3.2. Eigenfrequency Analysis

The first six modes of the BAW problem are presented in this part. The real part of the first six modes of BAW device is 221.4200 222.02, 222.97, 224.16, 225.51, and 226.96 MHz. The imaginary parts are three orders of magnitude less than the real parts with the values of 83.593, 84.982, 84.029, 90.701, 90.162, and 90.142 kHz. Detailed parameters are given in Table 11. To have a better understanding of given parameters they have been plotted in Figure 9. As shown, the coefficients have a fluctuation from one frequency to another. In participation factors of x and z, even frequencies (2, 4, 6, ...) have higher values and y participation factors have maximum in odd frequencies (1, 3, 5, ...).

Table 11. Eigenfrequency study of six modes of BAW resonator.

Description	Units	Value
Eigenfrequency	Hz	221.42 + 0.083574i
Participation factor, normalized, X-translation	-	−1.4177E−9−2.3951E−9i
Participation factor, normalized, Y-translation	-	2.3391E−8−2.3982E−9i
Participation factor, normalized, Z-rotation	–	1.3453E−12 + 1.4519E−13i
X-translation modal mass	kg	−3.7265E−18 + 6.7912E−18i
Y-translation modal mass	kg	5.4141E−16−1.1219E−16i
Z-rotation modal mass	kg·m^2	1.7886E−24 + 3.9064E−25i
Eigenfrequency	Hz	222.02 + 0.084982i
Participation factor, normalized, X-translation	-	−1.2376E−7−2.2132E−8i
Participation factor, normalized, Y-translation	-	−6.3621E−10 + 1.2395E−9i
Participation factor, normalized, Z-rotation	–	−7.0904E−12 + 1.7065E−11i
X-translation modal mass	kg	1.4827E−14 + 5.4782E−15i
Y-translation modal mass	kg	−1.1316E−18−1.5772E−18i
Z-rotation modal mass	kg·m^2	−2.4093E−22−2.4199E−22i
Eigenfrequency	Hz	222.97 + 0.084029i
Participation factor, normalized, X-translation	-	2.3988E−9−8.4602E−10i
Participation factor, normalized, Y-translation	-	−1.0829E−8−2.0163E−8i
Participation factor, normalized, Z-rotation	–	−1.1635E−13−1.7268E−13i
X-translation modal mass	kg	5.0385E−18−4.0589E−18i
Y-translation modal mass	kg	−2.8925E−16 + 4.3670E−16i
Z-rotation modal mass	kg·m^2	−1.6281E−26 + 4.0181E−26i
Eigenfrequency	Hz	224.16 + 0.090701i
Participation factor, normalized, X-translation	-	5.3732E−8−2.4727E−7i
Participation factor, normalized, Y-translation	-	6.2749E−10−1.0431E−9i
Participation factor, normalized, Z-rotation	–	−6.0091E−12−6.5558E−12i
X-translation modal mass	kg	−5.8254E−14−2.6572E−14i
Y-translation modal mass	kg	−6.9441E−19−1.3091E−18i
Z-rotation modal mass	kg·m^2	−6.8693E−24 + 7.8788E−23i
Eigenfrequency	Hz	225.51 + 0.090162i
Participation factor, normalized, X-translation	-	−2.9494E−9−1.3085E−9i
Participation factor, normalized, Y-translation	-	−3.4432E−9 + 5.0666E−8i
Participation factor, normalized, Z-rotation	–	6.7515E−13 + 6.2670E−13i
X-translation modal mass	kg	6.9867E−18 + 7.7184E−18i
Y-translation modal mass	kg	−2.5551E−15−3.4891E−16i
Z-rotation modal mass	kg·m^2	6.3072E−26 + 8.4624E−25i
Eigenfrequency	Hz	226.96 + 0.090142i
Participation factor, normalized, X-translation	-	3.8473E−8−1.7100E−8i
Participation factor, normalized, Y-translation	-	−1.0000E−9−9.8674E−10i
Participation factor, normalized, Z-rotation	–	−2.1903E−11 + 6.6730E−12i
X-translation modal mass	kg	1.1878E−15−1.3158E−15i
Y-translation (modal mass	kg	2.6426E−20 + 1.9736E−18i
Z-rotation modal mass	kg·m^2	4.3522E−22−2.9232E−22i

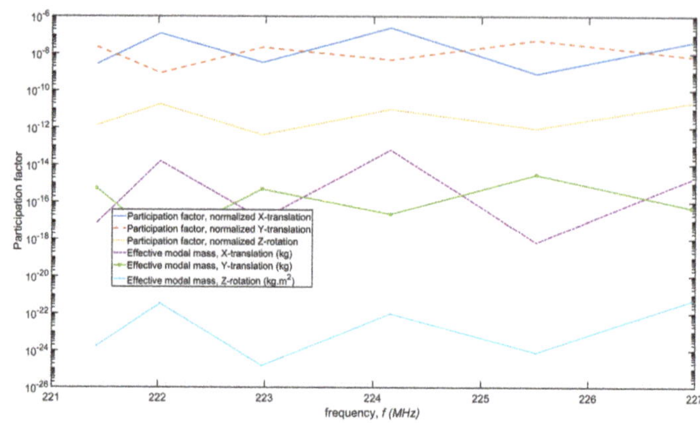

Figure 9. Eigenfrequency, participation factors, and effective modal masses of the first six modes of BAW piezoelectric in eigenfrequency evaluation.

Figures 10–15 show horizontal and vertical displacement of mode 1 to mode 3. Each of propagating modes and evanescent modes can also be visualized by its displacement field. As seen in Figure 10, the horizontal displacement of mode 1 has two peaks at the middle of terminals. Figure 11 presents four peaks in the vertical displacement of mode 1 where one of them is in opposite direction. Horizontal displacement of mode 2 as expected has four ($2 \times N$) peaks which are symmetric around the middle of piezoelectric which is visible in Figure 12. As illustrated in Figure 13, vertical displacement of mode 2 has four maximum and four minimum which are aligned in three parallel lines. The presence of propagating and evanescent modes is clear. The exhibition of six ($2 \times N$) simple maximum of mode 3 of horizontal displacement of mode 3 is demonstrated in Figure 14. As shown the eigenfunctions are aligned in three parallel lines and are symmetric conditions. Finally, the eigenfunctions of mode 3 of vertical displacement with six visible maximum are depicted in Figure 15. A detailed BAW design, general discussion on BAW resonators, and electric potential modes are discussed in [1,6]. In surface plotted figures including Figures 10–15 the color is automatically proportional to surface height.

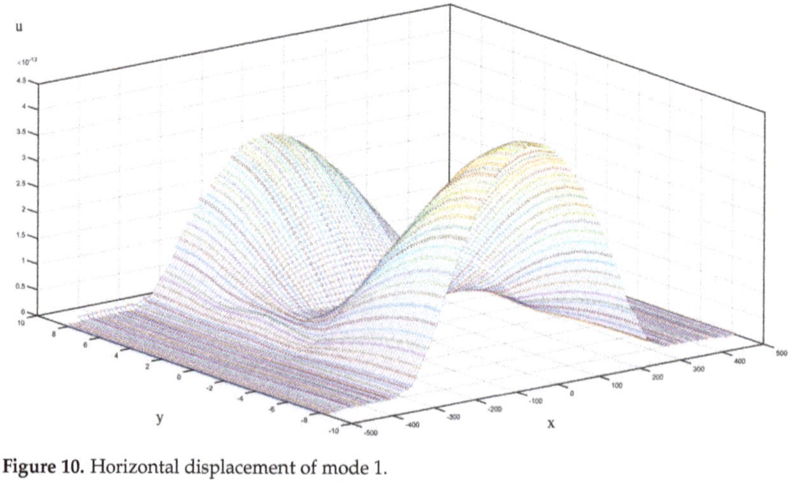

Figure 10. Horizontal displacement of mode 1.

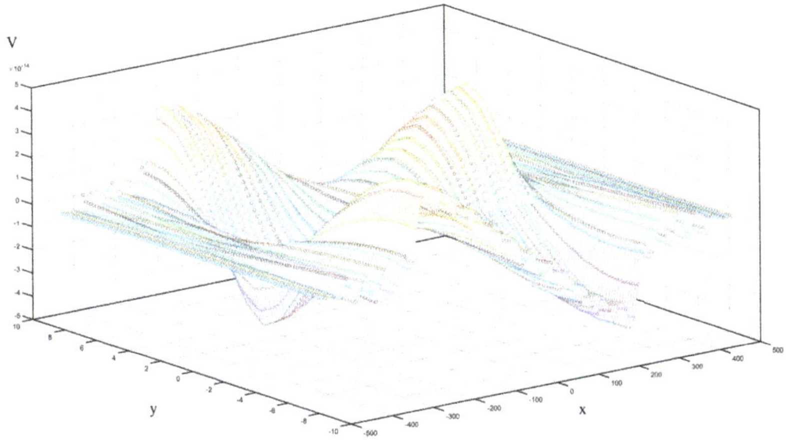

Figure 11. Vertical displacement of mode 1.

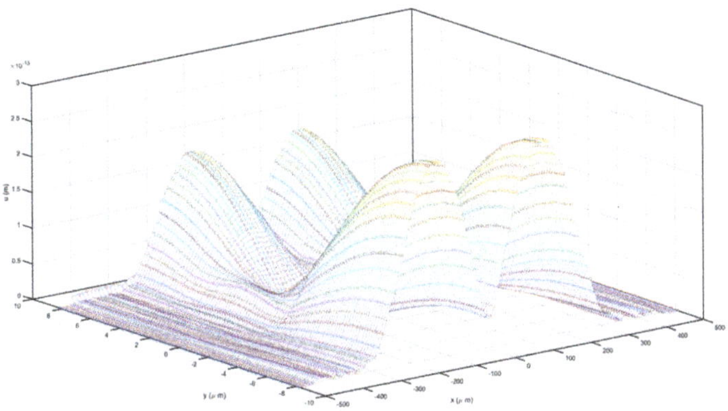

Figure 12. Horizontal displacement of mode 2.

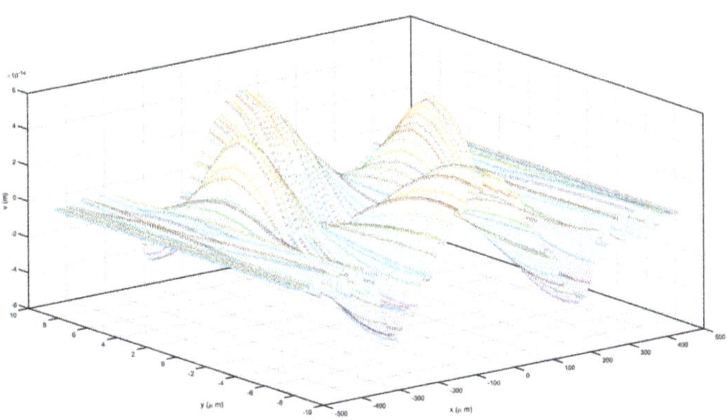

Figure 13. Vertical displacement of mode 2.

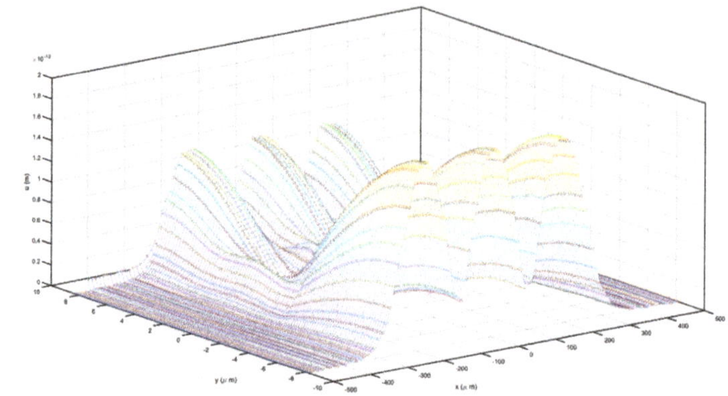

Figure 14. Horizontal displacement of mode 3.

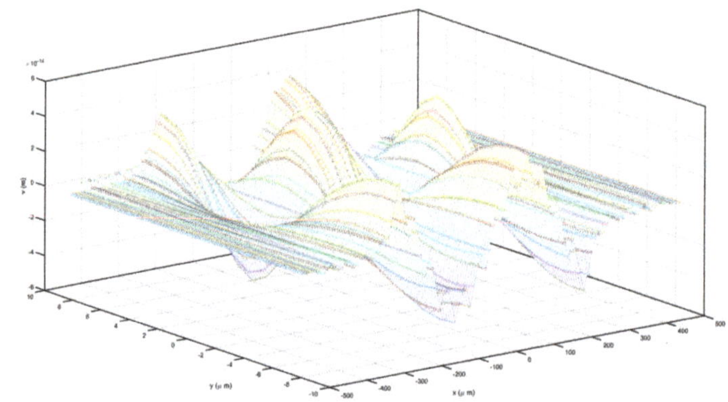

Figure 15. Vertical displacement of mode 3.

The damping ration, quality factor, and S_{11} are plotted in Figures 16–18 respectively. Damping ratio (ratio of imaginary of the frequency to its absolute value) and quality factor (ratio of absolute value to two times of the imaginary of the frequency) as a function of frequency are plotted in Figure 16. The quality factor for frequency as a function of frequency is depicted in Figure 17. The real part and the imaginary part of the scattering parameter as a function of frequency are illustrated in Figure 18 for the layer structure of a BAW resonator. Here the quality plot is against all selected intervals of frequency. The dispersion curve of imaginary (evanescent mode) and real (propagating mode) parts of the wavenumbers are figured in Figure 19 for a thin-film BAW composite resonator. The dispersion schematic includes both imaginary and real parts of the wavenumber. The plots actually are the contours of admittance $|Y_{11}|$ for real part of k, and contours of admittance $|Y_{22}|$ for imaginary the part of k. Figure 19 shows the calculated dispersion curve. The results for the the real and the imaginary parts of the wavenumber are combined by taking the positive *k*-axis with an optimum near $(6 \times 10^4, 2 \times 10^8)$ for the real part and the negative *k*-axis with a maximum near $(-2 \times 10^4, 2 \times 10^8)$ for the imaginary part. The horizontal axis of dispersion curve is k, wave number, with the unit of $\frac{1}{m}$ while the vertical axis is f, frequency, with aunit of $\frac{1}{s}$. They correspond to the lower branch for the TS_2 mode (thickness-shear), the upper branch for the TE_1 mode (thickness-extension), and the left branch for the evanescent modes.

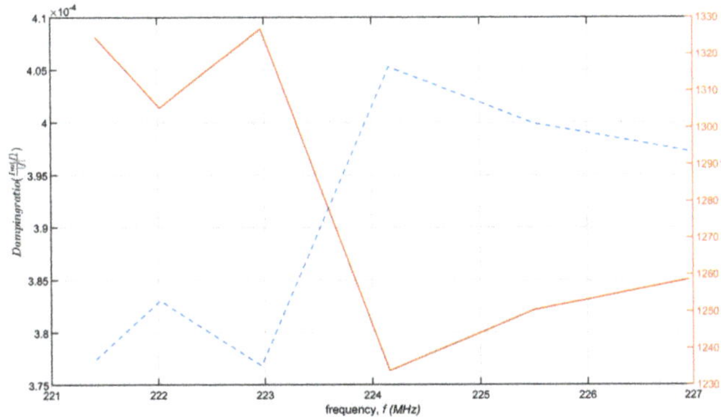

Figure 16. Damping ratio (blue dashed-line) and quality factor (red line) as a function of frequency.

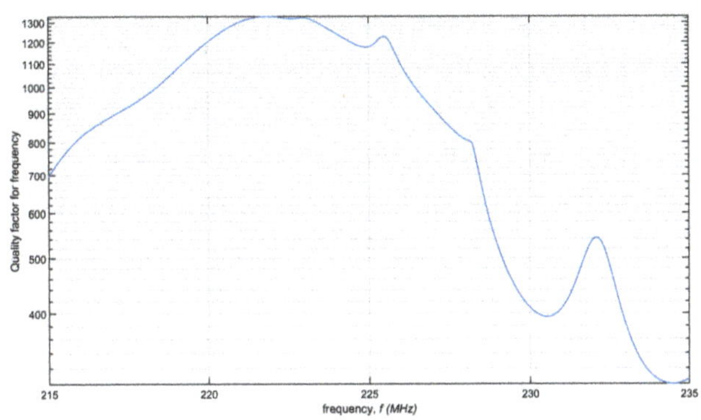

Figure 17. Quality factor for frequency as a function of frequency.

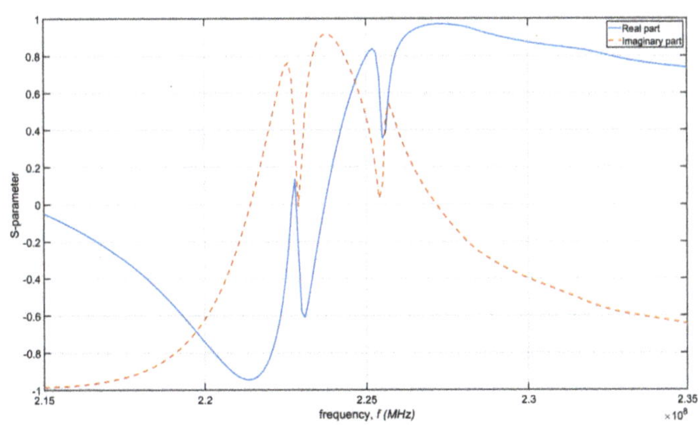

Figure 18. Real part (blue line) and imaginary part (red dashed-line) of scattering parameter as a function of frequency.

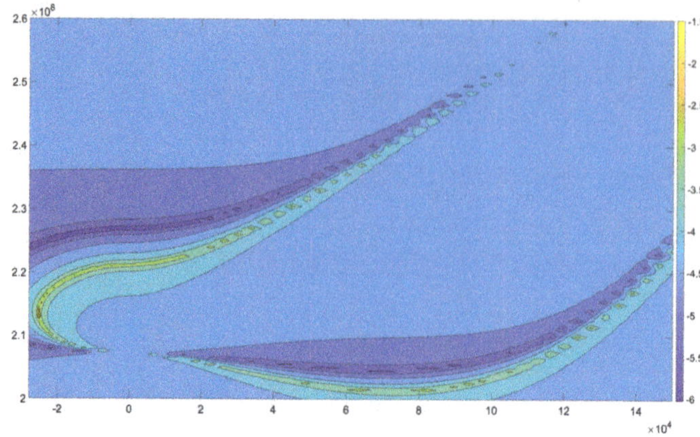

Figure 19. Dispersion curve for imaginary wave numbers (evanescent waves) and for real wave numbers (propagating waves).

Total power dissipation as a function of frequency is plotted in Figure 20 for the specific device structure thin-film BAW composite resonator considered here. As shown, the maximum value of heat generation happened in the first mode while is in accordance with maximum admittance (Y_{11}). This graph will be used in the next part as the input for heat generation.

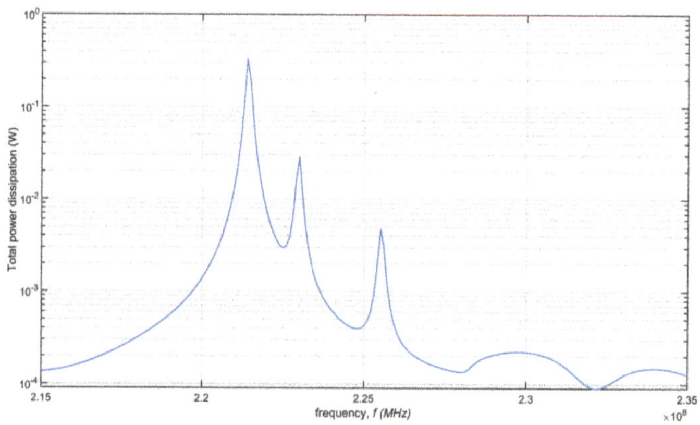

Figure 20. Total power dissipation as a function of frequency.

3.3. Fluid and Thermal Analysis

Profile development and contours of velocity magnitude distribution in the liquid part of the PCM cell are plotted in Figure 21. As shown, the liquid profile is developed fast near the initial condition as the absorbed heat in the PCM is converted to phase change in the liquid–solid boundary. Figure 21 does not have a color legend and is focused on showing the melting profile development. In the red zone, the velocity magnitude is maximum while in blue zones (near walls) the velocity magnitude is near zero. As time goes on, the melting process developed in the PCM box. As shown, the velocity is minimum near walls and is maximum at the middle of two flow channels created horizontally at the PCM cell. Detailed velocity distribution in PCM cell with cooling at the end of melting process is plotted in Figure 22. As shown in velocity vectors and magnitude in the PCM field, the maximum velocity happens at the middle of two flow channels with the horizontal direction. At the end of the flow stream, it turns vertically to turn inside by natural convection.

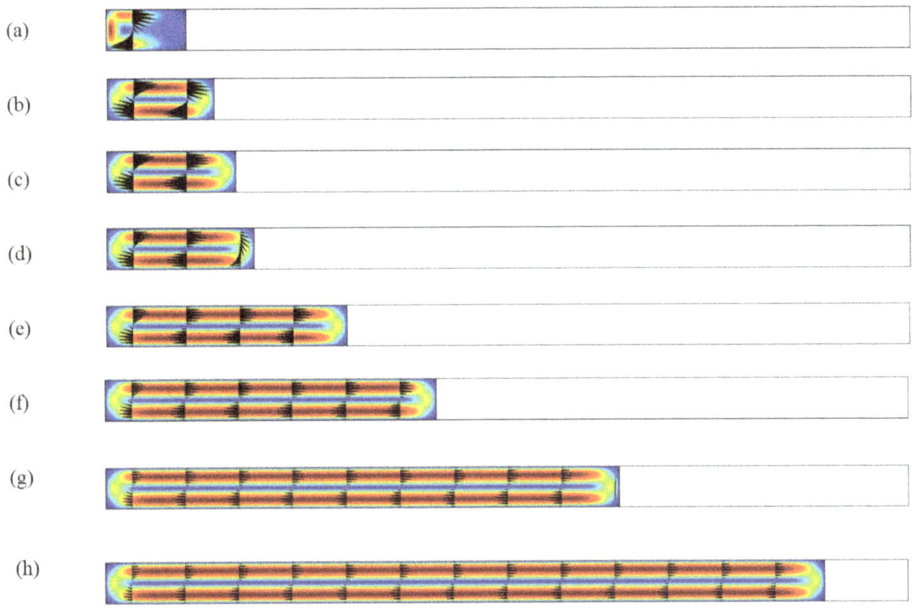

Figure 21. Profile development and contours of velocity magnitude in liquid part of PCM cell (**a**) t = 10, (**b**) t = 100, (**c**) t = 200, (**d**) t = 500, (**e**) t = 1000, (**f**) t = 2000, (**g**) t =5000, (**h**) t = 10,000.

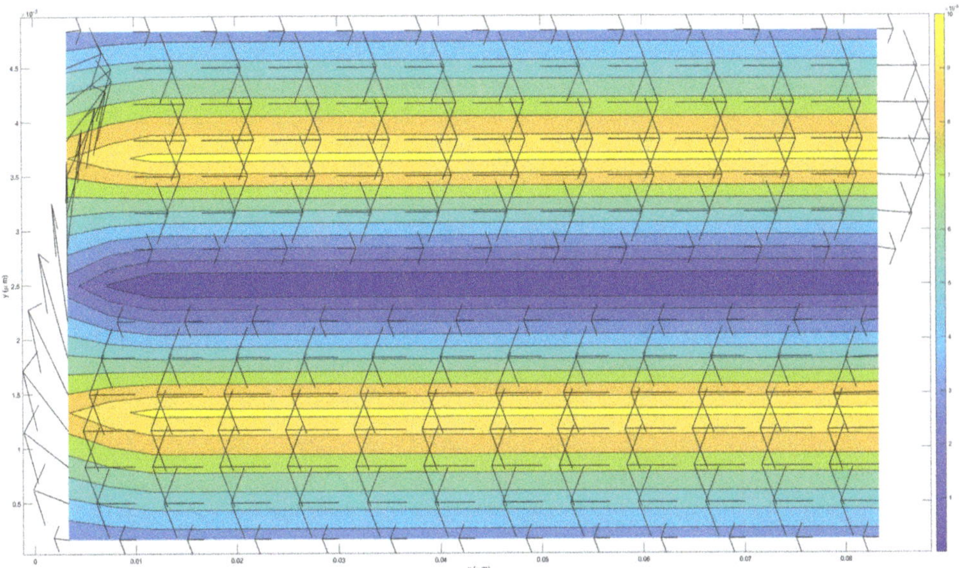

Figure 22. Velocity vectors and magnitude in PCM field.

The power distribution inside BAW slab for first and second mode are plotted in Figure 23. The distribution is axially and matches the analytic solution given in Appendix A.

The temperature distribution inside BAW slab for first and second mode are plotted in Figure 24. The distribution is axially and for the case of constant temperature BC at both sides.

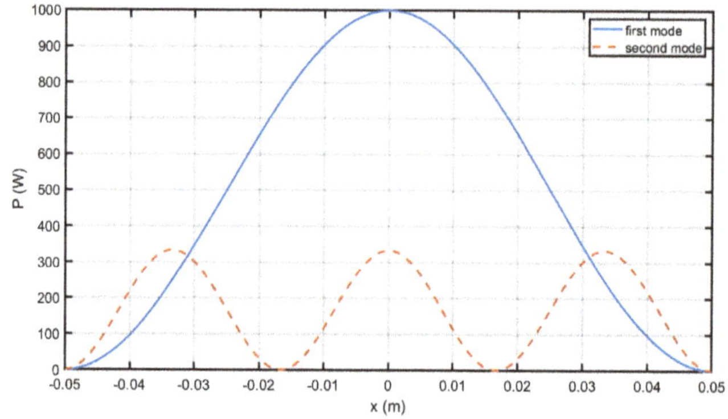

Figure 23. Power distribution inside BAW slab for first and second mode.

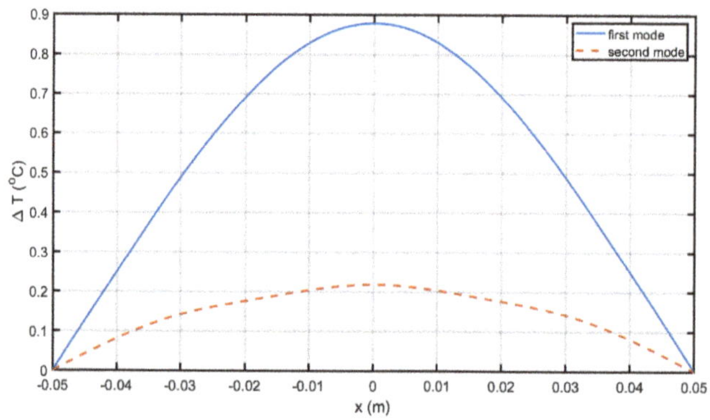

Figure 24. Temperature distribution inside BAW slab for first and second mode.

Temperature difference ($\Delta T = T - T_\infty$) contours are plotted in Figure 25. Figure 25 does not have a color legend and is focused on showing the melting profile development. In the red zone, the velocity magnitude is maximum while in blue zones (near walls) the velocity magnitude is near zero. The figure shows the temperature profiles through the time for various BC. The initial condition of PCM cell is temperature equals the environmental temperature ($T = T_\infty$). As time goes on, the cold zone is shortened through the volume, and the heated zone expands. If both sides are kept at environmental temperature the temperature profile inside the slab is parabolic. The case of constant temperature BC at both sides happens when the periodic structure of the BAW + PCM cell forced the melting temperature at both side until the end of phase change. The case of adiabatic BC at the left happens when each BAW + PCM cell is considered as an isolated domain from the other. Such conditions force the melting temperature on the right side until the end of phase change while adiabatic condition at the left of the cell. As shown, the periodic structure causes less temperature increase.

Temperature distribution inside BAW + PCM cell is illustrated in Figure 26. Time dependency of the temperature in the BAW + PCM cell shows that, by considering PCM, the thermal inertia happens in the system.

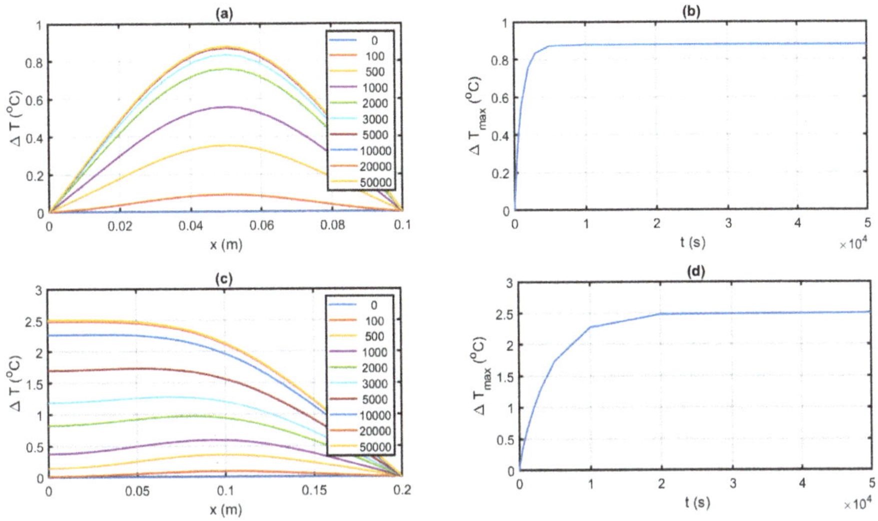

Figure 25. Temperature profiles inside liquid part of PCM cell (**a**) constant temperature BC at the both sides, (**b**) maximum temperature while constant temperature BC at the both sides, (**c**) adiabatic BC at the left, (**d**) maximum temperature while adiabatic BC at the left.

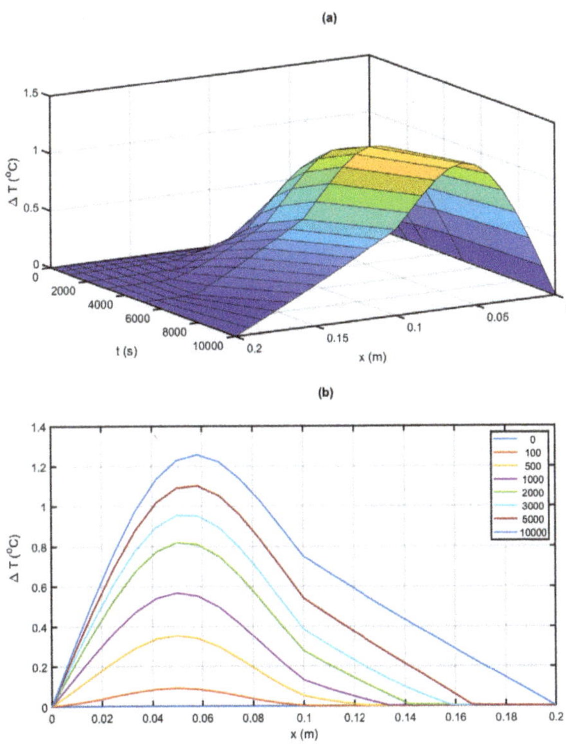

Figure 26. Temperature distribution inside BAW + PCM cell (**a**) profile plot (**b**) surface plot.

The effect of nanoparticle concentration on time dependency of maximum temperature is illustrated in Figure 27. By increasing the nanoparticle the rate of interface convection increased and faster phase change happens.

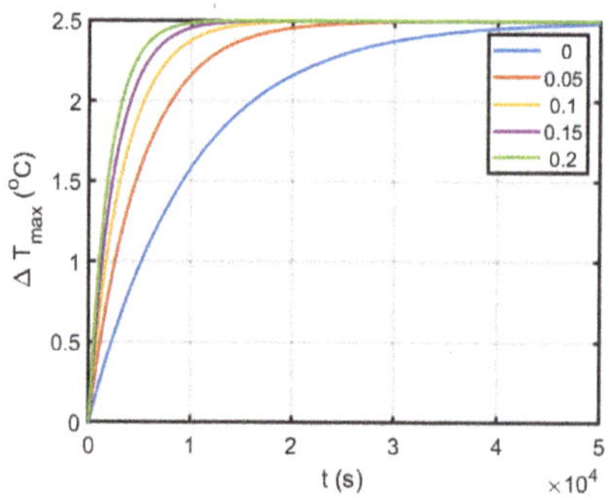

Figure 27. Effect of nanoparticle concentration on maximum temperature.

4. Conclusions

In this paper, a FEM code is introduced for the mathematical arrangement of the electro–elastic conditions that oversee the directly constrained vibrations of piezoelectric media. A consonant time reliance is accepted. Both of the methodologies, that of taking care of the field issue (consonant investigation) and that of taking care of the relating eigenvalue issue (modular examination), are portrayed. A FEM programming bundle has been made without any preparation. Significant angles fundamental to the proficient execution of FEM are clarified, for example, the memory of the executives and tackling the summed up piezoelectric eigenvalue issue. Calculations for diminishing the necessary PC memory through enhancement of the framework profile, just as Lanczos calculation for the arrangement of the eigenvalue issue are connected into the product from outer mathematical libraries. Current FEM programming is applied to nitty-gritty mathematical demonstrating of thin-film bulk acoustic wave (BAW) composite resonators. Correlation of results from 2D and full 3D recreations of a resonator is introduced. Specifically, 3D reproductions are utilized to explore the impact of the top terminal shape on the resonator electrical reaction. The legitimacy of the displaying strategy is shown by contrasting the recreated and estimated removal profiles at a few frequencies. The outcomes show that valuable data on the exhibition of the flimsy film resonators can be acquired even with generally coarse cross-sections and, subsequently, moderate computational assets. To highlight the advantages and disadvantages of using NEPCM for cooling application of BAW resonator it is good to notice that when comparing the current devised solution with other solutions from the scientific literature, the efficiency of the current design is more clear. While in common circuits the heat is removed from the semiconductor by conduction to support or with convection or radiative heat transfer modes to the environment, here the phase change plays the role of the heat sink with more thermal inertia. As well, the use of nanoparticles facilitates the heat removal, and also decreases the delay time and thermal inertia of the cooling system. As it would be useful to add information on further research of the authors related to the continuation of this research topic, future research could be conducted to consider the 3D design of a thin-film bulk acoustic wave composite resonators in conjunction with another electrical part to see the effectiveness of current study in an engineering case.

Funding: This research received no external funding.

Acknowledgments: The author want to acknowledge Humboldt officials for Research Fellowship Program for Experienced Researchers and Ing. Ulrich Nieken, head of Institut für Chemische Verfahrenstechnik at Universität Stuttgart, for hosting during writing of the paper.

Conflicts of Interest: The authors declare no conflict of interest.

Abbreviations

The following abbreviations are used in this manuscript:

2D	Two dimension
3D	Three dimension
BAW	Bulk acoustic waves
BC	Boundary conditions
FEM	Finite element method
IoT	Internet of Things
IDT	Interdigital transducers
RFFE	Radio frequency front-end
SAW	Surface acoustic waves

Nomenclature

T	Stress
S	Strain
E	Electric field
D	Electric displacement
c_E	Elasticity matrix (rank 4 tensor c_{ijkl})
e	Coupling matrix (rank 3 tensor e_{ijk})
ϵ_S	Permittivity matrix (rank 2 tensor ϵ_{ij})
C_{cp}	Constant pressure specific heat ($\text{Jkg}^{-1}\text{K}^{-1}$)
g	Gravity constant (ms^{-2})
Nu	Nusselt number
p	Pressure (Pa)
	Greek symbols
μ	Dynamic viscosity ($\text{kgs}^{-1}\text{m}^{-1}$)
ϕ	Electric potential
s	Solid

Appendix A. Analytical Solution in One Dimension

The current model as presented in Figure 5 simulates a uniform layer structure extending horizontally to infinity. The computed dispersion diagram of such an infinite plate can be used to optimize device designs. For this tutorial, one can take a thin slice from the center of the complete device and use the periodic boundary condition to extend it laterally to infinity. If we apply $\phi(z = h) = \phi_m e^{j\omega t}$ at the top electrode while bottom electrode is ground ($\phi(z = 0) = 0$) then from Laplace equation in electrostatic approximation

$$\frac{\partial^2 \phi}{\partial z^2} = 0 \tag{A1}$$

we have

$$\phi = \phi_m e^{j\omega t} \frac{z}{h} \tag{A2}$$

thus the electric field in z-direction is

$$E_3 = -\frac{\phi_m e^{j\omega t}}{h} \tag{A3}$$

if we differentiate from the first equation of piezoelectric

$$S_{11} = s_{11} T_{11} - d_{31} \frac{\partial \phi}{\partial z} \tag{A4}$$

respect to x direction, then

$$\frac{\partial S_{11}}{\partial x} = s_{11}\frac{\partial T_{11}}{\partial x} \quad \text{(A5)}$$

Considering the definition of strain from displacement field ($S_{11} = \frac{\partial u}{\partial x}$) the following equation for the differential of the stress respect to x direction is obtained

$$\frac{\partial T_{11}}{\partial x} = \frac{1}{s_{11}}\frac{\partial^2 u}{\partial x^2} \quad \text{(A6)}$$

substituting in right hand side of the dynamic equation

$$-\rho_s\omega^2 u = \frac{\partial T_{11}}{\partial x} \quad \text{(A7)}$$

one can obtain

$$-\rho_s\omega^2 u = \frac{1}{s_{11}}\frac{\partial^2 u}{\partial x^2} \quad \text{(A8)}$$

The above equation can solve by the free stress boundaries at the ends ($T_{11}(x = \pm L/2) = 0$)

$$S_{11}(x = \pm L/2) = \frac{\partial u(x = \pm L/2)}{\partial x} = -d_{31}\frac{\partial \phi}{\partial z} \quad \text{(A9)}$$

gives the displacement field as

$$u(x,t) = -\frac{d_{31}\phi_m e^{j\omega t}}{h\sqrt{\rho_s s_{11}}\omega}\frac{sin(\sqrt{\rho_s s_{11}}\omega x)}{cos(\sqrt{\rho_s s_{11}}\omega L/2)} \quad \text{(A10)}$$

Considering the definition of strain from the displacement field

$$S_{11}(x,t) = -\frac{d_{31}\phi_m e^{j\omega t}}{h}\frac{cos(\sqrt{\rho_s s_{11}}\omega x)}{cos(\sqrt{\rho_s s_{11}}\omega L/2)} \quad \text{(A11)}$$

Considering the definition of stress from the first piezoelectric equation

$$T_{11}(x,t) = -\frac{d_{31}\phi_m e^{j\omega t}}{h s_{11}}\left(\frac{cos(\sqrt{\rho_s s_{11}}\omega x)}{cos(\sqrt{\rho_s s_{11}}\omega L/2)} - 1\right) \quad \text{(A12)}$$

Considering the definition of electric displacement from the second piezoelectric equation

$$D_3 = d_{31}T_{11} - \varepsilon_{33}^T\frac{\partial \phi}{\partial z} \quad \text{(A13)}$$

The analytical solution of electric displacement is obtained

$$D_3 = -\frac{d_{31}^2\phi_m e^{j\omega t}}{h s_{11}}\left(\frac{cos(\sqrt{\rho_s s_{11}}\omega x)}{cos(\sqrt{\rho_s s_{11}}\omega L/2)} - 1\right) - \varepsilon_{33}^T\frac{\phi_m e^{j\omega t}}{h} \quad \text{(A14)}$$

or

$$D_3 = -\frac{\phi_m e^{j\omega t}}{h}\left[\frac{d_{31}^2}{s_{11}}\left(\frac{cos(\sqrt{\rho_s s_{11}}\omega x)}{cos(\sqrt{\rho_s s_{11}}\omega L/2)} - 1\right) + \varepsilon_{33}^T\right] \quad \text{(A15)}$$

If piezoelectric material coefficients are real when there will be no losses. Internal losses are attributed to the imaginary part of them, as complex piezoelectric coefficients are:

$$s_{11} = \bar{s}_{11} + js'_{11} \quad \text{(A16)}$$

$$d_{31} = \bar{d}_{31} + jd'_{31} \quad \text{(A17)}$$

$$\varepsilon^T_{33} = \bar{\varepsilon}^T_{33} + j\varepsilon^{T'}_{33} \tag{A18}$$

where $\bar{s}_{11}$, $\bar{d}_{31}$, $\bar{\varepsilon}^T_{33}$ denote real and s'_{11}, d'_{31}, $\varepsilon^{T'}_{33}$ denote imaginary parts, respectively. An expression for the power dissipation density in the slab is obtained by considering hysteresis loops. Integrating to determine the areas under the loops and averaging over time yields:

$$Q = \frac{-1}{T}\int_0^T D_3\frac{\partial E_3}{\partial t}dt + \frac{1}{T}\int_0^T T_{11}\frac{\partial S_{11}}{\partial t}dt \tag{A19}$$

where Q_E is

$$\begin{aligned} Q_E &= \frac{-1}{T}\int_0^T D_3\frac{\partial E_3}{\partial t}dt \\ &= \frac{-1}{T}\int_0^T -\frac{\phi_m e^{j\omega t}}{h}[\frac{d_{31}^2}{s_{11}}(\frac{cos(\sqrt{\rho_s s_{11}}\omega x)}{cos(\sqrt{\rho_s s_{11}}\omega L/2)} - 1) + \varepsilon^T_{33}] \\ &\quad \frac{\partial}{\partial t}(-\frac{\phi_m e^{j\omega t}}{h})dt \\ &= \frac{-\phi_m^2}{Th^2}\int_0^T j\omega e^{2j\omega t}[\frac{d_{31}^2}{s_{11}}(\frac{cos(\sqrt{\rho_s s_{11}}\omega x)}{cos(\sqrt{\rho_s s_{11}}\omega L/2)} - 1) + \varepsilon^T_{33}]dt \end{aligned} \tag{A20}$$

and Q_M is

$$\begin{aligned} Q_M &= \frac{1}{T}\int_0^T T_{11}\frac{\partial S_{11}}{\partial t}dt \\ &= \frac{1}{T}\int_0^T -\frac{d_{31}\phi_m e^{j\omega t}}{hs_{11}}(\frac{cos(\sqrt{\rho_s s_{11}}\omega x)}{cos(\sqrt{\rho_s s_{11}}\omega L/2)} - 1) \\ &\quad \frac{\partial}{\partial t}(-\frac{d_{31}\phi_m e^{j\omega t}}{h}\frac{cos(\sqrt{\rho_s s_{11}}\omega x)}{cos(\sqrt{\rho_s s_{11}}\omega L/2)})dt \\ &= \frac{\phi_m^2}{Th^2}\int_0^T e^{2j\omega t}\frac{d_{31}^2}{s_{11}}(\frac{cos(\sqrt{\rho_s s_{11}}\omega x)}{cos(\sqrt{\rho_s s_{11}}\omega L/2)} - 1) \\ &\quad j\omega\frac{cos(\sqrt{\rho_s s_{11}}\omega x)}{cos(\sqrt{\rho_s s_{11}}\omega L/2)}dt \end{aligned} \tag{A21}$$

if one replaces the complex piezoelectric coefficients in Q_E the following is obtained

$$\begin{aligned} Q_E &= \frac{-\phi_m^2}{Th^2}\int_0^T j\omega e^{2j\omega t}[\frac{(\bar{d}_{31} + jd'_{31})^2}{\bar{s}_{11} + js'_{11}} \\ &\quad (\frac{cos(\sqrt{\rho_s(\bar{s}_{11} + js'_{11})}\omega x)}{cos(\sqrt{\rho_s(\bar{s}_{11} + js'_{11})}\omega L/2)} - 1) + \bar{\varepsilon}^T_{33} + j\varepsilon^{T'}_{33}]dt \\ &= \frac{-\phi_m^2}{Th^2}\int_0^T j\omega(cos(2\omega t) + jsin(2\omega t))[\frac{(\bar{d}_{31} + jd'_{31})^2}{\bar{s}_{11} + js'_{11}} \\ &\quad (\frac{cos(\sqrt{\rho_s(\bar{s}_{11} + js'_{11})}\omega x)}{cos(\sqrt{\rho_s(\bar{s}_{11} + js'_{11})}\omega L/2)} - 1) + \bar{\varepsilon}^T_{33} + j\varepsilon^{T'}_{33}]dt \end{aligned} \tag{A22}$$

As

$$\begin{aligned} &\frac{1}{T}\int_0^T cos(2\omega t)dt = 0 \\ &\frac{1}{T}\int_0^T sin(2\omega t)dt = \frac{1}{4\pi} \end{aligned} \tag{A23}$$

The Q_E is

$$Q_E = \frac{\phi_m^2 j\omega}{h^2}[\frac{(\bar{d}_{31}+jd'_{31})^2}{\bar{s}_{11}+js'_{11}}(\frac{cos(\sqrt{\rho_s(\bar{s}_{11}+js'_{11})}\omega x)}{cos(\sqrt{\rho_s(\bar{s}_{11}+js'_{11})}\omega L/2)}-1) + \bar{\varepsilon}_{33}^T + j{\varepsilon^T}'_{33}] \tag{A24}$$

where the real part is

$$\bar{Q}_E = \frac{\phi_m^2\omega}{h^2}[{\varepsilon^T}'_{33} + imag(\frac{(\bar{d}_{31}+jd'_{31})^2}{\bar{s}_{11}+js'_{11}}\frac{cos(\sqrt{\rho_s(\bar{s}_{11}+js'_{11})}\omega x)}{cos(\sqrt{\rho_s(\bar{s}_{11}+js'_{11})}\omega L/2)})] \tag{A25}$$

Appendix B. ANSYS APDL Input Files

```
! ==== 3D bulk acoustic wave resonator model =======
!Parameters
xsubs=2e-4
ysubs=2e-4
zsubs=5e-7
xelectrobot=2e-4
yelectrobot=2e-4
zelectrobot=26e-8
xelectrotop=15e-5
yelectrotop=15e-5
zelectrotop=26e-8
xpiezo=2e-4
ypiezo=2e-4
zpiezo=1e-6
!Element type
/PREP7
ET,1,solid226,1001
ET,2,solid95
ET,3,solid95
MP,DENS,1,3.26e3
MP,PERX,1,8
MP,PERY,1,8
MP,PERZ,1,10
TB,PIEZ,1
TBDATA,1,0,0,-0.58
TBDATA,4,0,0,-0.58
TBDATA,7,0,0,1.53
TBDATA,10,0,0,0
TBDATA,13,0,-0.48,0
TBDATA,16,-0.48,0,0
TB,ANEL,1
TBDATA,1,3.45e7,1.25e7,1.2e7
TBDATA,7,3.45e7,1.2e7
TBDATA,12,3.95e7
TBDATA,16,1.1e7
TBDATA,19,1.18e7
TBDATA,21,1.18e7
DMPRAT, 0.0003
```

```
mp,damp,1,1e-8
MP,DENS,2,2181.5
MP,EX,2,17652e7
MP,EY,2,17652e7
MP,EZ,2,17652e7
MP,PRXY,2,0.22
MP,PRYZ,2,0.22
MP,PRXZ,2,0.22
MP,GXY,2,72345e6
MP,GYZ,2,72345e6
MP,GXZ,2,72345e6
MP,DENS,3,10096
MP,EX,3,23236e7
MP,EY,3,23236e7
MP,EZ,3,23236e7
MP,PRXY,3,0.362
MP,PRYZ,3,0.362
MP,PRXZ,3,0.362
MP,GXY,3,8530e7
MP,GYZ,3,8530e7
MP,GXZ,3,8530e7
! Mesh
BLOCK,0,xelectrobot/2,0,yelectrobot/2,0,zsubs
z0=zsubs+zelectrobot
BLOCK,0,xelectrobot/2,0,yelectrobot/2,zsubs,z0
layers = 1
*DO,i,1,layers
BLOCK,0,xpiezo/2,0,ypiezo/2,z0+(i-1)*zpiezo/layers, z0+(i)*zpiezo/layers
*ENDDO
z1=z0+zpiezo
h1=(ypiezo-yelectrotop)/2
h2=(xpiezo-xelectrotop)/2
BLOCK,h2,h2+xelectrotop/2,h1,h1+yelectrotop/2, z1,z1+zelectrotop
VGLUE,all
ALLSEL
VSEL,S,LOC,Z,0,zsubs
VATT,2, ,2,0
ALLSEL
VSEL,S,LOC,Z,zsubs,z0
VATT,3, ,3,0
ALLSEL
VSEL,S,LOC,Z,zsubs+zelectrobot,zsubs+zelectrobot+zpiezo
VATT,1, ,1,0
ALLSEL
VSEL,S,LOC,Z,z0+zpiezo,z0+zpiezo+zelectrotop
VATT,3, ,3,0
size=1e-2 ALLSEL
ESIZE,size
mshkey,0 MSHAPE,1,3D VMESH,all
! BC
NSEL,s,LOC,x,xpiezo/2
NSEL,R,LOC,y,ypiezo/2
D,ALL,Ux,0
D,ALL,Uy,0
```

```
NSEL,s,LOC,x,xpiezo/2
D,ALL,Ux,0
NSEL,sR,LOC,y,ypiezo/2
D,ALL,Uy,0
ALLSEL
NSEL,R,LOC,x,0,
D,ALL,Ux,0
D,ALL,Uy,0
D,ALL,Uz,0
NSEL,s,LOC,y,0,
D,ALL,Ux,0
D,ALL,Uy,0
D,ALL,Uz,0
NSEL,S,LOC,z,zsubs+zelectrobot
CM,electromasasup,NODE
CP,1,VOLT,ALL
*GET,electremasasupnumber,NODE,,NUM,MIN
NSEL,s,LOC,z,zsubs+zelectrobot+zpiezo
NSEL,R,LOC,x,h2,h2+xelectrotop/2
NSEL,R,LOC,y,h1,h1+yelectrotop/2
CM,electrodetop,node
CP,2,VOLT,ALL
*GET,electrodetopnumer,NODE,,NUM,MIN
ALLSEL
D,electromasasup,VOLT,0.0
D,electrodetop,VOLT,1.0
!solution
/SOLU
ANTYPE,HARMONIC
HARFRQ,1100E6,1250E6
NSUB,151
OUTRES,ALL,ALL
KBC,1
SOLVE
! plot
/post26
rforce,2,ELECTRODETOPNUMER,chrg,,Ladug
PI=3.141592
cfact,1,0,0,2*PI,0,0
prod,3,2,1,,current
quot,4,1,1,,volt
quot,5,4,3,,impedance
quot,6,3,4,,admittance
quot,7,4,5,,power
plvar,freq,power
```

References

1. Abdollahzadeh Jamalabadi, M.Y. Rational Design Calculation of Surface Acoustic Wave Gas Sensor. *Phys. J.* **2020**, *6*, 63–87.
2. Abdollahzadeh Jamalabadi, M.Y.; Park, J.H. Investigation of Property Variations on Freezing of PCM Containing Nanoparticles. *World Appl. Sci. J.* **2014**, *32*, 672–677.
3. Makkonen, T.; Pensala, T.; Vartiainen, J.; Knuuttila, J.V.; Kaitila, J.; Salomaa, M.M. Estimating materials parameters in thin-film BAW resonators using measured dispersion curves. *IEEE Trans. Ultrason. Ferroelectr. Freq. Control* **2004**, *51*, 42. [CrossRef]
4. Abdollahzadeh Jamalabadi, M.Y.; Park, J.H. Effects of Brownian motion on freezing of PCM containing nanoparticles. *Therm. Sci.* **2016**, *20*, 1533–1541. [CrossRef]

5. Abdollahzadeh Jamalabadi, M.Y. MD Simulation of Brownian Motion of Buckminsterfullerene Trapping in nano-Optical Tweezers. *Int. J. Opt. Appl.* **2015**, *5*, 161–167.
6. Abdollahzadeh Jamalabadi, M.Y. Use of Nanoparticle Enhanced Phase Change Material for Cooling of Surface Acoustic Wave Sensor. *Fluids* **2021**, *6*, 31. [CrossRef]
7. Singh, R.P.; Sze, J.Y.; Kaushik, S.C.; Rakshit, D.; Romagnoli, A. Thermal performance enhancement of eutectic PCM laden with functionalised graphene nanoplatelets for an efficient solar absorption cooling storage system. *J. Energy Storage* **2021**, *33*, 102092. [CrossRef]
8. Sarbu, I.; Sebarchievici, C. A comprehensive review of thermal energy storage. *Sustainability* **2018**, *10*, 191. [CrossRef]
9. Khan, Z.; Khan, Z.; Ghafoor, A. A review of performance enhancement of PCM based latent heat storage system within the context of materials, thermal stability and compatibility. *Energy Convers. Manag.* **2016**, *115*, 132–158. [CrossRef]
10. Jiang, Z.Y.; Qu, Z.G. Lithium-ion battery thermal management using heat pipe and phase change material during discharge—Charge cycle: A comprehensive numerical study. *Appl. Energy* **2019**, *242*, 378–392. [CrossRef]
11. Tariq, S.L.; Ali, H.M.; Akram, M.A.; Janjua, M.M.; Ahmadlouydarab, M. Nanoparticles enhanced phase change materials (NePCMs)-A recent review. *Appl. Therm. Eng.* **2020**, *176*, 115305. [CrossRef]
12. Raj, C.R.; Suresh, S.; Singh, V.K.; Bhavsar, R.R.; Chandrasekar, M.; Archita, V. Life cycle assessment of nanoalloy enhanced layered perovskite solid-solid phase change material till 10000 thermal cycles for energy storage applications. *J. Energy Storage* **2021**, *35*, 102220. [CrossRef]
13. Tiari, S.; Hockins, A.; Mahdavi, M. Numerical study of a latent heat thermal energy storage system enhanced by varying fin configurations. *Case Stud. Therm. Eng.* **2021**, *25*, 100999. [CrossRef]
14. Tiari, S.; Mahdavi, M. Computational study of a latent heat thermal energy storage system enhanced by highly conductive metal foams and heat pipes. *J. Anal. Calorim.* **2020**, *141*, 1741–1751. [CrossRef]
15. Li, W.Q.; Guo, S.J.; Tan, L.; Liu, L.L.; Ao, W. Heat transfer enhancement of nano-encapsulated phase change material (NEPCM) using metal foam for thermal energy storage. *Int. J. Heat Mass Transf.* **2021**, *166*, 120737. [CrossRef]
16. Dutkowski, K.; Kruzel, M.; Zajączkowski, B. Determining the heat of fusion and specific heat of microencapsulated phase change material slurry by thermal delay method. *Energies* **2021**, *14*, 179. [CrossRef]
17. Abdulateef, A.M.; Jaszczur, M.; Hassan, Q.; Anish, R.; Niyas, H.; Sopian, K.; Abdulateef, J. Enhancing the melting of phase change material using a fins–nanoparticle combination in a triplex tube heat exchanger. *J. Energy Storage* **2021**, *35*, 102227. [CrossRef]
18. Nie, C.; Liu, J.; Deng, S. Effect of geometric parameter and nanoparticles on PCM melting in a vertical shell-tube system. *Appl. Therm. Eng.* **2021**, *184*, 116290. [CrossRef]
19. Aqib, M.; Hussain, A.; Ali, H.M.; Naseer, A.; Jamil, F. Experimental case studies of the effect of Al_2O_3 and MWCNTs nanoparticles on heating and cooling of PCM. *Case Stud. Therm. Eng.* **2020**, *22*, 100753. [CrossRef]
20. Khatibi, M.; Nemati-Farouji, R.; Taheri, A.; Kazemian, A.; Ma, T.; Niazmand, H. Optimization and performance investigation of the solidification behavior of nano-enhanced phase change materials in triplex-tube and shell-and-tube energy storage units. *J. Energy Storage* **2021**, *33*, 102055. [CrossRef]
21. Maher, H.; Rocky, K.A.; Bassiouny, R.; Saha, B.B. Synthesis and thermal characterization of paraffin-based nanocomposites for thermal energy storage applications. Therm. *Sci. Eng. Prog.* **2021**, *22*, 100797.
22. Chen, G. *Nanoscale Energy Transport and Conversion: A Parallel Treatment of Electrons, Molecules, Phonons, and Photons*; Oxford University Press: Oxford, UK, 2005; ISBN 9780195159424.
23. Santhosh, S.; Satish, M.; Yadav, A.; Anish Madhavan, A. Thermal analysis of Fe_2O_3—Myristic acid nanocomposite for latent heat storage. *Mater. Today Proc.* **2021**, *52*, 687.
24. U Mekrisuh, K.; Giri, S.; Udayraj; Singh, D.; Rakshit, D. Optimal design of the phase change material based thermal energy storage systems: Efficacy of fins and/or nanoparticles for performance enhancement. *J. Energy Storage* **2021**, *33*, 102126. [CrossRef]
25. Hosseinzadeh, K.; Erfani Moghaddam, M.A.; Asadi, A.; Mogharrebi, A.R.; Jafari, B.; Hasani, M.R.; Ganji, D.D. Effect of two different fins (longitudinal-tree like) and hybrid nano-particles (MoS2-TiO2) on solidification process in triplex latent heat thermal energy storage system. *Alex. Eng. J.* **2021**, *60*, 1967–1979. [CrossRef]
26. Nakhchi, M.E.; Hatami, M.; Rahmati, M. A numerical study on the effects of nanoparticles and stair fins on performance improvement of phase change thermal energy storages. *Energy* **2021**, *215*, 119112. [CrossRef]
27. Li, F.; Almarashi, A.; Jafaryar, M.; Hajizadeh, M.R.; Chu, Y.-M. Melting process of nanoparticle enhanced PCM through storage cylinder incorporating fins. *Powder Technol.* **2021**, *381*, 551–560. [CrossRef]
28. Yu, Q.; Zhang, C.; Lu, Y.; Kong, Q.; Wei, H.; Yang, Y.; Gao, Q.; Wu, Y.; Sciacovelli, A. Comprehensive performance of composite phase change materials based on eutectic chloride with SiO_2 nanoparticles and expanded graphite for thermal energy storage system. *Renew. Energy* **2021**, *172*, 1120–1132. [CrossRef]
29. Fang, Y. A Comprehensive Study of Phase Change Materials (PCMs) for Building Wall Applications. Ph.D. Dissertation, University of Kansas, Lawrence, KS, USA, 2009.

magnetochemistry

Article

Size and Ion-Doping Effects on Magnetic, Optical, and Phonon Properties of $CuAlO_2$ Nanoparticles

Iliana Naumova Apostolova [1], Angel Todorov Apostolov [2] and Julia Mihailova Wesselinowa [3,*]

1 Faculty of Forest Industry, University of Forestry, 1756 Sofia, Bulgaria
2 Faculty of Hydrotechnics, University of Architecture, Civil Engineering and Geodesy, 1046 Sofia, Bulgaria
3 Department of Physics, University of Sofia, 1164 Sofia, Bulgaria
* Correspondence: julia@phys.uni-sofia.bg

Abstract: The magnetic, optical, and phonon properties of ion-doped $CuAlO_2$ nanoparticles on the Cu or Al site are theoretically investigated. The room temperature ferromagnetism in $CuAlO_2$ nanoparticles can be due to the surface, size, and doping effects. The magnetization increases with the decreasing nanoparticle size. The different radii of the transition metal ion and the host Cu ion lead to compressive strain, to the enhancment of the exchange interaction constants, and to increased magnetization M_s and Curie temperature T_C. By substitution with Mn or Cr on the Al site, tensile strain, a decrease in M_s, and an increase in dopants are observed. The size and ion-doping influence on the band-gap energy is also discussed. The phonon energy ω decreases, whereas the phonon damping γ increases with increasing temperature and decreasing NP size. They show a kink around $T_C \sim 400$ K. The behavior of ω and γ for different ion dopings is observed.

Keywords: $CuAlO_2$; ion doping; magnetization; band gap; phonon energy; microscopic model

Citation: Apostolova, I.N.; Apostolov, A.T.; Wesselinowa, J.M. Size and Ion-Doping Effects on Magnetic, Optical, and Phonon Properties of $CuAlO_2$ Nanoparticles. *Magnetochemistry* **2022**, *8*, 169. https://doi.org/10.3390/magnetochemistry8120169

Academic Editors: Cătălin-Daniel Constantinescu and Lucian Petrescu

Received: 17 October 2022
Accepted: 22 November 2022
Published: 25 November 2022

1. Introduction

Diluted magnetic semiconductors (DMS) play an important role in interdisciplinary materials science and future spintronics. Ferromagnetic DMS have been studied from first principles within mean field approximation [1,2]. Ferromagnetism in 3d transition metal (TM)-doped II–VI and III–V based DMS is predicted [3]. Kizaki et al. [4] have investigated the magnetism in a new DMS, namely $CuAlO_2$ (CAO), by the Korringa–Kohn–Rostoker method. Using density functional theory (DFT) calculations, Iordanidou et al. [5] have examined the hole doping effect on the magnetic and electronic properties of CAO. It is known that pure CAO is a p-type width band gap semiconductor with E_g of about 3.5 eV [6]. It has the hexagonal delafossite structure and the space group of R$\bar{3}$m. CAO can be applied as a DMS when doped with TM ions. Kizaki et al. [4,7] have investigated the electronic and magnetic properties under carrier doping treatment in TM-doped CAO on the Cu site (TM = Fe, Co, Mn, and Ni). The Curie temperature T_C of the doped examples increases with increasing dopant concentration. Moreover, Fe or Co doping on the Al site of CAO bulk and thin films increases the spontaneous magnetization [8–12].

The origin of ferromagnetism in DMS is not yet clear, even if various methods have been proposed. First-principal calculations showed that (Cu,Fe)AlO_2 can be a candidate of ferromagnetic DMS [13]. From the calculated density of states, it seems that the double exchange interaction is the dominant exchange mechanism in (Cu,Fe)AlO_2 [13]. It was recently found that the weak magnetism of these materials may derive from the polarised unpaired electrons around impurities [14]. The magnetic properties of Mn-doped CAO, Cu(Al,Mn)O_2 have been reported by Zhang et al. [15]. The magnetization decreases with increasing Mn concentration. The influence of Cd impurity at Cu and Al sites on the electronic properties of CAO from first-principles calculations is discussed in Ref. [16].

The TM doping on the Al site also influences the optical properties of CAO thin films and nanofibers [17–21]. Raman spectra of pure and ion-doped CAO bulk and NPs are studied in [18,22–26]. So, carrier concentration and ion doping together significantly affect the oxide-based DMS.

The aim of the present paper is to investigate the magnetic, optical, and phonon properties of ion-doped CAO nanoparticles (NPs) for the first time using a microscopic model and Greens's function theory in order to clarify the origin of room temperature ferromagnetism in these systems on microscopic level.

It should be noted that the most theoretical papers consider the magnetic properties of CAO NPs using DFT. The DFT is a very powerful tool in investigation of many body problems. However, DFT is mostly concerned with ground-state properties at zero temperature. In our approach, we are able to cover the whole temperature regime. It is a finite temperature analysis including the entire excitation spectrum. In particular, the method allows us to study the total phase diagram, which is based on the different excitation energies realized in the system. The disadvantage of our approach consists of the consideration of collective properties from the beginning. Our basic quantities are not the "naked" electrons but effective spins of the underlying quasi-particles. Whereas within DFT all of the parameters of the system can be—at least in principle—calculated, we are forced to use additional models to find out those parameters. We are convinced that both approaches, DFT and Green's function method, are appropriate and, to a certain extent, are an alternative in describing many body systems.

2. The Model and Green's Functions

The room temperature ferromagnetism in bulk CAO can be caused by ion doping and can be described by the Heisenberg hamiltonian H_d:

$$H_d = \sum_{i,j} xJ_{dij}(\mathbf{S}_j \cdot \mathbf{S}_j) - \sum_i D_i(S_i^z)^2 - g\mu_B h \sum_i S_i^z. \tag{1}$$

$\mathbf{S}_i$ is the Heisenberg spin-operator of the TM at site i. J_d is the exchange interaction between the TM ions. Kizaki et al. [4] found that J_d between the magnetic ions in the same Cu-plane are ferromagnetic, whereas those between the Cu-planes can be neglected. D is the single-ion anisotropy, h is an external magnetic field, and x is the doping concentration.

In the case of a CAO NP, there also appears a ferromagnetism from surface and size effects due to the uncompensated Al spins on the surface described by H_s as well as due to ion-doping effects (see H_d):

$$H_s = -\sum_{i,j}(1-x)J_{sij}(\mathbf{S}_j \cdot \mathbf{S}_j). \tag{2}$$

From the spin Green's function defined as $G_{ij} = \ll S_i^+; S_j^- \gg$ the magnetization M for arbitrary spin S is calculated:

$$M = \frac{1}{N^2}\sum_k \Big[(S+0.5)\coth[(S+0.5)\beta E_m(\mathbf{k})] - 0.5\coth(0.5\beta E_m(\mathbf{k})\Big], \tag{3}$$

where $\beta = 1/k_B T$, $E_m(\mathbf{k})$ is the spin-wave energy.

The spin–phonon and phonon–phonon interactions are described by:

$$H_{sp-ph} = \frac{1}{2}\sum_{i,j,k} F(i,j,k)Q_i S_j^z S_k^z - \frac{1}{4}\sum_{i,j,r,s} R(i,j,r,s)Q_i Q_j S_r^z S_s^z + h.c. \tag{4}$$

The normal coordinate Q_i can be expressed in terms of phonon creation a^+ and annihilation a operators, $Q_i = (2\omega_{0i})^{-1/2}(a_i + a_i^+)$. F and R are the spin–phonon coupling constants in the first and second order, respectively.

$$\begin{aligned} H_{ph-ph} &= \frac{1}{2!}\sum_i \omega_{0i} a_i a_i^+ + \frac{1}{3!}\sum_{i,j,r} B(i,j,r) Q_i Q_j Q_r \\ &+ \frac{1}{4!}\sum_{i,j,r,s} A(i,j,r,s) Q_i Q_j Q_r Q_s, \end{aligned} \tag{5}$$

where ω_{0i} is the frequency of the lattice mode.

From the poles of the phonon Green's function $\tilde{G}_{ij}(t) = \langle\langle a_i(t); a_j^+\rangle\rangle$ the phonon energies are observed:

$$\begin{aligned} \omega_{ij}^2 &= \omega_0^2 - 2\omega_0\Big(M_i M_j R_{ij}\delta_{ij} \\ &- \frac{1}{2N'}\sum_r A_{ijr}(2\bar{N}_r + 1) - B_{ij}\langle Q_{ij}\rangle\delta_{ij}\Big), \end{aligned} \tag{6}$$

$$\langle Q_{ij}\rangle = \frac{M_i M_j F_{ij}\delta_{ij} - \frac{1}{N'}\sum_r B_{ijr}(2\bar{N}_r + 1)}{\omega_0 - M_i M_j R_{ij}\delta_{ij} + \frac{1}{N'}\sum_r A_{ijr}(2\bar{N}_r + 1)}. \tag{7}$$

The phonon correlation function $\bar{N}_r = \langle a_r^+ a_r\rangle$ is obtained via the spectral theorem. The phonon damping $\gamma = \gamma_{sp-ph} + \gamma_{ph-ph}$ is also calculated taking into account anharmonic spin–phonon and phonon–phonon interactions.

The band-gap energy E_g of a CAO NP is defined by the difference between the valence and conduction bands:

$$E_g = \omega^+(\mathbf{k} = 0) - \omega^-(\mathbf{k} = \mathbf{k}_\sigma). \tag{8}$$

The electronic energies

$$\omega^{\pm}(k) = \epsilon_k - \frac{\sigma}{2} I\langle S^z\rangle \tag{9}$$

are observed from the Green's function $g(k,\sigma) = \ll c_{k,\sigma}; c_{k\sigma}^+ \gg$, $\sigma = \pm 1$, $c_{i\sigma}^+$ and $c_{i\sigma}$ are Fermi operators, and I-s-d is the interaction constant [27].

3. Numerical Results and Discussion

A CAO NP with a cubo-octahedral shape is defined by fixing the origin at a certain Al spin in the center of the particle and including all other spins within the particle into shells, which are numbered by $n = 1, \ldots, N$ from the central to the surface shell. For the numerical calculations, we use the following model parameters: J_d(Fe-Fe) = 195 meV, J_d(Co-Co) = 117 meV, J_d(Mn-Mn) = 78 meV, J_d(Ni-Ni) = 97.5 meV [4], J_s = 20 meV, $D = -0.068$ meV, $F = 23$ cm^{-1}, $R = -18$ cm^{-1}, $A = 6.61$ cm^{-1}, and $B = -2.94$ cm^{-1}. The exchange interaction $J_{ij} = J(r_i - r_j)$ depends on the distance between the spins, i.e., on the lattice parameters. It is inverse proportional to the lattice parameters. Therefore, the exchange interaction constant on the surface J_s and of the doped states J_d can be changed or can increase or decrease in dependence on the strain in the lattice. So, we take into account the direct connection between the microstructure and magnetic behavior.

Firstly, we will consider the spontaneous magnetization M_s in a pure CAO NP. It must be noted that bulk CAO is a p-type wide band gap semiconductor, and it is a nonmagnetic compound. However, due to the uncompensated surface spins of the Al^{3+} ions on the surface, a spontaneous magnetization M_s appears in the CAO NP, contrary to the bulk case, where $M_s = 0$. We obtain that J_s increases and the spontaneous magnetization M_s of a CAO NP increases with decreasing NP size. The result is presented in Figure 1. A finite magnetization M_s is also observed experimentally in pure CAO NPs [12,28] and in pure CAO thin films [29], due to the existence of point defects. It must be noted that we obtain similar behavior for the size dependence of the phase-transition temperature $T_C(N)$, T_C, which increases with decreasing NP size.

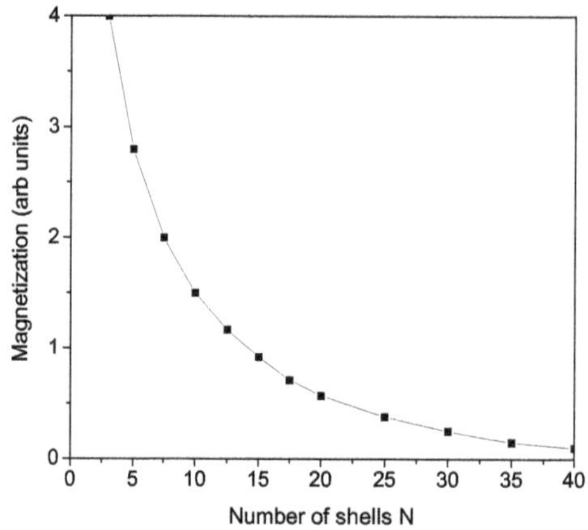

Figure 1. Size dependence of the spontaneous magnetization of a CAO NP for T = 300 K.

The origin of ferromagnetism in DMS is not yet clear. In order to investigate the observed room temperature ferromagnetism (RTFM) in ion-doped CAO-bulk and NPs, we have replaced some Cu-ions with Fe-ions, i.e., we consider $Cu_{1-x}Fe_xAlO_2$ NPs for x = 0–0.25. There are experimental data that show that in this concentration interval the samples are single-phase, whereas for larger x values, the second phase appears [4]. It must be noted that by the doping of TM on the Al site, the limit value of the ion doping concentration is smaller; it is around 0.05–0.1 [8,10,15,28,30], which is different for the different doping ions. The exchange interaction constant J depends on the distance between the spins, i.e., on the lattice parameter, on the different strain, on the lattice symmetry, and on the number of next neighbors. J is an inverse square function of the distance between two neighboring spins, i.e., $J \propto 1/(r_i - r_j)^2$. From the structural analysis, Cu^{1+} is replaced by Fe^{3+} or Fe^{2+}, which leads to cation vacancies increasing because of charge variation [4,7,9]. Through the different radii of the dopants, lattice defects can be introduced or intrinsic host-lattice defects can be activated when different ions, such as Fe, occupied the Cu sites. Chen et al. [9] have shown that Fe element exists in the ion form, with the valence of +2 and/or +3. Since the ionic radius of Fe^{2+} (0.92 $\dot{A}$) or Fe^{3+} (0.65 $\dot{A}$) is always smaller than Cu^{1+} (0.95 $\dot{A}$), the lattice parameters decrease with increasing the Fe dopants as supported by Chen et al. [9]. The reduction of the lattice constants in Fe-doped CAO nanostructures leads to a compressive strain, i.e., in our microscopic model, it leads to an increase in the exchange interaction constant of the doped states J_d with an increase in the Fe ion concentration because it is inverse proportional to the lattice parameters . So, we obtain an increase in the spontaneous magnetization M_s and the Curie temperature T_C. The ferromagnetic coupling between the doping Fe-ions contributes additively to this increase. Fe^{3+}-ion (Fe^{2+}-ion) has 5d (6d) electrons with the total spin of $S = 5/2$ ($S = 2$), whereas the Cu^{1+}-ion has $S = 0$. Thus, Fe^{3+} or Fe^{2+} substitution into Cu^{1+} induces an extra magnetic moment. The spontaneous magnetization M_s and the Curie temperature T_C in dependence on the Fe^{3+} doping concentration are calculated. We observe that both increase with increasing Fe doping concentration x. The result for $T_C(x)$ is demonstrated in Figure 2, curve 1. We would obtain a similar increase in the phase-transition temperature T_C, for example, for Ni-, Mn-, or Co-doped CAO, (Cu,TM)AlO_2 (Figure 2, curves 2–4), where the doping ions have a smaller ionic radius in comparison with that of the Cu ion. This behavior of the doping dependence of $T_C(x)$ is in good qualitative agreement with the experimental data [8–12].

It must be noted that for example, in Mn- or Cr-doped CAO on the Al site, where the ionic radius of Mn^{3+}(0.72 Å) or Cr^{3+}(0.63 Å) is larger than that of Al^{3+}(0.51 Å), the lattice parameters increase in agreement with the experimental data of Zhang et al. [15]. A tensile strain appears, i.e., J_d decreases, which leads to a decrease in M_s with increasing Mn or Cr ion doping concentration (see Figures 3 and 4, respectively). Similar behavior is reported for $Cu(Al,Mn)O_2$ by Zhang et al. [15]. Unfortunately, there are not experimental data for $M(x)$ of Cr-doped CAO. A decrease in the magnetization is observed additively with increasing Mn content x due to the antiferromagnetic coupling among the Mn spins (contrary to the case of Mn doping at the Cu site). By the substitution of the Al ion with a rare earth ion (RE) whose radius is larger than that of Al, tensile strain and a reduction of the spontaneous magnetization M_s with an increase in the RE ion doping concentration appears again. Unfortunately, there are not experimental data for RE doped CAO.

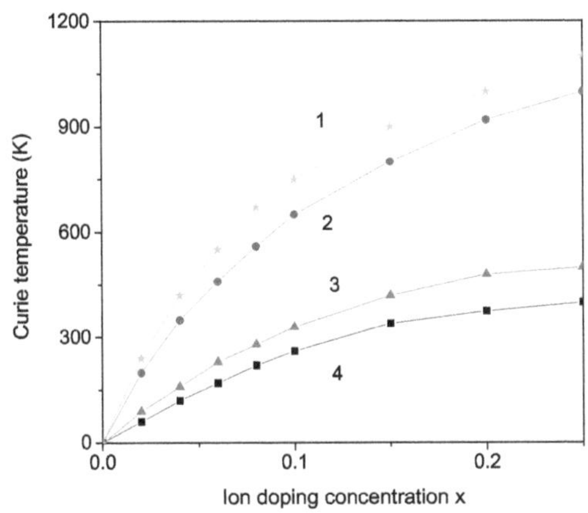

Figure 2. (Color online) The Curie temperature T_C as a function of the ion doping concentration x for different ions substituted the Cu ion: (1) Fe; (2) Co; (3) Mn; and (4) Ni.

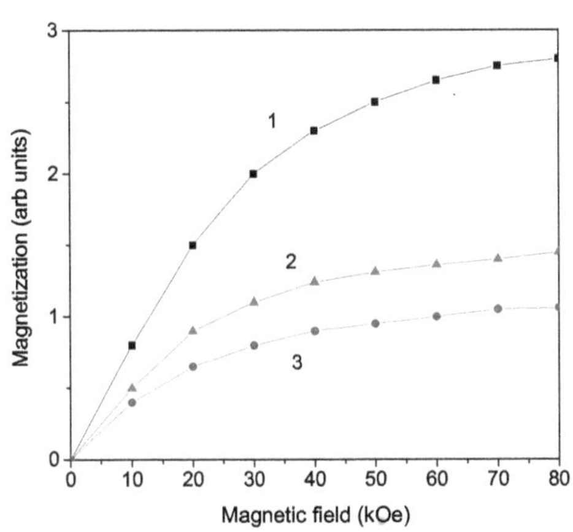

Figure 3. (Color online) Magnetic field dependence of the magnetization M_s of $Cu(Al,Mn)O_2$ for different Mn-ion doping concentrations x: (1) 0.01; (2) 0.03; and (3) 0.05.

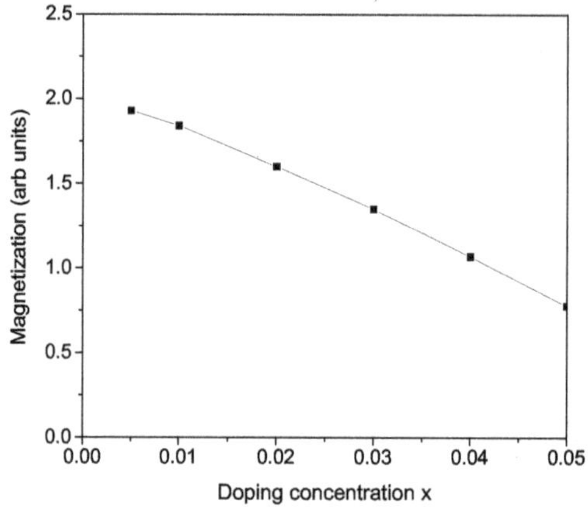

Figure 4. Dependence of the magnetization M_s of Cu(Al,Cr)O_2 on the doping Cr ion concentration.

From Equation (8), we have calculated the band-gap energy E_g for pure and ion-doped CAO NPs (see Figure 5). It must be noted that there is a broadening of the band gap E_g of pure CAO NPs with a decrease in the NP size. We now consider the case of a Cr^{3+}-doped CAO NP, $CuAl_{1-x}Cr_xO_2$. The lattice parameters increase with the increase in the Cr-doping concentration [17] because the ionic radius of the doping Cr ion (0.63 $\dot{A}$) is larger than that of the host Al ion (0.51 $\dot{A}$). A tensile strain appears, i.e., J_d decreases with increasing x, which leads to a decrease in the magnetization (see Figure 4) and an increase in E_g (curve 1). The observed behavior is in agreement with the experimental data of Jiang et al. [17]. We would also obtain a similar enhanced E_g within our model by doping with Sb^{3+} (0.76 $\dot{A}$) ions or Y^{3+} (1.04 $\dot{A}$) ions, which also causes a tensile strain, i.e., an increase in the lattice parameters as well as a decrease in the exchange interaction constant J_d and therefore reduced magnetization M_s and enhanced E_g, reported in [18,31]. Otherwise, by doping with the Mg ion, we observe the contrary result: a decrease in the band-gap energy (curve 2), in agreement with [28,30,32]. The optical absorption of the CAO increases due to the reduction in E_g by Mg doping. A smaller E_g is also reported by Bi and Eu doping at the Al site in CAO [21,33].

Finally, we will discuss the temperature, size, and ion-doping dependence of the phonon energy ω and damping γ in CAO NPs. The temperature dependence for the A_{1g} mode ω_0 = 767 cm^{-1} of the phonon energy ω and phonon damping γ is shown in Figure 6. We take into account the anharmonic spin–phonon interaction $R < 0$, as well as the three-phonon $A > 0$ and four-phonon $B < 0$ anharmonic interactions. Let us emphasize that in order to obtain the experimentally observed softening of ω, we have to chose $R < 0$. For $R > 0$, we would obtain a hardening, an increase in ω with T [34]. The damping γ is proportional to R^2 and so independent of the sign of R. γ corresponds to the full width at half-maximum of the Raman lines and is indirectly proportional to the phonon life time. The damping is a direct consequence of the anharmonic coupling. The phonon energy decreases, whereas the phonon damping increases with increasing temperature, which is in agreement with [22,26]. It can be seen from Figure 6 that around $T_c \sim 400$ K, we observe a kink due to the spin–phonon interaction R. For $R = 0$, the kink disappears. Moreover, the kink is size-dependent; it shifts to smaller T values with increasing NP size. In Figure 1, we have shown that with decreasing NP size spontaneous magnetization M_s with the corresponding T_C value appears. A similar anomaly in the phonon energy and damping of CAO thin films is observed experimentally between 423 and 573 K by Singh et al. [22] and around $T \sim 300$ K by Pellicer-Porres et al. [26]. The authors have proposed that these changes may be related to a negative thermal expansion (NTE) in the delafossite structure along the O-Cu-O linkage. A similar NTE is also reported in other AMO_2 (A = Cu or Ag; M = Al, Sc, In, or La) [35–37]. NTE means a volume contraction, i.e., there is a change in the structure, for example, by transitions between ferro- and paramagnetic phases.

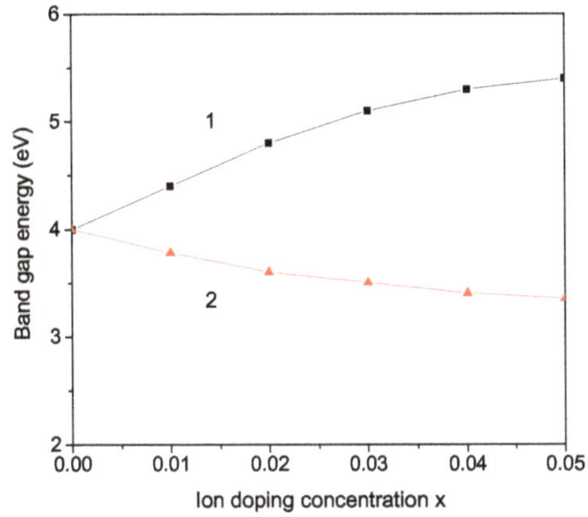

Figure 5. (Color online) Dependence of the band-gap energy E_g on the doping concentration x for a Cr-doped (1) and Mg-doped (2) substitution of the Al ion in CAO NP, $N = 10$, and $T = 300$ K.

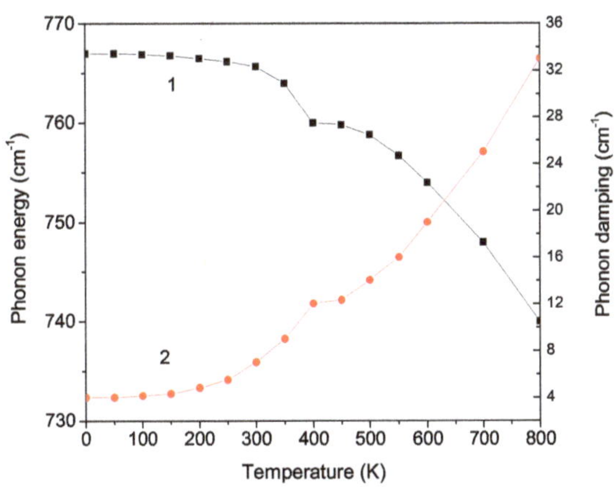

Figure 6. (Color online) Temperature dependence of the phonon energy ω (1) and phonon damping γ (2) of a CAO NP, $N = 10$.

The size dependence of the phonon energy ω and the phonon damping γ in a CAO NP is presented in Figure 7. It can be seen that the phonon energy decreases whereas the damping increases with decreasing nanoparticle size, which is in agreement with the experimental data of Yassin et al. [23].

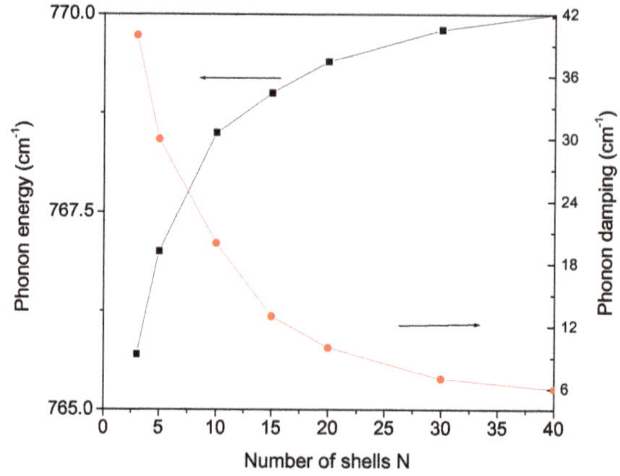

Figure 7. (Color online) Size dependence of the phonon energy ω and phonon damping γ of a CAO NP, $T = 300$ K.

The ion-doping also influences the phonon properties of CAO NPs. It may cause lattice distortions and unrelaxed strains. To show this, we will consider the Fe^{3+} doping at the Al^{3+} site. The ionic radius of the Fe ion (0.65 $\dot{A}$) is larger than that of the Al ion (0.51 $\dot{A}$), i.e., we observe tensile strain, an increase in the lattice parameters that leads to a decrease in J_d and R_d and to a decrease in the phonon energy with increasing Fe ions, see Figure 8, curve 1. This corresponds to a red-shift in the peak positions. The contribution to the damping of the ion doping is in addition to those of the bulk, surface, spin–phonon, and phonon–phonon interactions, so that γ increases with increasing Fe concentration (see Figure 8, curve 2). This means that the full-weight of the half maximum increases, i.e., the Raman lines are broadener. This is in agreement with the experimental results of Fe-doped CAO [38]. We obtain a similar decrease in ω and an increase in γ for Eu-doped CAO NP as reported by Liu et al. [21].

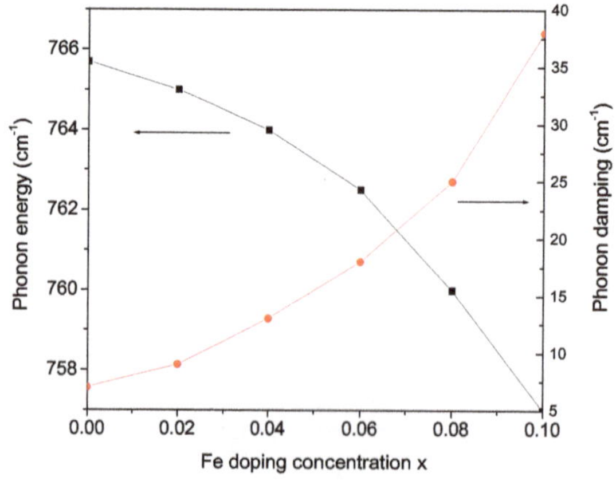

Figure 8. (Color online) Fe ion-doping dependence of the phonon energy ω and phonon damping γ of a $Cu(Fe,Al)O_2$ NP, $N = 10$, and $T = 300$ K.

4. Conclusions

To conclude, we have studied the size and ion-doping effects on the magnetic, optical, and phonon properties of transition metal (TM = Fe, Co, Mn, and Ni) ion-doped CAO NPs on the Cu and Al site using a microscopic model and Green's function theory. The RTFM in bulk CAO can be due to ion doping, whereas CAO NPs make an additive contribution in terms of surface and size effects . Due to surface defects, a magnetization is obtained that increases with decreasing NP size. The difference in the radii of the TM^{3+} and the host Cu^+ ions leads to a compressive strain, an increase in the exchange interaction constants, and increased spontaneous magnetization M_s and Curie temperature T_C. The doping TM^{3+} ion has an additional spin value compared to the zero spin value of the Cu^+ ion, which also enhances M_s. On the other hand, a tensile strain is observed by substitution with Mn^{3+} or Cr^{3+} on the Al^{3+} site, where the radius of the Mn or Cr ions is larger than that of the Al ion, which leads to a decrease in M_s with an increase in the ion-doping concentration. The band-gap energy increases with decreasing NP size and by doping with Cr or Sb ions and decreases by Mg ion doping. The phonon energy ω decreases, whereas the phonon damping γ increases, with increasing temperature. They show a kink around the phase-transition temperature $T_C \sim 400$ K, caused by the anharmonic spin–phonon interaction. Due to surface and size effects, ω decreases, whereas γ increases, with decreasing NP size. The phonon energy decreases, whereas the damping increases, by Fe and Eu doping on the Al site, i.e., a red shift and broadening of the Raman peaks is obtained. The observed results are in good qualitative agreement with the existing experimental data.

Let us emphasize that in order to take into account the carrier doping treatment in transition-metal-doped CAO on the Cu site, which plays an additive role in the effect of the different ionic radii, we have to expand our model and consider the s-d model by also taking into account the electron–phonon interaction, which is important for CAO [39]. This could be done in the next paper.

Author Contributions: All authors contributed equally to this work. All authors have read and agreed to the published version of the manuscript.

Funding: This research received no external funding.

Data Availability Statement: The raw data that support the findings of this study are available from the corresponding author upon reasonable request.

Conflicts of Interest: The authors declare no conflict of interest.

References

1. Dietl, T.; Ohno, H.; Matsukura, F.; Cibert J.; Ferrand, D. Zener Model Description of Ferromagnetism in Zinc-Blende Magnetic Semiconductors. *Science* **2000**, *287*, 1019. [CrossRef]
2. Sato, K.; Dederichs, P.H.; Katayama-Yoshida, H. Curie temperatures of III-V diluted magnetic semiconductors calculated from first principles. *Europhys. Lett.* **2003**, *61*, 403. [CrossRef]
3. Sato K.; Katayama-Yoshida, H. First principles materials design for semiconductor spintronics. *Semicond. Sci. Technol.* **2002**. *17*, 367. [CrossRef]
4. Kizaki, H.; Sato K.; Katayama-Yoshida, H. Materials Design of $CuAlO_2$-Based Dilute Magnetic Semiconductors by First-Principles Calculations and Monte Carlo Simulations. *Jpn. J. Appl. Phys.* **2008**, *47*, 6488. [CrossRef]
5. Iordanidou K.; Persson, C. Stoner Ferromagnetism in Hole-Doped $CuM^{IIIA}O_2$ with M^{IIIA} = Al, Ga, and In. *ACS Appl. Mater. Interfaces* **2021**, *13*, 29770. [CrossRef]
6. Kawazoe, H.; Yasukawa, M.; Hyodo, H.; Kurita, M.; Yanagi H.; Hosono, H. P-Type Electrical Conduction in Transparent Thin Films of $CuAlO_2$. *Nature* **1997**, *389*, 939. [CrossRef]
7. Kizaki, H.; Sato, K.; Katayama-Yoshida, H. Carrier concentration dependence of Curie temperature in $CuAlO_2$ based dilute magnetic semiconductor by first-principles calculations. *Phys. Stat. Sol. (c)* **2007**, *3*, 4135. [CrossRef]
8. Aziziha, M.; Byard, S.A.; Beesely, R.; Lewis, J.P.; Seehra, M.S.; Johnson, M.B. Magnetic properties of Fe-doped $CuAlO_2$ and role of impurities. *AIP Adv.* **2019**, *9*, 035030. [CrossRef]
9. Chen, C.; Dong, C.; Wang, B.; Huang, J.; Wang, Y. Synthesis and Room Temperature Ferromagnetism in Fe-Doped $CuAlO_2$ Semiconductor. *J. Wuhan Univ. Technol.-Mater. Sci. Ed.* **2013**, *28*, 500. [CrossRef]
10. Dong, C.J.; Yu, W.X.; Xu, M.; Cao, J.J.; Zhang, Y.; Chuai, Y.H.; Wang, Y.D. Evidence of room temperature ferromagnetism in Co-doped transparent $CuAlO_2$ semiconductor. *J. Alloys Compd.* **2012**, *512*, 195. [CrossRef]

11. Wang, Y.; Dong, C.; Chuai, Y.; Wang, Y. Room temperature ferromagnetism in Codoped $CuAlO_2$ nanofibers fabricated by electrospinning. *J. Wuhan Univ. Technol.-Mater. Sci.* **2015**, *30*, 1. [CrossRef]
12. Ray, N.; Gupta, V.; Sarma, L.; Kush, P.; Nag, J.; Sapra, S. Tuning the Electronic and Magnetic Properties of $CuAlO_2$ Nanocrystals Using Magnetic Dopants. *ACS Omega* **2018**, *3*, 509. [CrossRef]
13. Hidetoshi, K.; Kazunori, S.; Akira, Y.; Katayama-Yoshida, H. First-Principles Materials Design of $CuAlO_2$ Based Dilute Magnetic Semiconducting Oxide. *Jpn. J. Appl. Phys.* **2005**, *44*, L1187.
14. Sun, C.Q. Dominance of broken bonds and nonbonding electrons at the nanoscale. *Nanoscale* **2010**, *2*, 1930. [CrossRef] [PubMed]
15. Zhang, H.Y.; Li, P.G.; Chen, C.P.; Tu, Q.Y.; Tang, W.H. Magnetic properties of Mn-doped transparent $CuAlO_2$ semiconductor. *J. Alloys Compd.* **2005**, *396*, 40.
16. Lalic, M.V.; Mestnik-Filho, J.; Carbonari, A.W.; Saxena, R.N.; Moralles, M. Influence of Cd impurity on the electronic properties of $CuAlO_2$ delafossite: First-principles calculations. *J. Phys. Condens. Matter* **2002**, *14*, 5517. [CrossRef]
17. Jiang, H.F.; Zhu, X.B.; Lei, H.C.; Li, G.; Yang, Z.R.; Song, W.H.; Dai, J.M.; Sun,Y.P.; Fu, Y.K. Effect of Cr doping on the optical-electrical property of $CuAlO_2$ thin films derived by chemical solution deposition. *Thin Solid Films* **2011**, *519*, 2559.
18. Liu, Y.; Huang, Y.; Seo, H.J.; Wu, Y. Blueshift in near-band-edge emission in Y^{3+}-doped $CuAlO_2$ nanofibers. *Opt. Mater. Express* **2014**, *4*, 2602. [CrossRef]
19. Liu, R.; Li, Y.; Yao, B.; Ding, Z.; Jiang, Y.; Meng, L.; Deng, R.; Zhang, L.; Zhang, Z.; Zhao, H.; et al. Shallow Acceptor State in Mg-Doped $CuAlO_2$ and Its Effect on Electrical and Optical Properties: An Experimental and First-Principles Study. *ACS Appl. Mater. Interfaces* **2017**, *9*, 12608. [CrossRef]
20. Zhao, X.; Zhang, M.; Bai, P.; Hou, X.; Liu, F.; Yan, H. Optical, electrical, and structural properties of Fe-doped $CuAlO_2$ thin films. *Funct. Mater. Lett.* **2019**, *12*, 1850106. [CrossRef]
21. Liu, Y.; Gong, Y.; Mellott, N.P.; Wang, L.; Ye, H.; Wu, Y. Luminescence of delafossitetype $CuAlO_2$ fibers with Eu substitution for Al cations. *Sci. Technol. Adv. Mater.* **2016**, *17*, 200. [CrossRef] [PubMed]
22. Singh, M.K.; Dussan, S.; Sharma, G.L.; Katiyar, R.S. Raman scattering measurements of phonon anharmonicity in thin films. *J. Appl. Phys.* **2008**, *104*, 113503. [CrossRef]
23. Yassin, O.A.; Alamri, S.N.; Joraid, A.A. Effect of particle size and laser power on the Raman spectra of $CuAlO_2$ delafossite nanoparticles. *J. Phys. D Appl. Phys.* **2013**, *46*, 235301. [CrossRef]
24. Benreguia, N.; Barnabe, A.; Trari, M. Sol–gel synthesis and characterization of the delafossite $CuAlO_2$. *J. Sol-Gel Sci. Technol.* **2015**, *75*, 670. [CrossRef]
25. Byrne, D.; Cowley, A.; Bennettc, N.; McGlynn, E. The luminescent properties of $CuAlO_2$. *J. Mater. Chem. C* **2014**, *2*, 7859. [CrossRef]
26. Pellicer-Porres, J.; Martinez-Garcia, D.; Segura, A.; Rodriguez-Hernandez, P.; Munoz, A.; Chervin, J.C.; Garro, N.; Kim, D. Pressure and temperature dependence of the lattice dynamics of $CuAlO_2$ investigated by Raman scattering experiments and *ab initio* calculations. *Phys. Rev. B* **2006**, *74*, 184301. [CrossRef]
27. Wesselinowa, J.M.; Apostolov, A.T. Self-consistent theory of spin–phonon interactions in ferromagnetic semiconductors. *J. Phys. Condens. Matter* **1993**, *5*, 3555. [CrossRef]
28. Agrawal, S.; Parveen, A.; Azam, A. Influence of Mg on structural, electrical and magnetic properties of $CuAlO_2$ nanoparticles. *Mater. Lett.* **2016**, *168*, 125. [CrossRef]
29. Luo J.; Lin, Y.-J. Point defect-induced magnetic properties in $CuAlO_2$ films without magnetic impurities. *Appl. Phys. A* **2016**, *122*, 163. [CrossRef]
30. Dong, G.; Zhang, M.; Lan, W.; Dong, P.; Yan, H. Structural and physical properties of Mg-doped $CuAlO_2$ thin films. *Vacuum* **2008**, *82*, 1321. [CrossRef]
31. Ghosh, C.K.; Popuri, S.R.; Sarkar, D.; Chattopadhyay, K.K. Sb-doped $CuAlO_2$: Widening of band gap and nonlinear J-E characteristics. *J. Mater. Sci.* **2011**, *46*, 1613. [CrossRef]
32. Zou, Y.S.; Wang, H.P.; Zhang, S.L.; Lou, D.; Dong, Y.H.; Song, X.F.; Zeng, H.B. Structural, electrical and optical properties of Mg-doped $CuAlO_2$ films by pulsed laser deposition. *RSC Adv.* **2014**, *4*, 41294. [CrossRef]
33. Daichakomphu, N.; Klongratog, B.; Rodpun, P.; Pluengphon, P.; Harnwunggmoung, A.; Poo-arporn, Y.; Sakulkalavek, A.; Sakdanuphab, R. Improving the photo-thermoelectric performance of $CuAlO_2$ *via* doping with Bi. *Mater. Res. Bull.* **2021**, *144*, 111479. [CrossRef]
34. Wesselinowa, J.M.; Apostolov, A.T. Anharmonic effects in ferromagnetic semiconductors. *J. Phys. Condens. Matter* **1996**, *8*, 473. [CrossRef]
35. Li, J.; Sleight, A.; Jones, C.Y.; Toby, B. Trends in negative thermal expansion behavior for AMO_2 (A=Cu or Ag; M=Al, Sc, In, or La) compounds with the delafossite structure. *J. Solid State Chem.* **2005**, *178*, 285. [CrossRef]
36. Miller, W.; Smith, C.W.; Mackenzie, D.S.; Evans, K.E. Negative Thermal Expansion: A Review. *J. Mater. Sci.* **2009**, *44*, 5441. [CrossRef]

37. Salke, N.P.; Rao, R.; Achary, S.N.; Tyagi, A.K. Raman spectroscopic investigations on delafossite $CuLaO_2$ at high pressures. *J. Phys. Conf. Ser.* **2012**, *377*, 012020. [CrossRef]
38. Aziziha, M.; Akbarshahi, S.; Ghosh, S.; Pramanik, P.; Lewis, J.P.; Romero, A.H.; Thota, S.; Seehra, M.S.; Johnson, M.B. Phonon Dynamics in Anisotropic Dilute $CuAl_{1-x}Fe_xO_2$ Delafossite Alloys by a Weighted Dynamical Matrix Approach. *J. Phys. Chem. C* **2019**, *123*, 30604. [CrossRef]
39. Nakanishi, A.; Katayama-Yoshida, H. Chemical trend of superconducting transition temperature in hole-doped delafossite of $CuAlO_2$, $AgAlO_2$ and $AuAlO_2$. *Solid State Commun.* **2012**, *152*, 2078. [CrossRef]

magnetochemistry

MDPI

Article

Mesostructure and Magnetic Properties of SiO_2-Co Granular Film on Silicon Substrate

Natalia A. Grigoryeva [1,*], Victor Ukleev [2], Alexey A. Vorobiev [3], Alexander I. Stognij [4], Nikolay N. Novitskii [4], Leonid V. Lutsev [5] and Sergey V. Grigoriev [6]

1 Department of Physics, Saint-Petersburg State University, 198504 St. Petersburg, Russia
2 Laboratory for Neutron Scattering and Imaging, Paul Scherrer Institute, CH-5232 Villigen, Switzerland
3 Department of Physics and Astronomy, Uppsala University, 75120 Uppsala, Sweden
4 Scientific and Practical Materials Research Center of the National Academy of Sciences of Belarus, BY-220072 Minsk, Belarus
5 Ioffe Physical-Technical Institute of Russian Academy of Sciences, 194021 St. Petersburg, Russia
6 Petersburg Nuclear Physics Institute NRC "Kurchatov Institute", Gatchina, 188300 St. Petersburg, Russia
* Correspondence: n.a.grigorieva@yandex.ru

Abstract: Granular films SiO_2(Co) exhibit unusual magnetic and magnetotransport properties which are strongly dependent on the composition of the film and material of a substrate. For example, the injection magnetoresistance (IMR) coefficient reaches a giant (GIMR) value of 10^5% at room temperature in SiO_2(Co) films on an n-GaAs substrate. However, the IMR effect is negligible in the case of a similar granular film deposited on the n-Si substrate. In this report, the structural and magnetic properties of granular film SiO_2(Co) on Si substrate are studied with the aim to understand the cause of the difference in IMR coefficients for SiO_2(Co) thin film deposited on n-GaAs and on n-Si substrates. Investigations were carried out using complementary methods of Polarized Neutron Reflectometry, Grazing Incidence Small-Angle X-ray Scattering, X-ray Reflectometry, Scanning Electron Microscope, and SQUID magnetometry. It is shown that the interface layer between the granular film and Si substrate exhibits metallic rather than magnetic properties and eliminates the GIMR effect. This interface layer is associated with the Si diffusion to Co nanoparticles and the formation of the metallic cobalt silicides.

Keywords: granular film; nanoparticles; grazing-incidence small-angle X-ray scattering; polarized neutron reflectometry

PACS: 75.70.Cn; 75.75.–c; 75.50.Lk

Citation: Grigoryeva, N.A.; Ukleev, V.; Vorobiev, A.A.; Stognij, A.I.; Novitskii, N.N.; Lutsev, L.V.; Grigoriev, S.V. Mesostructure and Magnetic Properties of SiO_2-Co Granular Film on Silicon Substrate. *Magnetochemistry* **2022**, *8*, 167. https://doi.org/10.3390/magnetochemistry8120167

Academic Editors: Cătălin-Daniel Constantinescu and Lucian Petrescu

Received: 25 September 2022
Accepted: 15 November 2022
Published: 24 November 2022

1. Introduction

Granular films (GFs) are nanocomposites that consist of metallic magnetic nanoparticles incorporated in an insulating matrix. GFs with a very well-controlled nanostructure have promising technological applications in microwave devices, spintronics, medicine and biology [1–7]. Structural investigations of GFs play a crucial role and often facilitates the interpretation of their electrical and magnetic properties as well as their interconnection. A complex interplay of intrinsic properties can be caught through their structural features also accounting for finite-size effects, size distribution, surface effects, and nanoparticle interactions.

The effect of giant injection magnetoresistance (GIMR) has been observed in the GFs SiO_2(x at.% Co) deposited on GaAs substrate [8,9]. The GIMR phenomenon is a magnetic-field-induced suppression of the spin-polarized electron current from the GF into the semiconductor substrate. The IMR coefficient has maximum values for structures with Co concentrations in the range 54 ÷ 75 at.%. In the case of the opposite current direction (electrons drift from the semiconductor into the granular film) the magnetoresistance effect becomes less pronounced. The GIMR coefficient reaches 10^5 % in a narrow temperature range 240 K < T < 300 K at the magnetic field of H = 10 kOe applied in the sample plane and at the voltage U = 70 V applied perpendicular to the sample plane.

The extreme value of this effect at room temperature and the possibility of controlling the resistivity of heterostructure by varying the magnetic field could be used in the development of spintronic devices such as spin injectors, spin diodes, and transistors. Unlike crystalline silicon, the gallium arsenide substrate has a number of disadvantages in practical use: it is mechanically fragile and thermally unstable above 600 °C due to As evaporation, which imposes restrictions on the synthesis technology. The use of silicon substrates would eliminate these problems and provide additional advantages, such as its intrinsic carrier concentration being $n_i = 1.02 \times 10^{10}$ cm^{-3} and the energy gap being $E_g = 1.12$ eV (compared to GaAs – $n_i = 2.1 \times 10^6$ cm^{-3} and $E_g = 1.43$ eV, respectively). However, while the GIMR effect reaches 10^5 % in the case of the GF deposited on GaAs substrate, the GIMR is negligible in the case of a similar granular film deposited on n-Si substrate [8], and the intrinsic magnetoresistance of SiO_2(x at.% Co)/Si films has even negative values. In the present work we studied Au/SiO_2(Co x at.%)/Si heterostructures with cobalt content of x = 54, 60, 70, 75, 82 at.%. The aim of the study is to investigate the structural features and the associated magnetic properties of the heterostructures Au/GF/Si and to reveal the role of Si substrate in the negligible GIMR effect. For this purpose, we used the Polarized Neutron Reflectometry (PNR) technique, which we combined with complementary techniques of Grazing Incidence Small Angle X-ray Scattering (GISAXS), X-ray Reflectometry (XRR), Scanning Electron Microscope (SEM), and Superconducting Quantum Interference Device Magnetometry (SQUID).

The paper is organized in the following way. Section 2 gives details of the sample preparation and experimental results obtained via different methods to reveal structural and magnetic properties. Section 3 provides a discussion and interpretation of the results combined. A combination of different methods makes it possible to build a model explaining the negligible value of the GIMR effect, providing that almost all structural and magnetic properties of GF are practically the same for the films disposed on GaAs and Si substrates. It is shown that the interface layer between the GF and Si substrate shows metallic rather than magnetic properties and eliminates the GIMR effect. This interface layer is associated with the Si diffusion to Co nanoparticles and the formation of metallic cobalt silicides. Section 4 is the conclusion.

2. Experiment

2.1. Sample Preparation

The granular films SiO_2(x at.% Co) were synthesized by ion beam co-sputtering [10,11] of the composite cobalt-quartz target on the commercial n-Si substrate heated to 200 °C and misoriented by 2 degrees from (100) plane to (110) plane and with the thickness of 0.4 mm. The resistivity of the Si substrate was measured by the dc four-probe method at room temperature and was equal to 3.7 Ω·cm. The concentration of cobalt nanoparticles in SiO_2 was set by the ratio of cobalt and quartz target areas. The film was capped by 2 nm of the gold layer as a conductive contact for magnetoresistance measurements. The sputtering method and choice of components are described in ref. [12]. A detailed description of the characteristics of granular films is given in ref. [8].

The granular films SiO_2(x at.% Co) can be differentiated by their conductivity into three classes depending on the Co concentration.

(1) Class of GFs (SiO_2(x at.% Co), at x < 40 at.%) shows dielectric properties since isolated metal particles are dispersed in a dielectric continuum.

(2) Class of GFs (SiO_2(x at.% Co), at 40 at.% ≤ x ≤ 85 at.%) demonstrates semiconductor-like (intermediate) properties as metal particles with an increase in x become interconnected to form a maze structure. The electrical conductivity in this regime is due to percolation along the metallic maze and electron tunneling between isolated metal

particles. The concentration of Co x ≈ 40 at.% is a percolation threshold, below which only the tunneling process contributes to the conductivity of the granular metal.

(3) Class of GFs (SiO_2(x at.% Co), at x > 85 at.%) can be attributed to metals as the volume fraction of the metal x is large, metal grains merge and form a metallic continuum with dielectric inclusions.

For this study, the granular films SiO_2 (x at.% Co) with Co concentrations of x = 54, 60, 70, 75, 82 at.% deposited on Si substrate were chosen. A total of 5 batches of SiO_2 (x at.% Co)/Si heterostructures with the same concentrations were synthesized and studied to ensure the repeatability of synthesis results. Figure 1 demonstrates SEM images for SiO_2 (54 at.% Co)/Si (a), and for SiO_2 (82 at.% Co)/Si (b). Field Emission SEM (FE-SEM) Zeiss Leo 1530 capable of 2.5 nanometer resolution equipped with an In-Lens secondary electron detector has been used. The average size of cobalt nanoparticles has been obtained from a set of linear measurements using ImageJ software and equals about 75 ± 15 Å. The microscope achieves a resolution of approx. 1 nm.

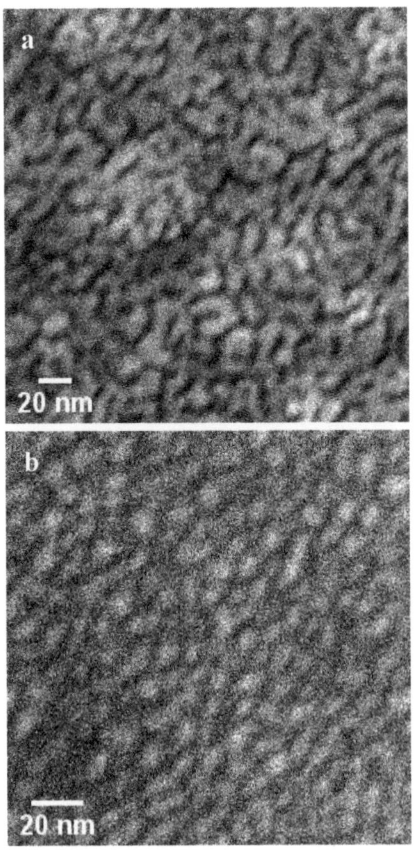

Figure 1. SiO_2(54 at.% Co)/Si (**a**) and SiO_2(82 at.% Co)/Si (**b**) images in the secondary electron mode with the InLens detector, mounted above the films. Light contrast corresponds to the SiO_2 matrix and dark areas correspond to Co inclusions.

2.2. *Grazing Incidence Small Angle X-ray Scattering and X-ray Reflectometry*

The Grazing Incidence Small Angle X-ray Scattering (GISAXS) and X-ray Reflectometry (XRR) experiments were carried out at ID10 beamline of European Synchrotron Radiation Facility (ESRF, Grenoble, France) [13]. The geometry of the GISAXS (Figure 2) is described by the grazing incidence angle α_i and two scattering angles α_f (out of sample plane) and φ (in-plane). Information on the electron density distribution in the z direction is obtained by measuring the scattering intensity as a function of the momentum transfer component Q_z perpendicular to the sample plane. Meanwhile, the lateral structure of the granular film, its surface and its interface with the silicon substrate can be studied by measuring intensity as a function of the component Q_y lying in the sample plane. A detailed description of the theory and methodology of the GISAXS technique can be found in the review [14–16]. The GISAXS experiment with the GF SiO_2(x at.% Co) on GaAs substrate can be found in [17]. The collimated X-ray beam of the size 1×0.1 mm^2 and wavelength λ = 0.95 Å, impinging the sample surface at the grazing angle α_i = 0.32°, was used for the GISAXS experiment. The scattering patterns were recorded with the two-dimensional position-sensitive detector MARCCD-133. The central part of the detector was protected from high-intensity direct and specular beams by a lead absorber.

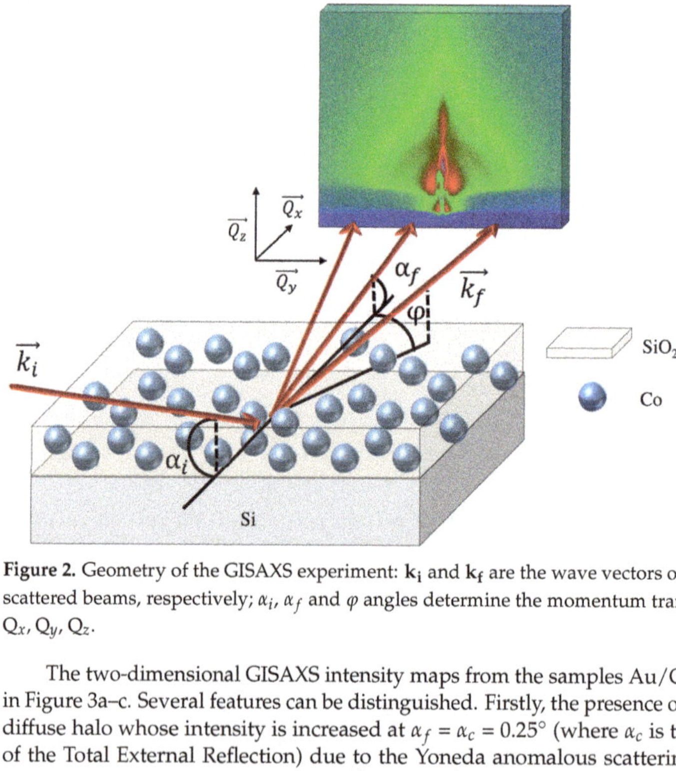

Figure 2. Geometry of the GISAXS experiment: $\mathbf{k_i}$ and $\mathbf{k_f}$ are the wave vectors of the incident and scattered beams, respectively; α_i, α_f and φ angles determine the momentum transfer components Q_x, Q_y, Q_z.

The two-dimensional GISAXS intensity maps from the samples Au/GF/Si are shown in Figure 3a–c. Several features can be distinguished. Firstly, the presence of a broad elliptic diffuse halo whose intensity is increased at $\alpha_f = \alpha_c$ = 0.25° (where α_c is the critical angle of the Total External Reflection) due to the Yoneda anomalous scattering and has zero intensity at α_f = 0, i.e., below the sample horizon. This broad elliptic diffuse halo is a typical manifestation of scattering from an isotropic three-dimensional system of scatterers (cobalt nanoparticles) located at a distance $l_1 = 2\pi/Q_1$ from each other, where Q_1 expressed in momentum transfer units is the radius of the diffuse halo.

The second feature of the scattering pattern is the presence of two symmetrical peaks at angles $\varphi = \pm 0.29°$. Such peaks correspond to the scattering from the nanoparticles spatially arranged within the monolayer at the interface with the substrate [14–16,18]. By taking the peaks coordinate Q_2 it is easy to calculate the average interparticle distance at the interface: $l_2 = 2\pi/Q_2$. To determine both l_1 and l_2 distances the cuts along φ direction of the two-dimensional intensity maps at α_f = 0.25° were taken (Figure 3d).

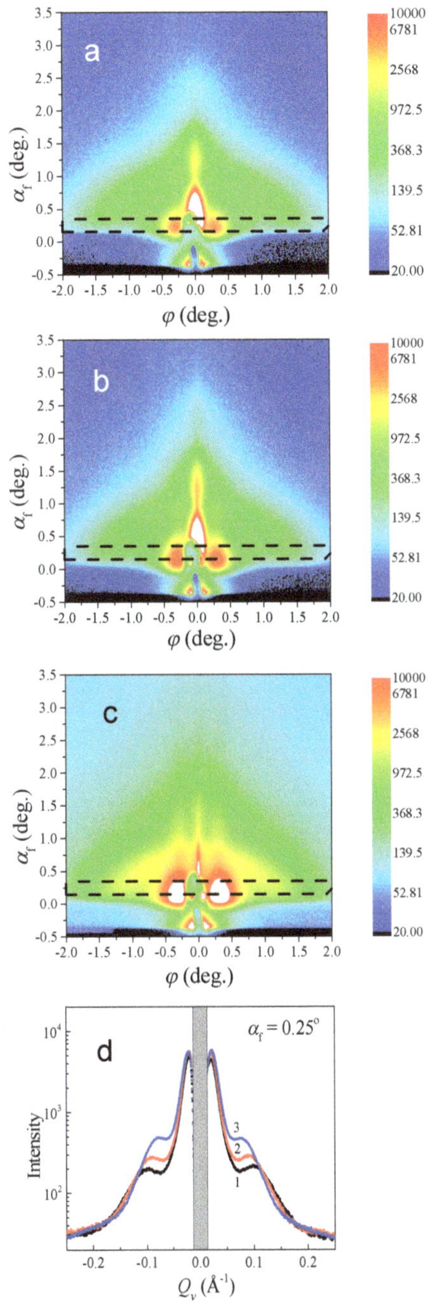

Figure 3. (**a**) The two-dimensional GISAXS intensity maps from the samples SiO_2(38 at.% Co)/Si (**a**), SiO_2(54 at.% Co)/Si (**b**), and SiO_2(82 at.% Co)/Si (**c**). The dashed horizontal lines indicate the integration regions for the one-dimensional profiles shown on (**d**) panel. (**d**) The scattering intensity profiles along ϕ direction at α_f = 0.25° for SiO_2(38 at.% Co)/Si—curve 1, SiO_2(54 at.% Co)/Si—curve 2, and SiO_2(82 at.% Co)/Si—curve 3. The gray area corresponds to the beamstop shadow.

In XRR experiments, the intensity of X-rays scattered in the direction $\alpha_i = \alpha_f$ and $\varphi = 0$ was measured as a function of the incidence angle α_i. The measurements were performed with a 1×0.1 mm^2 beam with a wavelength of 1.56 Å. To detect the reflected beam, we used an mBraun-50M linear position-sensitive detector, which made it possible not only to measure the desired signal but also to simultaneously determine the background, which was then subtracted. The specular reflection curve $R(Q_z)$ thus obtained is related to the one-dimensional in-depth distribution of the electron density of the sample $\rho(z)$ using the Parratt method [19]. The results of the XRR experiment are presented in Figure 4, where the intensity of the reflected beam is shown as a function of the momentum transfer for Au/SiO_2(60 at.% Co)/Si and Au/SiO_2(82 at.% Co)/Si.

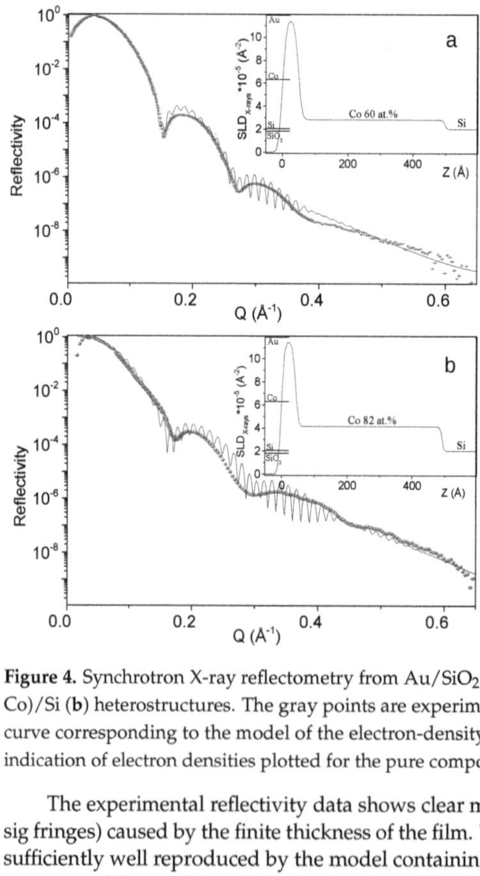

Figure 4. Synchrotron X-ray reflectometry from Au/SiO_2 (60 at.% Co)/Si (**a**) and Au/SiO_2 (82 at.% Co)/Si (**b**) heterostructures. The gray points are experimental data, the solid line is the theoretical curve corresponding to the model of the electron-density distribution shown in the inset with the indication of electron densities plotted for the pure components of the sample.

The experimental reflectivity data shows clear maxima with oscillating intensity (Kiessig fringes) caused by the finite thickness of the film. The experimental curves (Figure 4) are sufficiently well reproduced by the model containing two layers on a substrate Si. The parameters of the model include the depth profile of the electron scattering length density ($SLD_{X\text{-rays}}$), layer thickness (d) and interface (or surface) roughness (σ). The first layer corresponds to the technical layer of gold, the second layer corresponds to the granular SiO_2(x at.%Co) film with a thickness of 450 Å. The fitting parameters for each layer adjusted to minimize the value of reduced χ^2—a weighted measure of goodness of fit are presented in Table 1. The absence of fast oscillations corresponding to the total thickness of the bilayer Au/GF cannot be explained by the insufficient resolution of the instrument. Measurements were taken at a step of 0.02 degrees (50 steps/1 degree). This is enough to resolve a film with a thickness of about 450 Å (the oscillation period from 450 Å is 0.1 degrees). The reduction in the oscillation amplitude can be attributed to the interface roughness or, rather, to smeared $SLD_{X\text{-rays}}$ contrast (a contrast of the electron density) between a granular film and a substrate.

Table 1. Parameters of the model used to approximate the experimental X-ray reflectometry curves.

Layer	d (Å)	$SLD_{X\text{-rays}} \cdot 10^{-5}$ ($Å^{-2}$)	σ (Å)
		Au/SiO_2(60 at.% Co)/Si	
Au	48 ± 2	11.5 ± 0.1	9.0 ± 0.2
Granular film	450 ± 2	2.83 ± 0.05	9.0 ± 0.2
Si	–	2.024 ± 0.05	3.8 ± 0.2
		Au/SiO_2(82 at.% Co)/Si	
Au	43 ± 2	11.5 ± 0.1	7.0 ± 0.3
Granular film	450 ± 2	4.13 ± 0.05	7.0 ± 0.3
Si	–	2.024 ± 0.05	3.6 ± 0.2

2.3. Polarized Neutron Reflectometry

A detailed description of the polarized neutron reflectometry technique can be found elsewhere [20]. The experiment was carried out at the SuperADAM reflectometer at Institut Laue-Langevin (Grenoble, France) [21–23]. Typical geometry of the PNR experiment is shown in Figure 5.

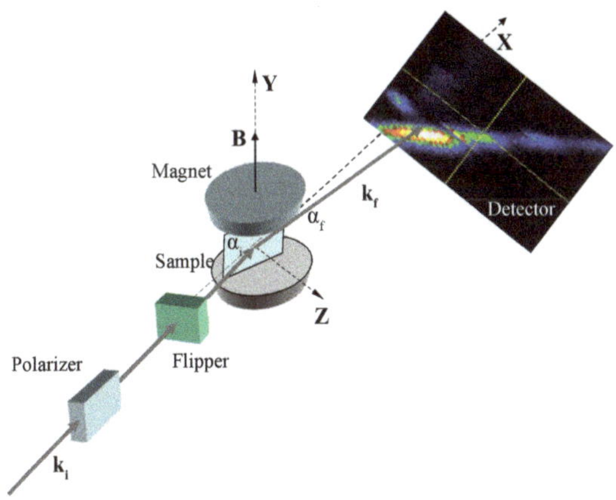

Figure 5. Typical geometry of the polarized neutron reflectometry experiment.

We exploited a neutron beam with the wavelength λ = 4.41 Å and the initial polarization P_0 = 0.98. The lateral components of neutron momentum transfer Q_x and Q_y were equal to zero, and the vertical component Q_z was equal to $4\pi\sin(\alpha_f)/\lambda$, where α_f is the angle of the scattered beam with wave vector $\mathbf{k_f}$. Intensities of the reflected beams with initial polarization $+\mathbf{P_0}$ and $-\mathbf{P_0}$, i.e., along and against the direction of the external magnetic field **B**, were measured one by one. The magnetic field from 0 to 320 mT was applied parallel to the film surface and perpendicular to the incident neutron beam with wave vector $\mathbf{k_i}$. The direction of $\mathbf{P_0}$ with respect to **B** was switched by a spin Flipper. Specularly reflected neutrons were detected by the ^{3}He position-sensitive detector. Measurements were performed at temperatures, where the GIMR effect is not detected (T = 120 K), reaches its maximum (T = 300 K) and again is not detected (T = 420 K) [8]. The data for every temperature were taken after the demagnetization process. No difference between reflectivity curves $R(Q_z, +\mathbf{P_0})$ and $R(Q_z, -\mathbf{P_0})$ was observed at B = 0 mT, while the application of the external field resulted in a pronounced splitting of two components (Figure 6).

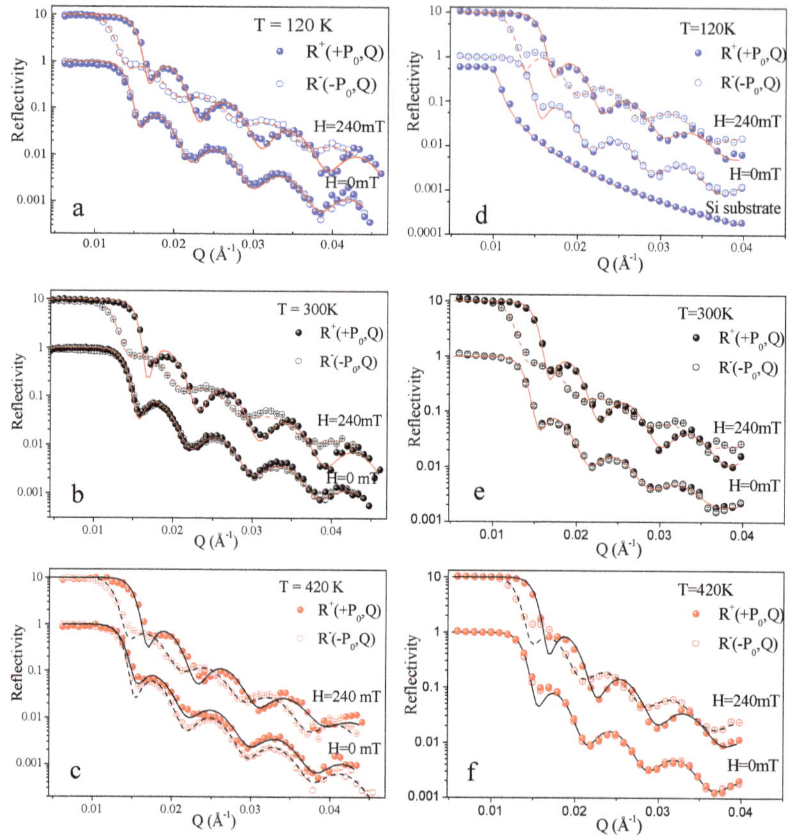

Figure 6. The experimental (symbols) and fitted model (lines) reflectivity curves for the samples Au/SiO_2(60 at.% Co)/Si (**a–c**) and Au/SiO_2(70 at.% Co)/Si (**d–f**) at fields B = 0 and B = 240 mT for T = 120 K (**a**,**d**), 300 K (**b**,**e**), and 420 K (**c**,**f**). Measured and fitted model curves at B = 240 mT are multiplied by 10 for convenience.

For fitting the Polarized Neutron Reflectometry experimental data the Parratt algorithm was used [19]. The following parameters were included in the fitting routine: the beam width, the sample length, the resolution, the background and the detector offset. The scattering length density (SLD) profile (Figure 7) represents the distribution of the scattering potential of the sample into the depth of the film (in **Z** direction Figure 5). The results for all studied samples are shown to coincide qualitatively with insignificant numerical variations associated with the concentration of cobalt in the granular film (Figure 7). All PNR curves at different temperatures and **B** were fitted with the same structural parameters: thickness (d), roughness (σ), and nuclear scattering length density (N_n). The magnetic scattering length density (N_m), which depends on the applied magnetic field, varied in the corresponding datasets (Tables 2 and 3).

From Figure 7, it is clear that a good quantitative agreement between model calculations and experimental data has been achieved. The difference in the granular film thickness, d, defined by PNR and XRR methods is related to different batches of samples.

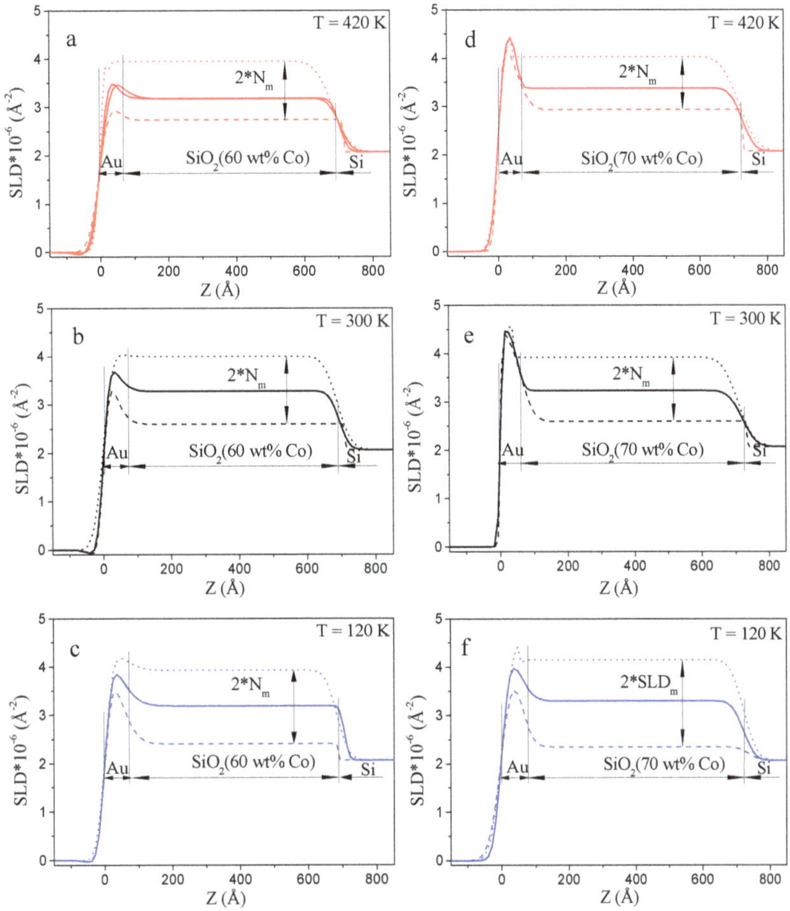

Figure 7. The SLD profiles ρ^{+} (dot line) and ρ^{-} (dashed line) of the samples Au/SiO_2 (60 at.% Co)/Si (**a–c**) and Au/SiO_2 (70 at.% Co)/Si (**d–f**) at B = 240 mT for T = 420 K (**a**,**c**), 300 K (**b**,**e**), and 120 K (**c**,**f**). The solid line corresponds to the SLD profile at H = 0 T for every polarization of neutrons.

Table 2. Parameters of the model used to approximate the PNR curves for Au/SiO_2 (60 at.% Co)/Si.

Layer	d (Å)	$N_n \cdot 10^{-6}$ (Å^{-2})	$N_m \cdot 10^{-7}$ (Å^{-2})	H (mT)	σ (Å)
			T = 120 K		
Au	45 ± 2	3.8 ± 0.2	–		11.0 ± 1.0
			0.33 ± 0.02	0	
			1.71 ± 0.02	20	
			3.25 ± 0.02	40	
GF	680 ± 2	3.29 ± 0.03	3.91 ± 0.02	60	10.0 ± 1.0
			5.02 ± 0.02	100	
			7.66 ± 0.02	240	
			8.82 ± 0.02	320	
Si		2.073 ± 0.02	–		40 ± 2

Table 2. *Cont.*

Layer	d (Å)	$N_n \cdot 10^{-6}$ (Å^{-2})	$N_m \cdot 10^{-7}$ (Å^{-2})	H (mT)	σ (Å)
		T = 300 K			
Au	45 ± 2	3.7 ± 0.2	-		11.0 ± 1.0
GF	680 ± 2	3.29 ± 0.03	0.13 ± 0.02	0	10.0 ± 1.0
			0.99 ± 0.02	20	
			1.83 ± 0.02	40	
			2.52 ± 0.02	60	
			3.62 ± 0.02	100	
			6.48 ± 0.02	240	
			7.19 ± 0.02	320	
Si		2.073 ± 0.02	-		50 ± 2
		T = 420 K			
Au	45 ± 2	3.5 ± 0.2	-		11.0 ± 1.0
GF	680 ± 2	3.38 ± 0.03	0.05 ± 0.02	0	10.0 ± 1.0
			1.10 ± 0.02	20	
			1.49 ± 0.02	40	
			2.21 ± 0.02	60	
			2.99 ± 0.02	100	
			5.05 ± 0.02	240	
			6.00 ± 0.02	320	
Si		2.073 ± 0.02	-		50 ± 2

Table 3. Parameters of the model used to approximate the PNR curves for Au/SiO_2 (70 at.% Co)/Si.

Layer	d (Å)	$N_n \cdot 10^{-6}$ (Å^{-2})	$N_m \cdot 10^{-7}$ (Å^{-2})	H (mT)	σ (Å)
		T = 120 K			
Au	45 ± 2	4.0 ± 0.2	-		11.0 ± 1.0
GF	685 ± 2	3.29 ± 0.03	0.06 ± 0.02	0	10.0 ± 1.0
			3.01 ± 0.02	20	
			4.37 ± 0.02	40	
			5.13 ± 0.02	60	
			6.53 ± 0.02	100	
			8.26 ± 0.02	240	
			8.76 ± 0.02	320	
Si		2.073 ± 0.02	-		50 ± 2
		T = 300 K			
Au	45 ± 2	3.7 ± 0.2	-		11.0 ± 1.0
GF	685 ± 2	3.29 ± 0.03	0.04 ± 0.02	0	10.0 ± 1.0
			0.65 ± 0.02	20	
			1.28 ± 0.02	40	
			1.92 ± 0.02	60	
			2.87 ± 0.02	100	
			5.70 ± 0.02	240	
			6.00 ± 0.02	320	
Si		2.073 ± 0.02	-		50 ± 2
		T = 420 K			
Au	45 ± 2	3.5 ± 0.2	-		11.0 ± 1.0
GF	685 ± 2	3.29 ± 0.03	0.00 ± 0.02	0	10.0 ± 1.0
			0.005 ± 0.02	20	
			0.02 ± 0.02	40	
			0.35 ± 0.02	60	
			1.39 ± 0.02	100	
			3.64 ± 0.02	240	
			3.84 ± 0.02	320	
Si		2.073 ± 0.02	-		50 ± 2

2.4. SQUID Magnetometry

Static magnetic properties of the samples were characterized by SQUID magnetometry (Quantum Design MPMS-5S) at the Institute of Condensed Matter Physics (Braunschweig, Germany). The magnetization curves M(H) were recorded up to 5 T. The magnetic field was applied parallel to the plane of the film. As an example, M(H) of the sample SiO_2 (75 at.% Co)/Si taken at T = 200 K and 300 K are shown in Figure 8 after subtraction of the diamagnetic contribution of the Si substrate and the SiO_2 matrix. A shift of the magnetic hysteresis loops along the field axis and a change of the coercive field as a function of temperature are observed. (insert (a) in Figure 8). In the fields region of 0.8 $\div$ 1.8 T there are specific hysteresis loops—"pockets" (inserts (b,c) in Figure 8).

The actual magnetic saturation value of the films is quite difficult to detect using SQUID techniques because the overall high-field signal includes the dominant diamagnetic contribution of the substrate, appearing due to its relatively large volume. If the granular film itself shows a small linear response in the high field regions, this is convoluted with an overwhelming diamagnetic signal and it is challenging to reliably subtract the latter contribution in order to separate the magnetic properties of the nanolayer itself. The saturation magnetization value of the SiO_2(Co) film obtained by SQUID magnetometry (Figure 8) $M_s \approx 60$ emu/cm^3 differs significantly from the saturation magnetization of bulk cobalt M_{Co} = 1422 emu/cm^3.

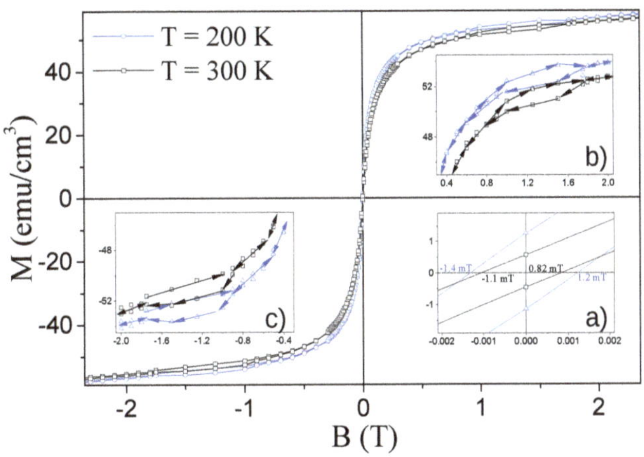

Figure 8. Magnetic field dependence of the SiO_2 (75 at.% Co)/Si sample magnetization at magnetic field from −5 T to 5 T. Inserts: (**a**)—region of the coercive fields, (**b**,**c**)—regions of the specific hysteresis loops.

3. Discussion

The GISAXS data from the samples Au/GF/Si has clearly shown two nanoparticle subsystems in GFs with different average interparticle distances l_1 and l_2 (Figure 3). The interparticle distance l_1 for the first subsystem increases with the increasing cobalt concentration: 64 $\pm$ 2 Å for SiO_2 (38 at.% Co)/Si, 70 $\pm$ 2 Å for SiO_2(54 at.% Co)/Si, and 88 $\pm$ 2 Å for SiO_2 (82 at.% Co)/Si (Figure 3d). It is clear that the determined interparticle distance l_1 for these concentrations of Co should be equal to the size of nanoparticles because of the percolation limit. As one can see, the value of l_1 obtained by GISAXS is in good agreement with the particle size observed by SEM on the top surface of the GF, where percolation is clearly visible (Figure 1).

In contrast, the second subsystem with a characteristic distance l_2 = 300 nm is not revealed by SEM measurements. This is due to the fact that it is most likely located at the burred GF/Si interface. A similar structure containing two subsystems of nanoparticles in the granular films SiO_2 (x at% Co) deposited on GaAs substrate was observed in [17,24].

There it was shown that the specific interface layer of cobalt nanoparticles is formed between the GF and GaAs substrate with an in-plane interparticle distance ~ 320 Å and the thickness ~70 Å [17,24].

The formation of such a specific layer on the GF/Semiconductor interface may be caused by a number of reasons. The deposition of cobalt on a substrate at the initial stage leads to the clustering of a thin cobalt film due to the dynamics of Co on Si, or GaAs [25–28]. The activation energy for clustering was found to be 0.3 ± 0.2 eV [25], indicating that a Co surface diffusion process dominates the kinetics. The surface energy of cobalt is about 2500–2900 erg/cm^2, while the surface energy of silicon corresponds to 1230 erg/cm^2 for the (111) plane, 1510 erg/cm^2 for the (110) plane, and 2130 erg/cm^2 for the plane (100). The surface energy of GaAs is 860 erg/cm^2 [29,30]. Another reason for the formation of a specific layer at the interface is presumably related to the process of heteroepitaxial growth of the GF by ion beam co-sputtering directly onto a semiconductor substrate, while the next layers are grown in the homoepitaxial regime [31].

However, the X-ray and Polarized neutron reflectometry data (Figures 4, 6 and 7) do not confirm the coexistence of two nanoparticle subsystems in GFs, or they are not sensitive enough to the existence of a specific layer at the GF/Semiconductor interface in SiO_2 (x at.% Co)/Si samples. In contrast to the GISAXS results, the film model which fits the XRR and PNR data in the best way consists of only two layers on the Si substrate: a granular film layer and the capping Au layer, i.e., there is no specific layer on the GF/Semiconductor interface. Why do the PNR method, which is highly sensitive to changes in magnetic properties, and the XRR method, which is highly sensitive to changes in electron density contrast, fail to detect an interface layer?

The next question deals with the study and interpretation of the magnetic properties of the granular films. The magnetic scattering length density N_m has the opposite sign for neutrons with initial polarization $+\mathbf{P}_0$ and $-\mathbf{P}_0$ (along and against **B**) and its value is directly proportional to the magnetization component **M** parallel to **B**:

$$M = \left(\frac{m_n \mu_n}{2\pi\hbar^2}\right)^{-1} \frac{N_m}{4\pi}, \tag{1}$$

where m_n and μ_n are the mass and magnetic moment of the neutron. Comparing SLD profiles $\rho^+ = N_n + N_m$ and $\rho^- = N_n - N_m$ corresponding to $R(Q_z, +\mathbf{P}_0)$ and $R(Q_z, -\mathbf{P}_0)$, respectively, one can obtain a value of the magnetization in every layer forming the sample film.

From the non-zero values of N_m at the SLD profiles (Figure 7) one can conclude that only GF is magnetized under the applied field and tends to saturate the magnetic moment at B ≥ 300 mT. Magnetizations of the GF in SiO_2(60 at.% Co)/Si and SiO_2 (70 at.% Co)/Si samples calculated from N_m according to Equation (1) are presented in Figure 9 as a function of the applied magnetic field for T = 120, 300 and 420 K. In Figure 9, the magnetic SLD N_m is normalized to magnetic SLD of bulk cobalt $N_m(Co) = 4.24 \times 10^{-6}$ Å^{-2}. From Figure 9, it follows that the GF is in the ferromagnetic state at all three temperatures. Its magnetization decreases with increasing temperature: at the maximum applied field B = 320 mT we found that $N_m/N_m(Co) = 0.21$ at T = 120 K while decreasing down to 0.17 at T = 300 K and to 0.14 at T = 420 K for SiO_2 (60 at.% Co) and decreasing down to 0.14 at T = 300 K and to 0.09 at T = 420 K for SiO_2 (70 at.% Co). Why is the magnetization, determined by the PNR method so much smaller than the magnetization of bulk Co? It is smaller by a factor of 5 to 10 at different temperatures.

Let us analyze the nuclear scattering length densities of neutrons and X-rays for all samples in order to answer these questions. The fit parameters in PNR and XRR models show excellent agreement between the measured and expected nuclear SLD values of Si (2.073×10^{-6}Å^{-2} and 2.024×10^{-5}Å^{-2}, respectively). The fits of the nuclear SLD values of Au in the PNR and the XRR experiments are slightly lower than the expected bulk values (4.5×10^{-6}Å^{-2} and 12.47×10^{-5}Å^{-2}, respectively). The reason could be the discontinuous layer of gold since its thickness is below 50 Å. A number of studies demonstrate changes in the physical and chemical properties of thin gold films associated with their island structure [32–34].

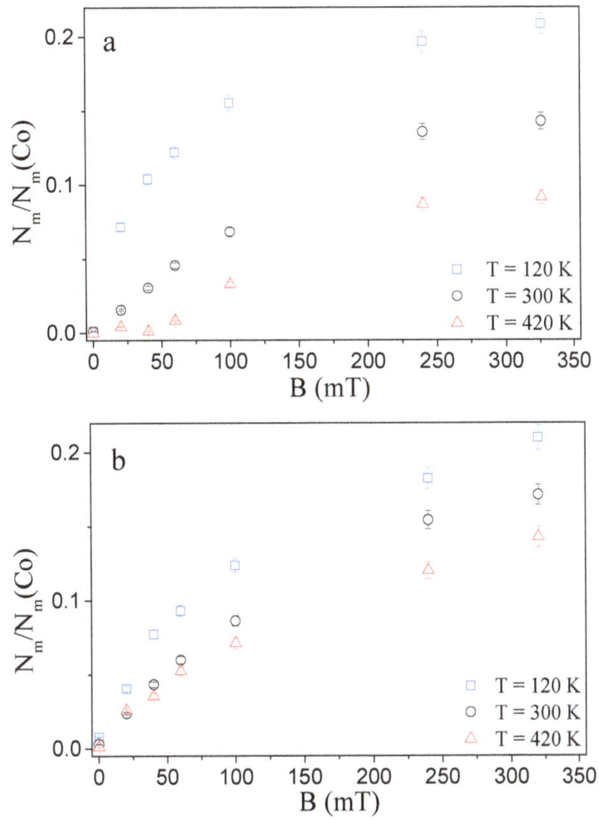

Figure 9. Magnetic field dependence of fitted magnetic SLD value N_m normalized to magnetic SLD of bulk cobalt N_m(Co) for the ferromagnetic films SiO_2 (70 at.% Co) (**a**) and SiO_2 (60 at.% Co) (**b**).

The measured nuclear SLD of the GF layer is substantially higher for neutrons and lower for X-rays as compared to expected values (Tables 1–4). Table 4 shows the calculated N_n and $SLD_{X\text{-}ray}$ values for a given ratio of Co/SiO_2 in GF under the assumption that cobalt can interact with oxygen from the silicon oxide dielectric matrix or oxidize in air. Unfortunately, there is no reliable way to determine the thickness of the oxide layer on the surface of cobalt nanoparticles in the present study. It is obvious that the cobalt grains must be partially oxidized for the nuclear SLD to be higher for neutrons0 or lower for X-rays, compared to the values for a given ratio of Co/SiO_2 in GF. All cobalt oxides are antiferromagnets, in contrast to pure cobalt (Table 5). Magnetization measurements (Figure 8) reveal that for SiO_2(x at.%Co)/Si heterostructures exchange biasing of the magnetic hysteresis loop along the field axis is observed. The exchange bias describes a phenomenon associated with interfacial coupling between ferromagnetic and antiferromagnetic materials [35]. In our investigation, the exchange biasing of the magnetic hysteresis loop confirms the oxidation of cobalt nanoparticles. The exchange biasing field, like the coercive field, changes as a function of temperature and is about 10 percent of the coercive field (insert (a) in Figure 8).

Another small but well-distinguished feature of the magnetization curve is the small hysteretic pockets in the field range from 0.8 to 1.6 T. They are well-detected both in the positive and negative halves of the magnetization curve. They were also detected for GF on the GaAs substrate [36]. An interesting feature of this hysteresis loop is counterintuitive large values of the magnetization when the field increases and smaller magnetization values when the field decreases. This feature can be referred to as the same biasing phenomenon of the Co (ferromagnetic) particles with the oxidized (antiferromagnetic) surface, to the so-called "core-shell" model. In this case, the magnetization of the Co particle is responsible for the large ferromagnetic response of the GF system, while the antiferromagnetic shell may show a spin flop in a certain field range, which may be energetically more favorable than the collinear spin arrangement. Considering its really small (4% only) contribution imposed on the top of the magnetization curve, we point out the biased (ferro-antiferro) magnetic nature of this phenomenon but do not go into any further details.

Table 4. The calculated N_n and $SLD_{X\text{-rays}}$ values for a given ratio of Co/SiO_2 in GF under the assumption that cobalt can be oxidized.

Ratio $xCo/(1\text{-}x)SiO_2$	xCo		xCoO		xCo_2O_3	
	N_n, 10^{-6} Å^{-2}	$SLD_{X\text{-rays}}$, 10^{-5} Å^{-2}	N_n, 10^{-6} Å^{-2}	$SLD_{X\text{-rays}}$, 10^{-5} Å^{-2}	N_n, 10^{-6} Å^{-2}	$SLD_{X\text{-rays}}$, 10^{-5} Å^{-2}
0.38/0.62	3.0	3.56	3.77	2.97	3.74	2.64
0.54/0.46	2.81	4.27	3.91	3.44	3.86	2.97
0.60/0.40	2.74	4.54	3.95	3.6	3.90	3.1
0.70/0.30	2.62	4.98	4.04	3.9	3.98	3.4
0.75/0.25	2.55	5.22	4.08	4.05	4.01	3.38
0.82/0.18	2.47	5.51	4.14	4.24	4.08	3.52

Table 5. The calculated N_n and $SLD_{X\text{-rays}}$ values are given with Density (D), Molecular weight (MW) and Number of molecules per Å^3 (n) for compounds (Comp) that may be formed in the granular film, or at the interfaces.

Comp	D, g/ccm	MW, g/mole	n	N_n, 10^{-6} Å^{-2}	$SLD_{X\text{-rays}}$, 10^{-5} Å^{-2}	Type of Magnetism at T = 300 K	Type of Conductivity at T = 300 K
Au	19.3	196.97	0.059	4.5	12.47	diamagnetic	metal
SiO_2	2.2	60.08	0.022	3.46	1.88	diamagnetic	insulator
CoO	6.45	74.93	0.0518	4.29	4.77	antiferromagnetic	semiconductor
Co_2O_3	5.18	165.86	0.0188	4.2	3.88	antiferromagnetic	semiconductor
Co_3O_4	6.073	240.80	0.0152	4.66	4.53	antiferromagnetic	semiconductor
Co	8.9	58.933	0.091	2.26	6.3	ferromagnetic	metal
CoSi	6.3	87.02	0.0436	2.88	4.77	diamagnetic	semimetal
Co_2Si	7.1	145.95	0.0293	2.67	5.23	diamagnetic	metal
$CoSi_2$	5.3	115.10	0.0277	2.97	4.15	diamagnetic	metal
Si	2.33	28.086	0.05	2.074	2.024	diamagnetic	semiconductor

Figure 10 demonstrates the neutron spin asymmetry calculated by:

$$SA = \frac{R(Q_z, +\mathbf{P}_0) - R(Q_z, -\mathbf{P}_0)}{R(Q_z, +\mathbf{P}_0) + R(Q_z, -\mathbf{P}_0)}, \tag{2}$$

As the neutron spin asymmetry is a very reliable method to study the intrinsic nano-magnetic features [37,38], it provides direct evidence that these magnetic properties originate from the composite thin film region and the substrate and surface play no detectable role. The spin asymmetry is purely related to the local magnetic moment in the sample and characterizes the in-depth distribution of the magnetization in GF directly. The observed peak SA occurs at $Q_z = 0.0136 \pm 0.004$ Å^{-1} and exceeds 80% (Figure 10a), which competes with that of standard magnetic alloys [39] and is large compared to the spin asymmetries typical of dilute magnetic systems [40]. The magnetic moment of the sample proportional to the SA peak area increases with the applied external field and saturates at the experimental field of 240 ÷ 320 mT (insert in Figure 10a). In contrast to SQUID magnetometry, PNR is unaffected by complications of diamagnetic or weak paramagnetic responses because they produce very low magnetizations per volume. Instead, the PNR signal is dominated by scattering effects from large local net magnetization, typically caused by a well-aligned collinear spin structure. Consequently, the small value of the neutron spin asymmetry at low fields (insert in Figure 10a) confirms the traits noted in the hysteresis from the SQUID data (Figure 8)—a low remanence and a low coercive field ($B_c < 1.4$ mT) which are attributed to either superparamagnetic behavior or a wandering anisotropy axis. This shows that, without a sufficient applied field, the magnetic moments in GF do not stay well aligned in any specific direction. The possible reason for this is that there are inhomogeneous magnetically ordered patches in the GF layer that experience a distribution of local spin anisotropies and consequently have a distribution of magnetization angles that are only co-aligned in a magnetic field and relax under zero field. In addition, the magnetic moment in GF changes moderately with temperature (insert in Figure 10b), indicating a strong magnetic exchange coupling.

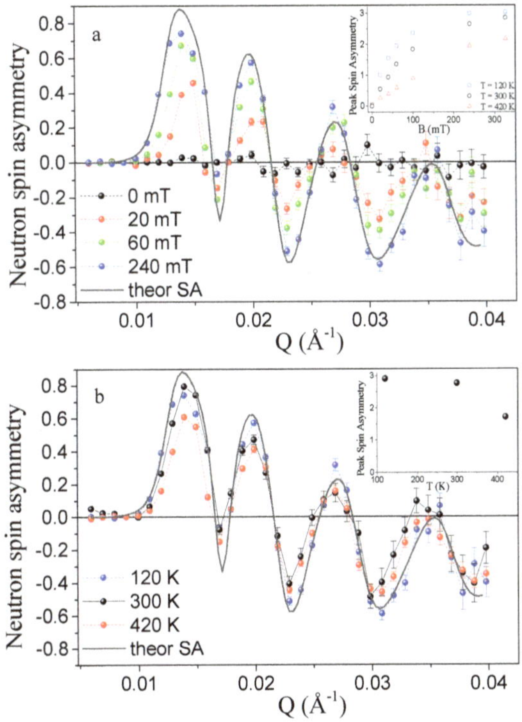

Figure 10. The neutron spin asymmetry, SA, for Au/SiO_2 (70 at.% Co)/Si fitted (gray line) to the chemical and magnetic model previously described in the text (Figure 7): (**a**)—under applied magnetic fields up to 320 mT, (**b**)—for different temperatures at B = 240 mT. Insert in panel (**a**)—the field-dependence of the integral intensity of the peak spin asymmetry for different temperatures. Insert in panel (**b**)—the temperature dependence of the integral intensity of the peak spin asymmetry at B = 240 mT. For clarity, data for (**a**) at B = 40, 100, and 320 mT are omitted.

As it is shown in Figure 10, the SA changes its sign upon the increase in the momentum transfer Q_z. We interpret this observation as the presence of the "magnetically dead" layer in the GF at its interface with the Si substrate. The thickness of this layer is difficult to determine exactly, but it should be of the order of (1/10) of the thickness of the granular film itself. A similar effect of the magnetically dead layers in the PNR was observed in the multilayered structures (see, for example, [41,42].)

Based on the above, it should be assumed that the oxidation of cobalt nanoparticles is the reason for such a low magnetization value of the granular film with cobalt concentrations from 54 at.% to 82 at.% compared to pure cobalt. The same difference of the magnetic SLD value normalized to magnetic SLD of bulk cobalt was observed for GF deposited on GaAs substrate in [43].

The answer to the question of why the highly sensitive PNR and XRR methods fail to detect an interface layer between GF and Si substrate will also allow us to answer the main question of this article: what is the reason for the dramatic difference in IMR coefficients for identical SiO_2(x at.% Co) granular films deposited on GaAs, or Si substrates.

It was noted above that Co due to the higher surface free energy as compared to Si would tend to grow as islands at the beginning of the deposition. The phase composition and magnetic properties of cobalt films deposited on the silicon substrate were studied in [36]. It was shown that as the film thickness increases, an interface cobalt silicide and an island film of a solid solution of silicon in cobalt are successively formed. The growth of metallic cobalt begins only after the deposition of a Co layer with a thickness of more than 7 Å. The calculated N_n values are given in Table 5 for chemical compounds that may be formed in the granular film, or at the interfaces due to boundary interlayer diffusion leading to the formation of an interface of variable binary composition—cobalt silicides. Table 5 shows that cobalt silicides and cobalt oxides have similar $SLD_{X\text{-rays}}$ values, which does not allow the GF and an additional interface layer to be distinguished in the Z direction (Figure 2) by XRR experiments. Wherein in the Y direction (Figure 2), an additional interface layer with large periodicity $l_2 = 300 \pm 10$ Å is detectable by GISAXS due to the high contrast of $SLD_{X\text{-rays}}$ between the multiphase cobalt granules and the SiO_2. Thus, both the XRR and PNR experimental results are well described by the model of a uniform granular cobalt film with a noticeably oxidized surface of the granules and a high surface roughness at the interface with the silicon substrate.

Increasing the silicon concentration to 30% in the Co-Si solid solution makes the large cobalt granules practically non-magnetic due to the charge transfer effect. Moreover, due to the formation of cobalt silicide at the GF/Si interface, the conductivity of the granular film will be determined by the chains "granule (Co)-metal (Co-Si)-semiconductor (Si)", and not by electron tunneling between ferromagnetic granules through the SiO_2 dielectric matrix into the Si semiconductor. This leads to a strong decrease in the magnetoresistive coefficient to a few percent in SiO_2 (x at.% Co) heterostructures on a Si substrate compared to the really giant IMR value in SiO_2 (71 at.% Co) on a GaAs substrate [8].

4. Conclusions

In the present paper, we studied the structure and magnetic properties of the maze structure with interconnected cobalt particles SiO_2 (x at.% Co) deposited on Si substrate at 54 at.% $\leq$ x $\leq$ 82 at.%). Investigations were carried out using Polarized Neutron Reflectometry, Grazing Incidence Small Angle X-ray Scattering, X-ray Reflectometry, Scanning Electron Microscopy, and Superconducting Quantum Interference Device Magnetometry.

The reason for the dramatic difference in IMR coefficients for SiO_2 (x at.% Co) heterostructures deposited on GaAs or Si substrates was established. Despite the fact that both granular films have very similar values for thickness, interparticle distances, and magnetizations, as well as the additional layer of Co at the interface granular film/substrate, there is a difference in the interface morphology for SiO_2 (x at.% Co)/Si. On the interface granular film/Si substrate there is boundary interlayer diffusion of Si atoms into Co, leading to the formation of an interface of variable binary composition - cobalt silicides. The formation of cobalt silicide at the GF/Si interface leads to the conductivity of the granular film determined by the electron movement via chains "ferromagnetic granule (Co)-metal (Co-Si)-semiconductor (Si)", and not by the electron tunneling between ferromagnetic granules through the SiO_2 dielectric matrix into the Si semiconductor. A high value of the GIMR effect in SiO_2 (x at.% Co) on GaAs [8,44] substrate is probably related to the spin-dependent potential barrier formed in the accumulation electron layer in the semiconductor near the interface. The action of the spin-dependent potential barrier is amplified by the avalanche process and by the electron accumulation in the quantum well in the semiconductor interface region induced by the backscattering process of injected electrons on exchange-split levels [44]. In the granular film SiO_2 (x at.% Co) on the Si substrate, the cobalt silicide does not allow the formation of the region at the GF/Si interface saturated with electrons capable of polarizing in the direction of the applied magnetic field. Therefore, further efforts to form a quantum well with exchange-split levels on the well top at the surface of Si are necessary to gain a high positive magnetoresistance in SiO_2 (x at.% Co)/Si heterostructure. For example, this can be achieved by introducing a buffer layer between the granular film and substrate [45,46].

Author Contributions: Conceptualization, N.A.G.; methodology, N.A.G.; validation, N.A.G. and S.V.G.; investigation, N.A.G., V.U., A.A.V.; resources, A.I.S., N.N.N. and L.V.L.; data curation, V.U. and A.A.V.; writing—original draft preparation, N.A.G. and V.U.; writing—review and editing, S.V.G.; visualization, A.A.V.; supervision, S.V.G.; project administration, L.V.L. All authors have read and agreed to the published version of the manuscript.

Funding: This work is financially supported by the Ministry of Science and Higher Education of the Russian Federation in the framework of Agreement No. 075-15-2022-830 (Prolongation of Agreement No. 075-15-2021-1358 from 12 October 2021).

Institutional Review Board Statement: Not applicable.

Informed Consent Statement: Not applicable.

Data Availability Statement: The data that support the findings of this study are available from the corresponding author upon reasonable request.

Acknowledgments: Authors would like to thank Institut Laue-Langevin, European Synchrotron Radiation Facility and Institute of Condensed Matter Physics of TU Braunschweig for the provided experimental opportunities. We would like to acknowledge M. Wolf, A.Mistonov and D.Menzel for the assistance with the neutron and magnetometry measurements. The authors are grateful to the staff of the Interdisciplinary Resource Center for Nanotechnology and the Center of X-ray diffraction studies at the Research park of Saint Petersburg State University for *preliminary research* of Au/SiO_2(Co)/Si heterostructures, as well as Saint Petersburg State University for financial support (Activity 6—grant for academic mobility 2019, ID:41160111).

Conflicts of Interest: The authors declare that they have no conflict of interest.

References

1. Gerber, A.; Milner, A.; Groisman, B.; Karpovsky, M.; Sulpice, A. Magnetoresistance of granular ferromagnet-insulator films. *Thin Solid Films* **1997**, *304*, 319–322. [CrossRef]
2. Stuart, H.R.; Hall, D.G. Island size effects in nanoparticle-enhanced photodetectors. *Appl. Phys. Lett.* **1998**, *73*, 3815–3817. [CrossRef]
3. Kodama, R. Magnetic Nanoparticle. *J. Magn. Magn. Mater.* **1999**, *200*, 359–372. [CrossRef]
4. Fujimori, H.; Ohnuma, S.; Kobayashi, N.; Masumoto, T. Spintronics in metal-insulator nanogranular magnetic thin films. *J. Magn. Magn. Mater.* **2006**, *304*, 32–35. [CrossRef]
5. Žutić, I.; Fabian, J.; Sarma, S.D. Spintronics: Fundamentals and applications. *Rev. Mod. Phys.* **2004**, *76*, 323.
6. Atwater, H.A.; Polman, A. Plasmonics for improved photovoltaic devices. *Nat. Mater.* **2010**, *9*, 205–213. [CrossRef]
7. Lambert, C.-H.; Mangin, S.; Varaprasad, B.C.S.; Takahashi, Y.; Hehn, M.; Cinchetti, M.; Malinowski, G.; Hono, K.; Fainman, Y.; Aeschlimann, M.; et al. All-optical control of ferromagnetic thin films and nanostructures. *Science* **2014**, *345*, 1337–1340. [CrossRef]
8. Lutsev, L.V.; Stognij, A.I.; Novitskii, N.N. Giant magnetoresistance in semiconductor/granular film heterostructures with cobalt nanoparticles. *Phys. Rev. B* **2009**, *80*, 184423. [CrossRef]
9. Lutsev, L. Potential barrier for spin-polarized electrons induced by the exchange interaction at the interface in the ferromagnet/semiconductor heterostructure. *J. Phys. Condens. Matter* **2006**, *18*, 5881. [CrossRef] [PubMed]
10. Miller, N.C.; Shirn, G.A. Co-sputtered Au-SiO_2 cermet films. *Appl. Phys. Lett.* **1967**, *10*, 86. [CrossRef]
11. Miller, N.C.; Hardiman, B.; Shirn, G.A. Transport properties, microstructure, and conduction model of cosputtered Au-SiO_2 cermet films. *J. Appl. Phys.* **1970**, *41*, 1850. [CrossRef]
12. Stognij, A.I.; Novitskii, N.N.; Stukalov, O.M. Separate cobalt-copper interface smoothening under the action of low-energy argon ion bombardment. *Tech. Phys. Lett.* **2003**, *29*, 43–46. [CrossRef]

13. Smilgies, D.M.; Boudet, N.; Struth, B.; Konovalov, O. Troika II: A versatile beamline for the study of liquid and solid interfaces. *J. Synchrotron Radiat.* **2005**, *12*, 329–339. [CrossRef]
14. Renaud, G.; Lazzari, R.; Leroy, F. Probing surface and interface morphology with Grazing Incidence Small Angle X-ray Scattering. *Surf. Sci. Rep.* **2009**, *64*, 255–380. [CrossRef]
15. Renaud, G.; Lazzari, R.; Revenant, C.; Barbier, A.; Noblet, M.; Ulrich, O.; Leroy, F.; Jupille, J.; Borensztein, Y.; Henry, C.R.; et al. Real-time monitoring of growing nanoparticles. *Science* **2003**, *300*, 1416–1419. [CrossRef] [PubMed]
16. Lazzari, R.; Renaud, G.; Revenant, C.; Jupille, J.; Borensztein, Y. Adhesion of growing nanoparticles at a glance: Surface differential reflectivity spectroscopy and grazing incidence small angle X-ray scattering. *Phys. Rev. B* **2009**, *79*, 125428. [CrossRef]
17. Grigoreva, N.; Vorobev, A.; Ukleev, V.; Dyadkina, E.; Lutsev, L.; Stognij, A.; Novitskii, N.; Grigorev, S. Investigation of the SiO_2 (Co)/GaAs heterostructures using the surface scattering of synchrotron radiation. *JETP Lett.* **2010**, *92*, 767–773. [CrossRef]
18. Schwartzkopf, M.; Buet, A.; Körstgens, V.; Metwalli, E.; Schlage, K.; Benecke, G.; Perlich, J.; Rawolle, M.; Rothkirch, A.; Heidmann, B.; et al. From atoms to layers: in situ gold cluster growth kinetics during sputter deposition. *Nanoscale* **2013**, *5*, 5053–5062. [CrossRef] [PubMed]
19. Parratt, L.G. Surface Studies of Solids by Total Reflection of X-rays. *Phys. Rev.* **1954**, *95*, 359. [CrossRef]
20. Ankner, J.; Felcher, G. Polarized-neutron reflectometry. *J. Magn. Magn. Mater.* **1999**, *200*, 741–754. [CrossRef]
21. Wolff, M.; Zhernenkov, K.; Zabel, H. Neutron reflectometery with ADAM at the ILL: Present status and future perspectives. *Thin Solid Films* **2007**, *515*, 5712–5715. [CrossRef]
22. Devishvili, A.; Zhernenkov, K.; Dennison, A.J.; Toperverg, B.; Wol, M.; Hjörvarsson, B.; Zabel, H. SuperADAM: Upgraded polarized neutron reflectometer at the Institut Laue-Langevin. *Rev. Sci. Instruments* **2013**, *84*, 025112. [CrossRef]
23. Vorobiev, A.; Devishvilli, A.; Palsson, G.; Rundlöf, H.; Johansson, N.; Olsson, A.; Dennison, A.; Wollf, M.; Giroud, B.; Aguettaz, O.; et al. Recent upgrade of the polarized neutron re?ectometer Super ADAM. *Neutron News* **2015**, *26*, 25–26. [CrossRef]
24. Ukleev, V.A.; Grigoryeva, N.A.; Dyadkina, E.A.; Vorobiev, A.A.; Lott, D.; Lutsev, L.V.; Stognij, A.I.; Novitskiy, N.N.; Mistonov, A.A.; Menzel, D.; et al. Magnetic properties of the SiO_2(Co)/GaAs interface: Polarized neutron reflectometry and SQUID magnetometry. *Phys. Rev. B* **2012**, *86*, 134424. [CrossRef]
25. Ottavian, G.; Tu, K.N.; Psaras, P.; Nobili, C. In situ resistivity measurement of cobalt silicide formation. *J. Appl. Phys.* **1987**, *62*, 2290–2294. [CrossRef]
26. Cabral, C., Jr.; Barmak, K.; Gupta, J.; Clevenger, L.A.; Arcot, B.; Smith, D.A.; Harper, J.M.E. Role of stress relief in the hexagonaclos-packed to face-centered-cubic phase transformation in cobalt thin films. *J. Vac. Sci. Technol.* **1993**, *A11*, 1435–1440. [CrossRef]
27. Colgan, E.G., Jr.; Cabral, C.; Kotecki, D.E. Activation energy for CoSi and $CoSi_2$ formation measured during rapid thermal annealing. *J. Appl. Phys.* **1995**, *77*, 614–619. [CrossRef]
28. Lowes, T.D.; Zinke-Allmang, M. Cobalt grain growth on clean Si(100) Surfaces. *Scanning Microsc.* **1998**, *12*, 119–129.
29. Jaccodine, R.J. Surface energy of germanium and silicon. *J. Electrochem. Soc.* **1963**, *110*, 524. [CrossRef]
30. Messmer, C.; Bilello, J.C. The surface energy of Si, GaAs, and GaP. *J. Appl. Phys.* **1981**, *52*, 4623. [CrossRef]
31. Zinke-Allmang, M. Phase separation on solid surfaces: Nucleation, coarsening and coalescence kinetics. *Thin Solid Films* **1999**, *346*, 1–68. [CrossRef]
32. Loncaric, M.; Sancho-Parramon, J.; Zorc, H. Optical properties of gold island films—A spectroscopic ellipsometry study. *Thin Solid Films* **2011**, *519*, 2946–2950. [CrossRef]
33. Qian, H.; Xiao, Y.; Lepage, D.; Chen, L.; Liu, Z. Quantum electrostatic model for optical properties of nanoscale gold films. *Nanophotonics* **2015**, *4*, 413–418. [CrossRef]
34. Yakubovsky, D.I.; Arsenin, A.V.; Stebunov, Y.V.; Fedyanin, D.Y.; Volkov, V.S. Optical constants and structural properties of thin gold films. *Opt. Express* **2017**, *25*, 25574. [CrossRef]
35. Meiklejohn, W.H.; Bean, C.P. New Magnetic Anisotropy. *Phys. Rev.* **1957**, *105*, 904. [CrossRef]
36. Gomoyunova, M.V.; Grebenyuk, G.S.; Pronin, I.I. Formation of ultrathin magnetic cobalt films on the Si(111)7 × 7 surface. *Tech. Phys.* **2011**, *56*, 865–868. [CrossRef]
37. Katmis, F.; Lauter, V.; Nogueira, F.S.; Assaf, B.A.; Jamer, M.E.; Wei, P.; Satpati, B.; Freeland, J.W.; Eremin, I.; Heiman, D.; et al. A high-temperature ferromagnetic topological insulating phase by proximity coupling. *Nature* **2016**, *533*, 513–516. [CrossRef]
38. Mogi, M.; Nakajima, T.; Ukleev, V.; Tsukazaki, A.; Yoshimi, R.; Kawamura, M.; Takahashi, K.S.; Hanashima, T.; Kakurai, K.; Arima, T.H.; et al. Large Anomalous Hall Effect in Topological Insulators with Proximitized Ferromagnetic Insulators. *Phys. Rev. Lett.* **2019**, *123*, 016804. [CrossRef] [PubMed]
39. Cortie, D.L.; Lin, K.W.; Shueh, C.; Hsu, H.F.; Wang, X.L.; James, M.; Fritzsche, H.; Bruck, S.; Klose, F. Exchange bias in a nanocrystalline hematite/permalloy thin film investigated with polarized neutron reflectometry. *Phys. Rev. B* **2012**, *86*, 054408. [CrossRef]
40. Saoudi, M.; Fritzsche, H.; Nieuwenhuys, G.J.; Hesselberth, M.B.S. Size Effect in the Spin Glass Magnetization of Thin AuFe Films as Studied by Polarized Neutron Reflectometry. *Phys. Rev. Lett.* **2008**, *100*, 057204. [CrossRef]
41. Ankner, J.F.; Majrzak, C.F.; Homma, H. Magnetic dead layer in Fe/Si multilayer: Profile refinement of polarized neutron reflectivity data. *J. Appl. Phys.* **1993**, *73*, 6436. [CrossRef]
42. Dyadkina, E.A.; Grigoriev, S.V.; Lott, D.; Sitnikov, A.V.; Kalinin, Y.E. Study of the $[(Co_{45}Fe_{45}Zr_{10})_x(Al_2O_3)_{100-x}$/a-Si:H$]_m$ multilayer nanostructure by polarized neutron reflectometry. *Phys. B* **2011**, *406*, 2397–2400. [CrossRef]

43. Dyadkina, E.A.; Grigoryeva, N.A.; Vorobiev, A.A.; Grigoriev, S.V.; Lutsev, L.V.; Zhernenkov, K.; Wolff, M.; Lott, D.; Stognij, A.I.; Novitskii, N.N.; et al. Polarized neutron reflectometry from the interface of the heterostructures SiO_2(Co)/Si and SiO_2(Co)/GaAs. *Phys. B* **2009**, *404*, 2547–2549. [CrossRef]
44. Lutsev,L. *Magnetoresistance: Types, Roles and Research*; Gonzalez, R., Ed.; Nova Science Publishers, Incorporated: New York, NY, USA, 2017; pp. 65–121.
45. Fan, Z.; Li, P.; Zhang, L.; Mi, W.; Jiang, E.; Bai, H. Electrical transport properties and room-temperature positive magnetoresistance of Fe_3O_4/a-C/n-Si junctions. *Thin Solid Films* **2012**, *520*, 3641–3646. [CrossRef]
46. Volkov, N.; Tarasov, A.; Eremin, E.; Baron, F.; Varnakov, S.; Ovchinnikov, S. Extremely large magnetoresistance induced by optical irradiation in the Fe/SiO_2/p-Si hybrid structure with Schottky barrier. *J. Appl. Phys.* **2013**, *114*, 093903. [CrossRef]

magnetochemistry

MDPI

Article

Study of the Effects of Er Doping on the Physical Properties of CdSe Thin Films

Yuliana de Jesús Acosta-Silva [1,*], Luis A. Godínez [2], Manuel Toledano-Ayala [1], Rosendo Lozada-Morales [3], Orlando Zelaya-Angel [1] and Arturo Méndez-López [1,*]

1 División de Investigación y Posgrado, Facultad de Ingeniería, Universidad Autónoma de Querétaro, Centro Universitario, Queretaro 76010, Mexico
2 Centro de Investigación en Química para la Economía Circular, CIQEC, Facultad de Química, Universidad Autónoma de Querétaro, Centro Universitario, Queretaro 76010, Mexico
3 Facultad de Ciencias Físico-Matemáticas, Benemérita Universidad Autónoma de Puebla, Puebla 72570, Mexico
* Correspondence: yuliana.dejesus.acosta@uaq.mx (Y.d.J.A.-S.); arturo.mendez@uaq.mx (A.M.-L.)

Abstract: Erbium-doped cadmium selenide thin films grown on 7059 Corning glass by means of a chemical bath at 80 °C were prepared. Doping was performed by adding an aqueous $Er(NO3)_33{\cdot}H_2O$ dilution to the CdSe growth solution. The volume of Er doping solution was varied to obtain different Er concentration (x at%). Thus, in the $Cd_{1-x}Er_xSe$ samples, the x values obtained were in the 0.0–7.8 at% interval. The set of the CdSe:Er thin films synthesized in the hexagonal wurtzite (WZ) crystalline phase are characterized by lattice parameters (a and c) that increase until x = 2.4% and that subsequently decrease as the concentration of x increases. Therefore, in the primitive unit cell volume (UC), the same effect was observed. Physical parameters such as nanocrystal size, direct band gap (E_g), and optical longitudinal vibrational phonon on the other hand, shift in an opposite way to that of UC as a function of x. All the samples exhibit photoluminescence (PL) emission which consists of a single broad band in the $1.3 \leq h\nu \leq 2.5$ eV range ($954 \geq \lambda \geq 496$ nm), where the maximum of the PL-band shift depends on x in the same way as the former parameters. The PL band intensity shows a singular behavior since it increases as x augments but exhibits a strong decreasing trend in the intermediate region of the x range. Dark d.c. conductivity experiences a high increase with the lower x value, however, it gradually decreases as x increases, which suggests that the Er^{3+} ions are not only located in Cd^{2+} sites, but also in interstitial sites and at the surface. Different physical properties are correlated among them and discussed considering information from similar reports in the literature.

Keywords: CdSe nanocrystalline films; Er doping; CdSe:Er; CdSe photoluminescence; softening-hardening of phonons

Citation: Acosta-Silva, Y.d.J.; Godínez, L.A.; Toledano-Ayala, M.; Lozada-Morales, R.; Zelaya-Angel, O.; Méndez-López, A. Study of the Effects of Er Doping on the Physical Properties of CdSe Thin Films. *Magnetochemistry* **2023**, *9*, 107. https://doi.org/10.3390/magnetochemistry9040107

Academic Editors: Cătălin-Daniel Constantinescu and Lucian Petrescu

Received: 22 December 2022
Revised: 7 April 2023
Accepted: 11 April 2023
Published: 14 April 2023

1. Introduction

The wide implementation of thin films in various applications, prepared with compound semiconductors of the IIA-VIB group is due to their excellent characteristics; therefore, these compounds have attracted the attention of several research groups around the world [1,2]. Due to the compatibility in the synthesis of stoichiometric and non-stoichiometric compounds, the cadmoselite or cadmium selenide (CdSe) is a particularly attractive metal selenide from the II–VI group [3]. Its high absorption coefficient, appropriate band gap of 1.74 eV (which is close to the ideal band gap desired for an appropriate absorber layer), high electron affinity, elevated photosensitivity, and n-type conductivity, makes it one of the prominent competitors among Cd-based chalcogenides of the II–VI group [4–6]. Cadmium selenide is therefore an interesting semiconductor material that has been used as a model substrate to design, prepare and pursue novel applications such as semiconductor nanostructures [7], luminescent nanocrystals [8], phase transformations [9], vibrational modes [10] and colloidal luminescent monolayers [11], among others. In most cases, CdSe exhibits n-type conductivity, which is dependent on its growing dynamics and on the development of vacancies. The n-type conductivity of CdSe films is caused by the Se vacancies, but the p-type conductivity may be caused by an Se surplus [6]. The CdSe material primarily exhibits cubic and hexagonal phases of crystallization and a combination of the cubic and hexagonal phases could arise during crystallization [6,12]. Several reports on the synthesis, properties and potential applications of CdSe have been published [13–15] and in the specific field of synthesis, new techniques to prepare chemically grown nanostructured CdSe films have been discovered. Among these, thermolysis of Cd and Se reagents [16], green methods [17], reverse micelle assisted hydrothermal [18] and rapid microwave synthesis [19], stand out as approaches to synthesize materials that can be used to prepare composite structures such as polymerized CdSe nanocrystals (NCs) [20], organic-inorganic hybrid CdSe NCs [21], CdSe-polymer fibers [22] and biocompatible CdSe-ZnS NCs [23]. These novel CdSe arrangements have been explored for applications in semiconductor nanostructure technology [24,25], food science [26] biologic images [27], agriculture [28] and DNA studies [29], among others.

The intrinsic CdSe nanobelts (NBs) exhibits an extremely high resistivity of 1×10^{13} Ω cm and thus leads to its low conductivity [30]. In order to further expand its usage, it is necessary to adjust the carrier concentration and then mediate its bandgap. Several researchers intentionally introduced defects by doping and thus changed carrier concentrations to enhance its electro-optical properties [31–36]. Up to now, doping has resulted in optical, electronic and transport control as well as in spintronic properties of the semiconductor materials [37]. For example, Mg-doped CdSe can modulate its optical bandgap and makes possible light response in the whole visible range [38]. Sahu et al. doped CdSe nanocrystals with Ag impurities and investigated their optical and electrical properties. They observed that doping leads not only to dramatic changes but surprising complexity. The addition of just a few Ag atoms per nanocrystal causes a large enhancement in the fluorescence. While Ag was expected to be a substitutional acceptor, nonmonotonic trends in the fluorescence and Fermi level suggest that Ag changes from an interstitial (n-type) to a substitutional (p-type) impurity with increased doping [8].

In the case of rare earth (RE) element doping, interesting results have been obtained using CdSe as the semiconductor substrate. While CdSe:Y3+ for example, has been used for luminescence enhancement [39] and CdSe:Gd for potential optical and MR imaging [40], CdSeEu has been found to be an efficient photocatalytic material [41]. The morphological, structural, optical and other properties of a host lattice can be improved by gradual and systematic addition of various dopants [42,43]. On the other hand, erbium is another rare earth element used as impurity in optical amplifiers because of its high capacity to supply both high gain and high at-saturation output. The Er^{3+} ion is also an active lasing element. Therefore, Er^{3+} has become one of the most interesting topics of research due to ${}^4I_{9/2}\rightarrow{}^4I_{15/2}$ and ${}^4I_{11/2}\rightarrow{}^4I_{13/2}$ transitions [44], and these characteristic properties open up a new line of research for these materials due to the intra Er^{3+} 4*f* shell transition from its first excited state (${}^4I_{3/2}$) to the ground state (${}^4I_{5/2}$) [45]. In this context, the aim of the present work is to study some of the morphological, structural and optical properties of Er^{3+}-doped CdSe thin films in order to obtain crystalline films. CdSe thin films were grown using the chemical bath (CB) method. In this way, Er^{3+} ions were efficiently incorporated into the semiconductor CdSe hexagonal wurtzite (WZ) crystalline phase with no important distortion of the lattice structure. Er^{3+} ions in the CdSe hexagonal wurtzite lattice are expected to occupy Cd^{2+} sites to produce an n-type CdSe doped substrate and in this work, the Er^{3+} doping values fall within the 0.0–7.8 at% interval. The WZ unit cell parameters, crystal size, vibrational modes, energy band gap, luminescence spectra and dark conductivity experience changes depending, as expected, on the Er concentration. The variation in these parameters and properties are discussed on the basis of the correlation of the experimental data obtained and of the information from similar works in the literature.

2. Experimental

Nanocrystalline CdSe:Er thin films were deposited on Corning 7059 glass substrates at 80 ± 1 °C. The glass substrates were cleaned by immersion in water for a couple of hours, and successive ultrasonic treatment using trichloroethylene, acetone, and deionized distilled water. Finally, the glass substrates were dried under a stream of pure N_2. The solutions employed in the growth of the CdSe:Er layers were: $CdCl_2$ 0.02 M, KOH 0.05 M, NH_4NO_3 1.5 M, $CSe(NH_2)_2$ 0.02 M, and $Er(NO_3)_3{\cdot}3H_2O$ 0.5 M. The total volume of the growing solution, V_T = 100 mL, consists of the volume solution V_{CdSe} for CdSe plus the volume solution, V_{Er}, containing the doping Er chemical: $V_{CdSe} + V_{Er} = V_T = 100$ mL. $V_R = (V_{Er}/V_T)100\%$ on the other hand, corresponds to the relative volume, and different V_R values were chosen within the 1.0% to 25% interval in order to obtain eight different doping levels. The growing time was one hour for all the film depositions.

The average thickness (τ) of the films correspond to 200 ± 10 nm as measured with a KLA TENCO P-15 profilometer. Optical transmittance spectra were obtained using a UNICAM 8700 spectrophotometer. X-ray diffraction (XRD) patterns were measured using a Rigaku D/max-2100 diffractometer (CuKα radiation, λ = 0.15406 nm). High resolution (HR-TEM) images were registered in a Transmission Electron Microscope (TEM) Jeol 2010 working at 200 KeV, in which thin layers of the CdSe films were detached from the substrate and supported on TEM-grids. The selected area for electron diffraction (SAED) was performed at the camera length, $L\frac{1}{4}$-20 cm. Measurements of the atomic concentration of elements were obtained by means of energy dispersion spectroscopy (EDS) utilizing a Voyager II X-ray quantitative microanalysis in an1100/1110 EDX system from Noran Instruments. Photoluminescence (PL) measurements were obtained using a 0.5 m Jovin-Ivon double spectrograph, employing as the excitation source the 441.6 nm (2.808 eV) line of an Omnichrome 2056 He-Cd laser with 35 mW of power. Raman spectra were obtained by means of a microRaman spectrometer (Jobin Yvon, model Labram) using the 632.8 nm laser line from a He–Ne laser. Finally, dark electrical conductivity measurements were obtained in vacuum using the two points method and sign of carriers (n) by thermoelectric power. The essential drawback of the usual two-point method arises from the approximation of the intrinsic drain voltage by the external applied drain bias. This means that high series resistances can induce significant errors in the lifetime evaluation [46]. All the characterization measurements were carried out at room temperature (RT).

3. Results and Discussion

The CdSe:Er thin films prepared as described in the experimental section, appear bright and smooth to the naked eye, as can be seen from inspection of insets (i) and (ii) of Figure 1, which correspond to a doped (V_R = 15%) and undoped samples, respectively. Figure 1 also exhibits the Er concentration measured by means of EDS, where each data point is an average value of ten measurements. As it can be observed in Figure 1, the EDS data (in at%) vs. V_R plot, roughly follows a linear relationship. In the $Cd_{1\text{-}x}Er_xSe$ doped samples, the Er concentration (x) varied within the $0 \leq x \leq 7.8$ at% range.

Nanocrystallites corresponding to the $Cd_{1-x}Er_xSe$ WZ structure for x = 4.6 at% can be seen in the TEM image of Figure 1b. The crystalline [002] planes of the WZ phase can also be observed, and the (002) interplanar spacing (IS) was determined as 0.35 ± 0.01 nm, which can be associated to the standard IS value 0.3505 nm of the [002] planes in WZ-CdSe (JCPDS CAS No. 08-0459). The inset of Figure 1b exhibits the first diffraction rings of the WZ-CdSe structure as well as the SAED pattern of the sample where the nanocrystalline character of the film is confirmed.

The hexagonal wurtzite structure of the CdSe samples in the crystalline phase obtained from the diffractograms of Figure 2A can be seen for two representative samples prepared with different doping levels (x = 5.5 and 0.6 at%). The asterisks in the diffractograms mark where the XRD reflections of Er_2O_3 should be present [47,48]; however, most of the peaks showed low intensity. Er_2O_3 has been reported to appear as another phase when CdSe nanocrystals are obtained from molecular precursors [49] and when GaNEr films are prepared using MOCVD [50]. As it was observed for the CB method and in the preparation of Si by different methods [51], in these synthesis techniques, oxygen is captured by Er^{3+} ions and the resulting Er_2O_3 oxide has for instance, been observed in Er doped ZnO NCs prepared using the sol-gel method [48].

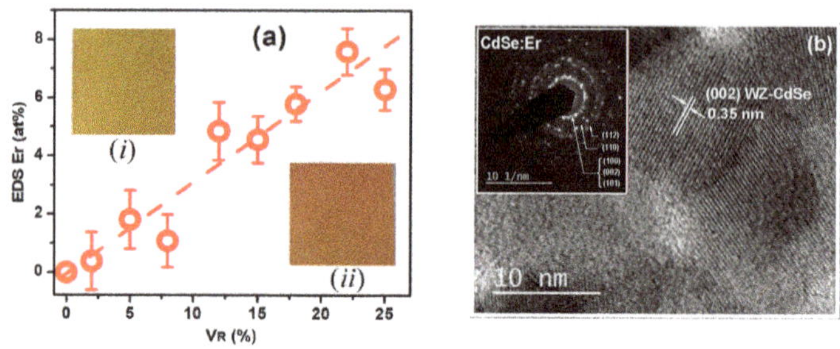

Figure 1. (**a**) EDS as a function of V_R. Dashed line represents a linear fitting. (i) and (ii) insets illustrate images of the V_R = 15% doped and undoped $Cd_{1-x}Er_xS$ samples, respectively. (**b**) TEM image of sample for x = 4.6 at%, where the (002) planes of wurtzite phase of CdSe are represented. The inset shows a SAED pattern of the WZ crystalline structure.

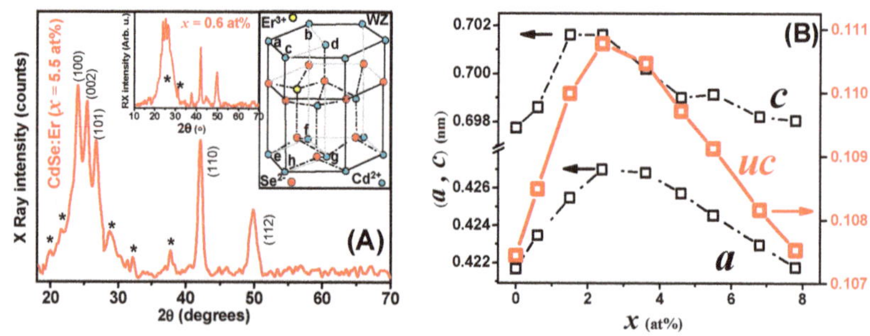

Figure 2. (**A**) XRD pattern of the CdSe:Er film for x = 5.5 at%. The left inset shows the pattern for x = 0.6 at%. The right inset illustrates the unit cell of WZ-CdSe lattice. (**B**) (left axis) lattice parameters a and c versus x, (right axis) primitive unit cell volume versus x.

The size of the CdSe:Er nanoparticles, observed by HRTEM coincides with the size of the crystallites observed by XRD [52]. In our case, the sizes of the crystallite (determined by XRD) using the Monshi–Scherrer equation [53] are comparable to the size of the particle sizes (observed by TEM). The observed nanoparticles are in the 8–15 nm size range (Figure 1b).

The XRD patterns revealed the CdSe and Er_2O_3 phases of the doped films and the right inset in Figure 2A shows the hexagonal WZ unit cell (UC) of CdSe, where the yellow circle represents the Er^{3+} ion in a Cd^{2+} site. The ab.gh parallelepiped represents the primitive UC with side "a" and height "c", which correspond to the WZ lattice parameters. Figure 2B on the other hand, displays the a and c values of the UC as a function of the doping level, x. The corresponding data in the left axis reveals that for both parameters, there is an initial increasing tendency that reaches a maximum at around x~2.4 at%, and that is followed by a decreasing trend. As expected, the UC volume (right axis graph) follows a similar volcano shaped curve.

It is evident from the XRD pattern's peaks that all films are polycrystalline, as well as from the spreading diffraction peak and the small shift in the (002) crystal plane. The integration of the Er^{3+} ion into the crystal structure of the CdSe film causes a shift in this peak that rises with Er^{3+}. The (100) and (002) planes are positioned in the CdSe:Er film's strongest reflection peaks. We believe that this behavior is related to Er^{3+} ion inclusion in the crystal lattice, which results in preferred orientation in certain crystal planes. The size of the ionic radii of Er^{3+} should be related with the lowest energy that favors preferential orientation in these crystal planes. The WZ phase is indicated by strong peaks on the CdSe:Er film, which indicates a better crystallinity.

The interesting relationship of the UC dimension data and the doping level of the semiconductor material can be interpreted considering the work of Hsu et al. [54] who reported a study of Er:GaSe. These authors suggested that the two acceptor levels were originated either from the substitution (SBSs) of one Er^{3+} for one pair of Ga^{2+} or interstitial (INSs) insertion of one Er^{3+} at interlayer sites. In the case of the material under study, it is possible to note that since the ionic radii are Cd^{2+}: 0.95Å, Er^{3+}: 0.89Å and Se^{2-}: 2.025Å, we can assume that Er^{3+} ions occupy Cd^{2+} sites (similar cation-size). However, with a further increase in the Er dopant concentration, the dopant ions are probably incorporated at the interstitial sites (several papers have reported this change) [55–59]. Based on this, it is possible to suggest that according to the oxidation state of the Er^{+3} ion, this chemical species is incorporated into the crystal lattice of CdSe replacing a Cd^{2+} ion, thus providing an extra hole due to its deficiency in valence electrons (similar investigations have been reported for CdSe:Er films) [60].

Raman spectra of Er-doped CdSe nanocrystalline thin films in 150–700 cm^{-1} intervals can be seen in Figure 3a. An inspection of the corresponding data shows an increase in the phonon signals as x increases. These observations are similar to those reported by Portillo et al. [60]. A Raman spectrum of a representative CdSe:Er sample is exhibited in Figure 3a, where three overtones are observed in addition to the optical longitudinal A1LO mode at 205 cm^{-1} of the WZ CdSe structure. A Raman phonon mode for the monoclinic structure of Er_2O_3 at ~560 cm^{-1} can also be seen in the CdSe:Er spectrum of Figure 3b [61]. The 1LO phonon mode shows an asymmetric band, which reflects the presence of a surface mode [62]. In nanocrystals (NCs) this is mainly due to a high surface to volume ratio [63], surface-ions-bond contraction and quantum confinement [64]. Deconvolution of these signals separates the 1LO and the surface phonon modes (SPM), A and B, as shown in the inset of Figure 3c. The lower inset shows the 1LO mode in a magnified scale, where the frequency position as a function of x is clearly observed. This relationship resembles an inverse function of the CdSe UC versus x (see Figure 2B), in which lattice expansion implies phonon softening, and lattice contraction and phonon hardening. In the inset of Figure 3c on the other hand, it is possible to observe vibration displacements of the 1LO mode, which reflect bond-stretching vibrations [65]. As it was previously indicated, contraction or expansion of the CdSe lattice depends on the concentration of Er^{3+} impurities in the interstitial sites of the CdSe lattice. The B-data trend in Figure 3c indicates that the SPM shifts to lower frequencies when the doping increases; a fact that suggests that disorder augments almost monotonically as x increases [66,67].

Figure 4a shows the Tauc method data used to calculate the CdSe direct energy band gap, E_g, in which the $(\alpha h\nu)^2$ *versus* hν plot was employed [68]. Here, α is the optical absorption coefficient, which was determined from the optical transmittance (T_o) ($\alpha = [\mathrm{Log}\{100/T_o\}]/[\tau \mathrm{Log} e]$), h$\nu$ is the photon energy; and only four samples are exhibited for clarity. The inset in this figure shows T_o spectra for the same four samples chosen for E_g calculation. Figure 4b displays E_g as function of x for the entire $0.0 \leq x \leq 7.8$ at% range. The E_g dependence on x is in agreement with the UC vs. x plot, because, in general, if UC increases, E_g is reduced, and vice versa. However, this is not in agreement with the average crystal size (ACS) as function of x, since it can be readily seen that as ACS decreases, E_g also decreases. This is contrary to the quantum confinement effect since ACS should be within the strong quantum confinement regime (CdSe Bohr radius: 5.3 nm) [69]. This anomalous behavior can be explained considering the modifications in the lattice produced by the doping, which in turn can result in distortion, vacancies, interstitials, stress, expansion, contraction and non-stoichiometry effects, which can break the crystal symmetry and modify the physical properties of materials [70–73], particularly in the case of lanthanide doping [74]. The ACS dependence on x can be explained by surface tension increase that reduces crystal size when Er^{3+} bonds at the NC surface are formed [75,76]. It is also interesting to note that at $x \cong 4.6$ at%, the effect of Er^{3+} on the NCs volume becomes important, and the ACS starts to increase. It can also be observed that the crystal size (Figure 4b green) behaves inversely to the unit cell size (Figure 2B red). We have obtained different values of E_g shown in Figure 4b red. E_g decreases with increasing erbium doping up to $x = 4.6\%$, then it increases for $x = 5.5$ to 7.8%. The decrease in E_g may be due to the erbium electron localized states that create new states near the conduction band, leading to a reduction in the band gap of pure CdSe (similar observations were reported for ZnO) [77]. The variation in band gap depends also on particle size (Figure 4b green) and lattice parameters (Figure 2B). The decrease in particle size with increasing doping concentration leads to a rising of the band gap as the quantum refinement effect occurs [78]. The Bras' effective mass model [79] can also explain our results. The energy gap (E_g) of the nanoparticles can be expressed, according to Bras' effective mass model, as a function of particle size, as shown by the following expression:

$$E_g = E_{bulk} + \frac{\hbar^2\pi^2}{2R^2}\left[\frac{1}{m_e} + \frac{1}{m_h}\right] - \frac{1.8e^2}{\varepsilon R} \tag{1}$$

where E_g is the measured band gap of the nanoparticles, E_{bulk} is the band gap of the bulk semiconductor, $\hbar$ is the second Plank's constant, R is the radius of the nanoparticle, m_e is the effective mass of the electron, m_h is the effective mass of the hole, e is the charge of electron, and ε is the electric permittivity of the material. The lattice parameters (a and c) and the unit cell volume (V) of nanoparticles can be calculated with the following equation:

$$\frac{1}{d^2} = \frac{4}{3}\left(\frac{h^2 + hk + k^2}{a^2}\right) + \frac{l^2}{c^2} \tag{2}$$

$$V = \frac{\sqrt{3}}{2}a^2c \tag{3}$$

where d is the interplanar distance and (h, k, l) are the Miller indices. The ratio a, c and the unit cell volume (V) for pure and Er-doped samples are shown in Figure 2B. It can be noticed that a, c and V are affected by erbium doping with an increase in their values up to $x = 2.4\%$ with respect to the pure sample. This behavior of a, c and V may be attributed to changes in the ionic size of Er^{3+} (0.89Å) substituting Cd^{2+} (0.95Å) in CdSe lattice. a, c and V decrease for $x > 2.4\%$ and in the end, they approach the values of pure CdSe. This cell contraction may be explained by the larger Er^{3+} ions leaving the CdSe lattice at high doping concentrations ($x > 2.4\%$) to form the Er_2O_3 impurity [80,81]. The decrease in the lattice parameter c may also be attributed to the formation of defects in the CdSe lattice as a result of increasing the number of dopant (Er^{3+}) ions, Dakhel et al. reported similar behavior when performing a different doping experiment [82].

Figure 5a shows the dark d.c. conductivity (σ) dependences (presented on a semi-logarithmic scale) of $Cd_{1-x}Er_xSe$, in the 10^{-7} to 10^1 Ω^{-1} cm^{-1} range, versus the inverse of the absolute temperature (T) times k (1/kT). Here, k corresponds to the Boltzmann constant. A metallic behavior of log (σ) vs. 1/kT is observed for $x = 0.6$, 1.5 and 2.4 at%, since σ remains essentially constant in the $300 \geq T \geq 35$ K range (which is $35 \leq 1/kT \leq 110$ $(eV)^{-1}$), and decreases for $T \leq 35$ K. For $3.6 \leq x \leq 7.8$ at%, the behavior is that of a semiconductor with shallow donor levels, where the free carrier density decreases as x increases. The metallic dependence of log (σ) vs. T has been observed in ITO films grown by reactive plasma [83], and has been attributed to disorder in the lattice. Grenberg and coworkers [84] also reported a similar behavior at low temperatures for ZnO nanocrystals (NCs), which was obtained by changing the contact area and the center-to-center distance between NCs. A strong decrease in free carriers' lifetime has been reported when the Er concentration increases, due to the formation of NCs [85,86].

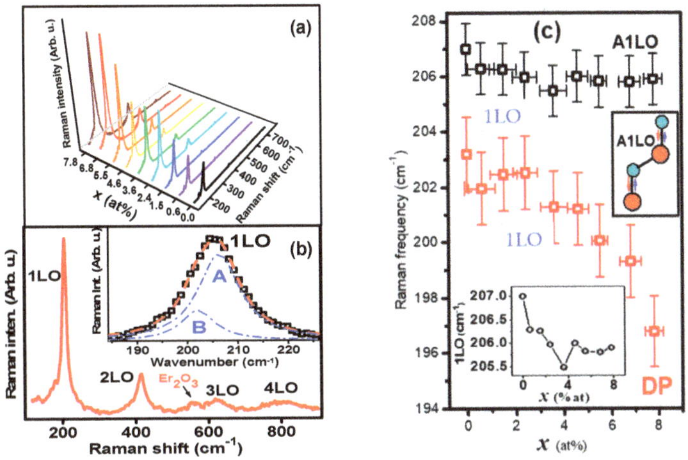

Figure 3. (**a**) Phonon Raman spectra of the series of $Cd_{1-x}Er_xSe$ samples. (**b**) Raman spectrum of a representative CdSe:Er sample where the 1LO phonon mode and up to the third overtone are observed, a Raman signal from Er_2O_3 is revealed in the spectrum. The inset illustrates the deconvolution of the band at ~205 cm^{-1} in the 1LO mode (A) and the surface mode (B). (**c**) The 1LO (A) and the surface (B) modes positions against x. The low inset exhibits the 1LO mode position vs. x in a magnified scale. The right inset illustrates the A1LO mode oscillation.

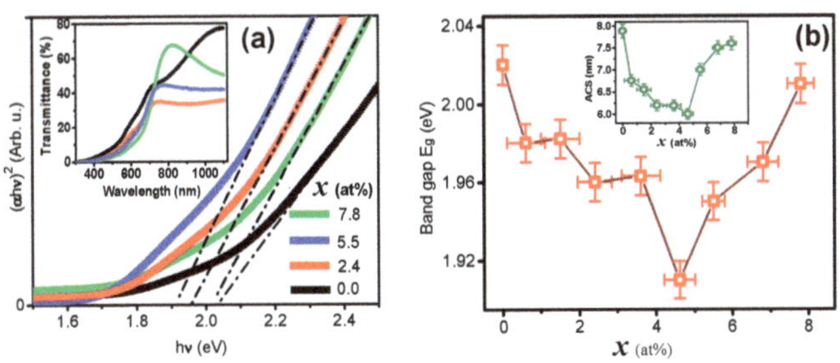

Figure 4. (**a**) Tauc method to calculate the direct energy band gap (E_g) of the $Cd_{1-x}Er_xSe$ for four values of x. The inset illustrates the optical transmittance (T_o) spectra for the same samples chosen for E_g calculation. (**b**) Direct band gap as function of x, the inset depicts the average crystal size (ACS) versus x plot.

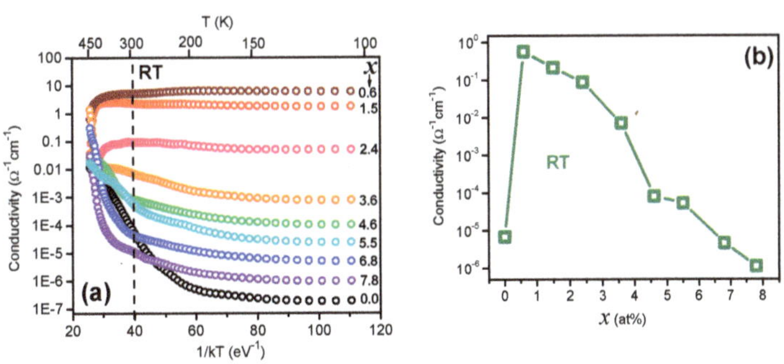

Figure 5. (**a**) Dark d.c. conductivity (σ) vs. inverse of absolute temperature (T). (**b**) Room temperature (RT) dark d.c. conductivity σ as a function of x.

In the case of the CdSe:Er NCs of this work, the incorporation of Er^{3+} produces an effective doping at low Er concentrations. When x rises, the formation of Er_2O_3 NCs (previously observed in XRD and Raman analysis) can produce barriers between the CdSe NCs which scatter the free carriers that, in turn, generate non-radiative transitions. σ at RT versus x is displayed in Figure 5b (see also dotted line in Figure 5a). The corresponding plot reveals that for low x values, σ increases almost five orders of magnitude. When Er concentration increases up to $x = 7.8$ at%, σ diminishes to values lower than those of the undoped sample. In this context, doping ZnO with Er^{3+} [87] and Ho^{3+} [88] shows that σ decreases when the impurity concentration increases. In the first case, the decrease is attributed to impurities scattering of free carriers and in the second case, to trapping at grain boundaries. Another effect of the Er^{3+} ion incorporation is the decrease in the crystallite size, which leads to higher grain boundary scattering, decreasing the mobility as well as the conductivity. As described by Callaway, when the grain size is decreased, the scattering rate of phonons become higher. Hence, it will lower the value of conductivity, which is favorable for various materials. However, as conductivity is reduced, the value of grain boundary scattering increases [89]. Morais et al., in their experiment of Er doped SnO_2 thin films, indicate that by analogy, the overall effect of increasing the Er concentration is to decrease the film conductivity since an increase in the electric charge compensation, and decrease in the crystallite size, result in a decrease in the mobility since the grain boundary scattering is increased [90].

Photoluminescence (PL) measurements on the materials under study were carried out using as excitation source the 2.808 eV (441.6 nm) line of a He-Cd laser. Figure 6a displays the spectra for the $Cd_{1-x}Er_xSe$ films registered for all the films in the $0.0 \leq x \leq 7.8$ at% range. All the samples exhibit a single PL broad band in the 1.4–2.5 eV interval (496–886 nm). Very narrow PL signals from Er^{3+} intra-atomic transitions [91] are observed in all spectra except for the undoped sample ($x = 0$). Several authors [92,93] report PL emission of rare earths doped and undoped CdSe NCs with a broad band centered in the 2.06–2.25 eV (550–600 nm) interval. The CdSe:Er samples of this work show the same PL band whose position depends on the NC size, doping type, and crystal defects. Other than the Er intra-atomic level transitions, no PL signals from Er_2O_3 oxide (see Figure 2A) could be observed. Figure 6b exhibits the PL band position (PLBP) as a function of the doping level x. Inspection of this data reveals that PLBP decreases in the $0.0 \leq x \leq 4$ at% interval and increases for the remaining values. PLPB follows a tendency similar to that of the band gap (see Figure 4b), since PLBP is defined by an exciton emission [93]. The PLBP intensity versus x plot (right axis) does not show a gradual increase in the intensity. This result is not clear. It seems that PL intensity emission is reduced for smaller NPs and increases for larger sizes (see inset of inset of Figure 4b). As it was previously indicated, the crystalline quality of CdSe:Er films improves as x increases; a fact that suggests that defects on the NCs surface are responsible for the intensity abatement of PL spectra. The non-radiative transitions at the surface are increased when NC size decreases, along the $0.0 \leq x \leq 4.6$ at% interval (see inset of Figure 4b). It is interesting note that a strong decrease in PL emissions has been observed in Er-doped GaAs due to the presence of ErAs precipitates in the GaAs matrix; giving place to Schottky barriers at the Er_2O_3-GaAs interface, which in turn form high n- and p-type depletion regions of high resistivity and very short carrier lifetimes [86]. Doping Er^{3+} in Hf-Al-silicate forms Er_2O_3 close the surface of planar waveguides, where the carriers' lifetime decreases when the Er concentration increases [85]. In the case of the CdSe:Er samples under study, while lifetime and PL decreases for low Er concentrations, NC increases for higher x values (x~4.6 at%) where radiative transitions surpass non-radiative ones because the surface/volume ratio increases. As a consequence, the carriers' lifetime also increases [94]. High density of Er_2O_3 NCs on the other hand, reduces σ and enhances PL signals.

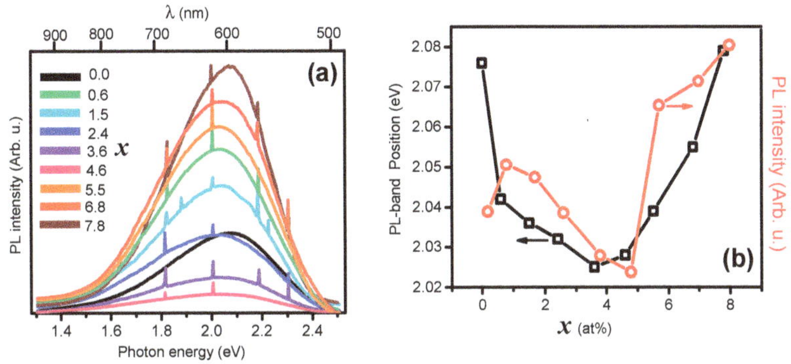

Figure 6. (**a**) Photoluminescence spectra of all the set of $Cd_{1-x}Er_xSe$ samples. (**b**) Energy position of the PL band maximum (left axis) and PL band intensity (right axis) versus x.

4. Conclusions

In summary, Er incorporation in CdSe nanocrystals provokes drastic changes in the $Cd_{1-x}Er_xSe$ lattice structure, depending on the level of Er doping. For low x values, Er^{3+} ions are sequentially incorporated as follows: (i) replacement of Cd^{2+} ions, (ii) to interstitial sites, and (iii) to NCs surface. In the first (i) case, the unit cell volume tends to increase, and in the (ii) and (iii) cases, it tends to decrease. Er^{3+} in Cd^{2+} sites induce crystalline quality of CdSe as suggested from XRD and Raman measurements, thus Er^{3+} reduces the structural disorder but increases the chemical one. When the lattice expands, the 1LO phonon softens, and as expected, hardens if the lattice contracts. The surface mode softens continuously when x increases, which indicates that chemical disorder increases. The CdSe energy band gap decreases when UC increases and increases in the opposite case, which is a property of all semiconductors. Er_2O_3 NCs grow as a secondary phase, which forms barriers at the oxide-semiconductor interface, which gradually reduces the carriers transfer through the interface when x increases and, consequently, the film dark conductivity decreases. On the other hand, the Er_2O_3-CdSe band gap coupling enhances the photoluminescence, via conversion in the oxide, of CdSe. The findings of this work can be useful for the design of CdSe-based devices.

Author Contributions: Y.d.J.A.-S.: experimental, investigation, methodology, formal analysis; L.A.G.: formal analysis, resources; M.T.-A.: investigation, formal analysis; R.L.-M.: resources, validation; O.Z.-A.: Validation, Writing; A.M.-L.: Validation, visualization, Writing—Review & Editing. All authors have read and agreed to the published version of the manuscript.

Funding: This research received no external funding.

Institutional Review Board Statement: Not applicable.

Informed Consent Statement: Not applicable.

Data Availability Statement: The datasets used and/or analyzed during the current study are available from the corresponding author on reasonable request.

Conflicts of Interest: The authors declare no conflict of interest.

References

1. Babu, N.S.; Khadar, M.A. Electrical properties of grain size tuned CdSe nanocrystal films for practical applications. *Sol. Energy Mater. Sol. Cells* **2018**, *178*, 106–114. [CrossRef]
2. Hassen, M.; Riahi, R.; Laatar, F.; Ezzaouia, H. Optical and surface properties of CdSe thin films prepared by sol-gel spin coating method. *Surf. Interfaces* **2019**, *18*, 100408. [CrossRef]
3. Ghobadi, N.; Sohrabi, P.; Hatami, H.R. Correlation between the photocatalytic activity of CdSe nanostructured thin films with optical band gap and Urbach energy. *Chem. Phys.* **2020**, *538*, 110911. [CrossRef]

4. Dai, G.; Chen, Y.; Wan, Q.; Zhang, Q.; Pan, A.; Zou, B. Fabrication and optical waveguide of Sn-catalyzed CdSe microstructures. *Solid State Commun.* **2013**, *167*, 31–35. [CrossRef]
5. Shyju, T.; Anandhi, S.; Indirajith, R.; Gopalakrishnan, R. Solvothermal synthesis, deposition and characterization of cadmium selenide (CdSe) thin films by thermal evaporation technique. *J. Cryst. Growth* **2011**, *337*, 38–45. [CrossRef]
6. Bayramoglu, H.; Peksoz, A. Electronic energy levels and electrochemical properties of co-electrodeposited CdSe thin films. *Mater. Sci. Semicond. Process.* **2019**, *90*, 13–19. [CrossRef]
7. Smith, A.M.; Nie, S. Semiconductor Nanocrystals: Structure, Properties, and Band Gap Engineering. *Acc. Chem. Res.* **2010**, *43*, 190–200. [CrossRef]
8. Sahu, A.; Kang, M.S.; Kompch, A.; Notthoff, C.; Wills, A.W.; Deng, D.; Winterer, M.; Frisbie, C.D.; Norris, D.J. Electronic Impurity Doping in CdSe Nanocrystals. *Nano Lett.* **2012**, *12*, 2587–2594. [CrossRef]
9. Tolbert, S.H.; Alivisatos, A.P. Size Dependence of a First Order Solid-Solid Phase Transition: The Wurtzite to Rock Salt Transformation in CdSe Nanocrystals. *Science* **1994**, *265*, 373–376. [CrossRef]
10. Trallero-Giner, C.; Debernardi, A.; Cardona, M.; Proupin, E.M.; Ekimov, A.I. Optical vibrons in CdSe dots and dispersion relation of the bulk material. *Phys. Rev. B* **1998**, *57*, 4664–4669. [CrossRef]
11. Delikanli, S.; Yu, G.; Yeltik, A.; Bose, S.; Erdem, T.; Yu, J.; Erdem, O.; Sharma, M.; Sharma, V.; Quliyeva, U.; et al. Ultrathin Highly Luminescent Two-Monolayer Colloidal CdSe Nanoplatelets. *Adv. Funct. Mater.* **2019**, *29*, 1901028. [CrossRef]
12. Alasvand, A.; Kafashan, H. Comprehensive physical studies on nanostructured Zn-doped CdSe thin films. *J. Alloys Compd.* **2019**, *789*, 108–118. [CrossRef]
13. Garg, N. A Brief Study on Characteristics, Properties, and Applications of CdSe. *Proc. ICEMIT* **2017**, *2*, 43–60.
14. Yu, J.; Chen, R. Optical properties and applications of two-dimensional CdSe nanoplatelets. *Infomat* **2020**, *2*, 905–927. [CrossRef]
15. Zhao, L.; Hu, L.; Fang, X. Growth and Device Application of CdSe Nanostructures. *Adv. Funct. Mater.* **2012**, *22*, 1551–1566. [CrossRef]
16. Trindade, T.; O'Brien, P.; Zhang, X.-M. Synthesis of CdS and CdSe Nanocrystallites Using a Novel Single-Molecule Precursors Approach. *Chem. Mater.* **1997**, *9*, 523–530. [CrossRef]
17. Ahmed, M.; Guleria, A.; Rath, M.C.; Singh, A.K.; Adhikari, S.; Sarkar, S.K. Facile and green synthesis of CdSe quantum dots in protein matrix: Tuning of morphology and optical properties. *J. Nanosci. Nanotechnol.* **2014**, *14*, 5730–5742. [CrossRef] [PubMed]
18. Xi, L.; Lam, Y.M.; Xu, Y.P.; Li, L.-J. Synthesis and characterization of one-dimensional CdSe by a novel reverse micelle assisted hydrothermal method. *J. Colloid Interface Sci.* **2008**, *320*, 491–500. [CrossRef]
19. Thomas, D.; Lee, H.O.; Santiago, K.C.; Pelzer, M.; Kuti, A.; Jenrette, E.; Bahoura, M. Rapid Microwave Synthesis of Tunable Cadmium Selenide (CdSe) Quantum Dots for Optoelectronic Applications. *J. Nanomater.* **2020**, *2020*, 5056875. [CrossRef]
20. Skaff, H.; Emrick, T. Reversible Addition Fragmentation Chain Transfer (RAFT) Polymerization from Unprotected Cadmium Selenide Nanoparticles. *Angew. Chem. Int. Ed.* **2004**, *43*, 5383–5386. [CrossRef]
21. Li, J.; Hao, X.; Wang, J.; Cui, X.; Li, X.; Wei, S.; Lu, J. Layered Inorganic/Organic Hybrid (CdSe) n·Monoamine Nanobelts: Controllable Solvothermal Synthesis, Multiple Stage Amine De-Intercalation Transformation, and Two-Dimensional Exciton Quantum Confinement Effect. *Inorg. Chem.* **2018**, *57*, 10781–10790. [CrossRef]
22. Miethe, J.F.; Schlosser, J.A.; Eckert, G.; Lübkemanna, F.; Bigall, N.C. Electronic transport in CdSe-polymer fibres. *J. Mater. Chem. C* **2018**, *6*, 10916–10923. [CrossRef]
23. Yildiz, I.; Callan, B.; Cruickshank, S.F.; Callan, J.F. Biocompatible CdSe-ZnS Core-Shell Quantum Dots Coated with Hydrophilic Polythiols. *Langmuir* **2009**, *25*, 7090–7096. [CrossRef]
24. Huang, C.; Mao, J.; Chen, X.M.; Yang, J.; Du, X.W. Laser-activated gold catalysts for liquid-phase growth of cadmium selenide nanowires. *Chem. Commun.* **2014**, *51*, 2145–2148. [CrossRef] [PubMed]
25. Rajan, P.I.; Vijaya, J.J.; Jesudoss, S.K.; Kaviyarasu, K.; Lee, S.-C.; Kennedy, L.J.; Jothiramalingam, R.; Al-Lohedan, H.A.; Abdullah, M.M. Investigation on preferably oriented abnormal growth of CdSe nanorods along (0002) plane synthesized by henna leaf extract-mediated green synthesis. *R. Soc. Open Sci.* **2018**, *5*, 171430. [CrossRef] [PubMed]
26. Bonilla, J.C.; Bozkurt, F.; Ansari, S.; Sozer, N.; Kokini, J.L. Applications of quantum dots in Food Science and biology. *Trends Food Sci. Technol.* **2016**, *53*, 75–89. [CrossRef]
27. Almeida-Silva, A.C.; Peres-Freschi, A.P.; Mendonça-Rodrigues, C.; França-Matias, B.; Prado-Maia, L.; Goulart, L.R.; Oliveira-Dantas, N. Biological analysis and imaging applications of CdSe/CdS_xSe_{1-x}/CdS core–shell magic-sized quantum dot. *Nanomed. Nanotechnol. Biol. Med.* **2016**, *12*, 1421–1430. [CrossRef]
28. Ung, T.D.T.; Tran, T.K.C.; Pham, T.N.; Nguyen, D.N.; Dinh, D.K.; Nguyen, Q.L. CdTe and CdSe quantum dots: Synthesis, characterizations and applications in agriculture. *Adv. Nat. Sci. Nanosci. Nanotechnol.* **2012**, *3*, 043001. [CrossRef]
29. Dunpall, R.; Nejo, A.A.; Pullabhotla, V.S.R.; Opoku, A.R.; Revaprasadu, N.; Shonhai, A. An in Vitro Assessment of the Interaction of Cadmium Selenide Quantum Dots With DNA, Iron, and Blood Platelets. *IUBMB Life* **2012**, *64*, 995–1002. [CrossRef]
30. Sathyalatha, K.; Uthanna, S.; Reddy, P. Electrical and photoconducting properties of vacuum evaporated pure and silver-doped CdSe films. *Thin Solid Films* **1989**, *174*, 233–238. [CrossRef]
31. Walukiewicz, W. Intrinsic limitations to the doping of wide-gap semiconductors. *Phys. B Condens. Matter* **2001**, *302–303*, 123–134. [CrossRef]
32. Majid, A.; Arshad, H.; Murtaza, S. Synthesis and characterization of Cr doped CdSe nanoparticles. *Superlattices Microstruct.* **2015**, *85*, 620–623. [CrossRef]

33. Whitham, P.J.; Knowles, K.E.; Reid, P.J.; Gamelin, D.R. Photoluminescence Blinking and Reversible Electron Trapping in Copper-Doped CdSe Nanocrystals. *Nano Lett.* **2015**, *15*, 4045–4051. [CrossRef] [PubMed]
34. Proshchenko, V.; Dahnovsky, Y. Magnetic effects in Mn-doped CdSe nanocrystals. *Phys. Status Solidi B* **2015**, *252*, 2275–2279. [CrossRef]
35. Al-Kabbi, A.S.; Sharma, K.; Saini, G.; Tripathi, S. Effect of doping on transport properties of nanocrystalline CdSe thin film. *Thin Solid Films* **2015**, *586*, 1–7. [CrossRef]
36. Sharma, K.; Al-Kabbi, A.S.; Saini, G.; Tripathi, S. Influence of Zn doping on structural, optical and electrical properties of nanocrystalline CdSe thin films. *J. Alloys Compd.* **2015**, *651*, 42–48. [CrossRef]
37. Zhang, K.-C.; Li, Y.-F.; Liu, Y.; Chi, F. Density-functional study on the robust ferromagnetism in rare-earth element Yb-doped SnO 2. *J. Magn. Magn. Mater.* **2014**, *360*, 165–168. [CrossRef]
38. Lee, W.; Kwak, W.-C.; Min, S.K.; Lee, J.-C.; Chae, W.-S.; Sung, Y.-M.; Han, S.-H. Spectral broadening in quantum dots-sensitized photoelectrochemical solar cells based on CdSe and Mg-doped CdSe nanocrystals. *Electrochem. Commun.* **2008**, *10*, 1699–1702. [CrossRef]
39. Martín-Rodríguez, R.; Geitenbeek, R.; Meijerink, A. Incorporation and Luminescence of Yb^{3+} in CdSe Nanocrystals. *J. Am. Chem. Soc.* **2013**, *135*, 13668–13671. [CrossRef]
40. Li, I.-F.; Yeh, C.-S. Synthesis of Gd doped CdSe nanoparticles for potential optical and MR imaging applications. *J. Mater. Chem.* **2010**, *20*, 2079–2081. [CrossRef]
41. Hanifehpour, Y.; Joo, S.W. Synthesis, characterization and sonophotocatalytic degradation of an azo dye on Europium doped cadmium selenide nanoparticles. *Nanochemistry Res.* **2018**, *3*, 178–188.
42. Tomás, S.A.; Lozada-Morales, R.; Portillo, O.; Lima-Lima, H.; Palomino-Merino, R.; Zelaya, O. Characterization of chemical bath deposited CdS thin films doped with methylene blue and Er^{3+}. *Eur. Phys. J. Spéc. Top.* **2008**, *153*, 299–302. [CrossRef]
43. Moreno, O.P.; Pérez, R.G.; Portillo, M.C.; Téllez, G.H.; Rosas, E.R.; Cruz, S.C.A. Moreno Rodríguez, Synthesis, morphological, optical and structural properties of PbSSe2−nanocrystals. *Optik* **2016**, *127*, 8341–8349. [CrossRef]
44. Dammak, M.; Zhang, D.-L. Spectra and energy levels of Er^{3+} in Er_2O_3 powder. *J. Alloys Compd.* **2006**, *407*, 8–15. [CrossRef]
45. Al-Kuhaili, M.F.; Durrani, S.M.A. Optical properties of erbium oxide thin films deposited by electron beam evaporation. *Thin Solid Films* **2007**, *515*, 2885–2890. [CrossRef]
46. Ionescu, A.M.; Munteanu, D.; Chovet, A.; Rusu, A.; Steriu, D. The intrinsic pseudo-MOSFET technique. In Proceedings of the 1997 International Semiconductor Conference 20th Edition, Sinaia, Romania, 7–11 October 1997; Volume 1, pp. 217–220.
47. Bakhsh, A.; Maqsood, A. Sintering effects on structure, morphology, and electrical properties of sol-gel synthesized, nano-crystalline erbium oxide. *Electron. Mater. Lett.* **2012**, *8*, 605–608. [CrossRef]
48. Bhatia, S.; Verma, N. Erbium-doped nanoparticles/films for enhancing percentage photodegradation of direct red-31 dye. *J. Mater. Sci. Mater. Electron.* **2018**, *29*, 14960–14970. [CrossRef]
49. Hegazy, M.A.; El-Hameed, A.M. A Characterization of CdSe-nanocrystals used in semiconductors for aerospace applications: Production and optical properties. *NRIAG J. Astron. Geophys.* **2014**, *3*, 82–87. [CrossRef]
50. Alajlouni, S.; Sun, Z.Y.; Li, J.; Zavada, J.M.; Lin, J.Y.; Jiang, H.X. Refractive index of erbium doped GaN thin films. *Appl. Phys. Lett.* **2014**, *105*, 081104. [CrossRef]
51. Kenyon, A.J. Erbium in silicon. *Semicond. Sci. Technol.* **2005**, *20*, R65–R84. [CrossRef]
52. Yu, J.; Zhang, C.; Pang, G.; Sun, X.W.; Chen, R. Effect of Lateral Size and Surface Passivation on the Near-Band-Edge Excitonic Emission from Quasi-Two-Dimensional CdSe Nanoplatelets. *ACS Appl. Mater. Interfaces* **2019**, *11*, 41821–41827. [CrossRef]
53. Nasiri, S.; Hosseinnezhad, M.; Rabiei, M.; Palevicius, A.; Janusas, G. The effect of calcination temperature on the photophysical and mechanical properties of copper iodide (5 mol%)–doped hydroxyapatite. *Opt. Mater.* **2021**, *121*, 111559. [CrossRef]
54. Hsu, Y.-K.; Chen, C.-W.; Huang, J.Y.; Pan, C.-L.; Zhang, J.-Y.; Chang, C.-S. Erbium doped GaSe crystal for mid-IR applications. *Opt. Express* **2006**, *14*, 5484–5491. [CrossRef] [PubMed]
55. Chen, X.; Shangguan, W. Hydrogen production from water splitting on CdS-based photocatalysts using solar light. *Front. Energy* **2013**, *7*, 111–118. [CrossRef]
56. Cheng, L.; Xiang, Q.; Liao, Y.; Zhang, H. CdS-Based photocatalysts. *Energy Environ. Sci.* **2018**, *11*, 1362–1391. [CrossRef]
57. Su, J.; Zhang, T.; Li, Y.; Chen, Y.; Liu, M. Photocatalytic Activities of Copper Doped Cadmium Sulfide Microspheres Prepared by a Facile Ultrasonic Spray-Pyrolysis Method. *Molecules* **2016**, *21*, 735. [CrossRef]
58. Liu, M.; Du, Y.; Ma, L.; Jing, D.; Guo, L. Manganese doped cadmium sulfide nanocrystal for hydrogen production from water under visible light. *Int. J. Hydrogen Energy* **2012**, *37*, 730–736. [CrossRef]
59. Wang, H.; Chen, W.; Zhang, J.; Huang, C.; Mao, L. Nickel nanoparticles modified CdS—A potential photocatalyst for hydrogen production through water splitting under visible light irradiation. *Int. J. Hydrogen Energy* **2015**, *40*, 340–345. [CrossRef]
60. Portillo-Moreno, O.; Gutiérrez-Pérez, R.; Palomino-Merino, R.; Chávez-Portillo, M.; Márquez-Specia, M.N.; Hernández-Torres, M.E.; Gracia-Jiménez, M.; Cerna, J.R.; Zamora-Tototzintle, M. Near-infrared-to-visible upconverting luminescence of Er^{3+}-doped CdSe nanocrystals grown by chemical bath. *Optik* **2017**, *138*, 229–239. [CrossRef]
61. Yan, D.; Wu, P.; Zhang, S.P.; Liang, L.; Yang, F.; Pei, Y.L.; Chen, S. Assignments of the Raman modes of monoclinic erbium oxide. *J. Appl. Phys.* **2013**, *114*, 193502. [CrossRef]
62. Ferrari, A.C. Raman spectroscopy of graphene and graphite: Disorder, electron–phonon coupling, doping and nonadiabatic effects. *Solid State Commun.* **2007**, *143*, 47–57. [CrossRef]

63. Lin, C.; Kelley, D.F.; Rico, M.; Kelley, A.M. The "Surface Optical" Phonon in CdSe Nanocrystals. *ACS Nano* **2014**, *8*, 3928–3938. [CrossRef] [PubMed]
64. Gomonnai, A.V.; Azhniuk, Y.M.; Yukhymchuk, V.O.; Kranjčec, M.; Lopushansky, V.V. Confinement-, surface- and disorder-related effects in the resonant Raman spectra of nanometric $CdS_{1-x}Se_x$ crystals. *Phys. Status Solidi B* **2003**, *239*, 490–499. [CrossRef]
65. Dzhagan, V.; Lokteva, I.; Himcinschi, C.; Jin, X.; Kolny-Olesiak, J.; Zahn, D.R.T. Phonon Raman spectra of colloidal CdTe nanocrystals: Effect of size, non-stoichiometry and ligand exchange. *Nanoscale Res. Lett.* **2011**, *6*, 79. [CrossRef]
66. Dıaz-Reyes, J.; Galvan-Arellano, M.; Mendoza-Alvarez, J.G.; Arias-Ceron, J.S.; Herrera-Perez, J.L.; Lopez-Cruz, E. Characterization of highly doped $Ga_{0.86}In_{0.14}As_{0.13}Sb_{0.87}$ grown by liquid phase epitaxy. *Rev. Mex. Física* **2017**, *63*, 55–64.
67. Dzhagan, V.M.; Azhniuk, Y.M.; Milekhin, A.G.; Zahn, D.R.T. Vibrational spectroscopy of compound semiconductor nanocrystals. *J. Phys. D Appl. Phys.* **2018**, *51*, 503001. [CrossRef]
68. Nasiri, S.; Dashti, A.; Hosseinnezhad, M.; Rabiei, M.; Palevicius, A.; Doustmohammadi, A.; Janusas, G. Mechanochromic and thermally activated delayed fluorescence dyes obtained from D–A–D′ type, consisted of xanthen and carbazole derivatives as an emitter layer in organic light emitting diodes. *Chem. Eng. J.* **2021**, *430*, 131877. [CrossRef]
69. Venkatram, N.; Sathyavathi, R.; Rao, D.N. Size dependent multiphoton absorption and refraction of CdSe nanoparticles. *Opt. Express* **2007**, *15*, 12258–12263. [CrossRef]
70. Gautam, S.K.; Gautam, N.; Singh, R.G.; Ojha, S.; Shukla, D.K.; Singh, F. Anomalous behavior of B_{1g} mode in highly transparent anatase nano-crystalline Nb-doped Titanium Dioxide (NTO) thin films. *AIP Adv.* **2015**, *5*, 127212. [CrossRef]
71. Choudhury, B.; Choudhury, A. Lattice distortion and corresponding changes in optical properties of CeO_2 nanoparticles on Nd doping. *Curr. Appl. Phys.* **2013**, *13*, 217–223. [CrossRef]
72. Wang, T.; Layek, A.; Hosein, I.D.; Chirmanov, V.; Radovanovic, P.V. Correlation between native defects and dopants in colloidal lanthanide-doped Ga_2O_3nanocrystals: A path to enhance functionality and control optical properties. *J. Mater. Chem. C* **2013**, *2*, 3212–3222. [CrossRef]
73. Sercel, P.C.; Shabaev, A.; Efros, A.L. Photoluminescence Enhancement through Symmetry Breaking Induced by Defects in Nanocrystals. *Nano Lett.* **2017**, *17*, 4820–4830. [CrossRef] [PubMed]
74. Zeng, S.; Ren, G.; Xu, C.; Yang, Q. Modifying crystal phase, shape, size, optical and magnetic properties of monodispersed multifunctional $NaYbF_4$nanocrystals through lanthanide doping. *CrystEngComm* **2011**, *13*, 4276–4281. [CrossRef]
75. Bezkrovnyi, O.; Małecka, M.A.; Lisiecki, R.; Ostroushko, V.; Thomas, A.G.; Gorantlad, S.; Kepinski, L. The effect of Eu doping on the growth, structure and red-ox activity of ceria nanocubes. *CrystEngComm* **2018**, *20*, 1698–1704. [CrossRef]
76. Wang, F.; Han, Y.; Lim, C.S.; Lu, Y.; Wang, J.; Xu, J.; Chen, H.; Zhang, C.; Hong, M.; Liu, X. Simultaneous phase and size control of upconversion nanocrystals through lanthanide doping. *Nature* **2010**, *463*, 1061–1065. [CrossRef] [PubMed]
77. Anandan, S.; Miyauchi, M. Ce-doped ZnO ($Ce_xZn_{1-x}O$) becomes an efficient visible-light-sensitive photocatalyst by co-catalyst (Cu^{2+}) grafting. *Phys. Chem. Chem. Phys.* **2011**, *13*, 14937–14945. [CrossRef] [PubMed]
78. He, H.Y.; Fei, J.; Lu, J. Sm-doping effect on optical and electrical properties of ZnO films. *J. Nanostructure Chem.* **2015**, *5*, 169–175. [CrossRef]
79. Zhang, F.; Jin, Q.; Chan, S.W. Ceria nanoparticles: Size, size distribution, and shape. *J. Appl. Phys.* **2004**, *95*, 4319–4326. [CrossRef]
80. Lang, J.; Wang, J.; Zhang, Q.; Xu, S.; Han, D.; Yang, J.; Han, Q.; Yang, L.; Sui, Y.; Li, X.; et al. Synthesis and photoluminescence characterizations of the Er^{3+}-doped ZnO nanosheets with irregular porous microstructure. *Mater. Sci. Semicond. Process.* **2016**, *41*, 32–37. [CrossRef]
81. Chen, W.-B.; Liu, X.-C.; Li, F.; Chen, H.-M.; Zhou, R.-W.; Shi, E.-W. Influence of oxygen partial pressure on the microstructural and magnetic properties of Er-doped ZnO thin films. *AIP Adv.* **2015**, *5*, 067105. [CrossRef]
82. Dakhel, A.A.; El-Hilo, M. Ferromagnetic nanocrystalline Gd-doped ZnO powder synthesized by coprecipitation. *J. Appl. Phys.* **2010**, *107*, 123905. [CrossRef]
83. Shi, J.; Shen, L.; Meng, F.; Liu, Z. Structural, electrical and optical properties of highly crystalline indium tin oxide films fabricated by RPD at room temperature. *Mater. Lett.* **2016**, *182*, 32–35. [CrossRef]
84. Greenberg, B.L.; Robinson, Z.L.; Ayino, Y.; Held, J.T.; Peterson, T.A.; Mkhoyan, K.A.; Pribiag, V.S.; Aydil, E.S.; Kortshagen, U.R. Metal-insulator transition in a semiconductor nanocrystal network. *Sci. Adv.* **2019**, *5*, eaaw1462. [CrossRef] [PubMed]
85. Almeida, R.M.; Marques, A.C.; Cabeça, R.; Zampedri, L.; Chiasera, A.; Ferrari, M. Photoluminescence of Erbium-Doped Silicate Sol-Gel Planar Waveguides. *J. Sol-Gel Sci. Technol.* **2004**, *31*, 317–322. [CrossRef]
86. Sethi, S.; Bhattacharya, P.K. Characteristics and Device Applications of Erbium Doped III-V Semiconductors Grown by Molecular Beam Epitaxy. *J. Electron. Mater.* **1996**, *25*, 467–477. [CrossRef]
87. Pradhan, A.K.; Douglas, L.; Mustafa, H.; Mundle, R.; Hunter, D.; Bonner, C.E. Pulsed-laser deposited Er:ZnO films for 1.54 μm emission. *Appl. Phys. Lett.* **2007**, *90*, 072108. [CrossRef]
88. Popa, M.; Schmerber, G.; Toloman, D.; Gabor, M.S.; Mesaros, A.; Petrişor, T. Magnetic and Electrical Properties of Undoped and Holmium Doped ZnO Thin Films Grown by Sol-Gel Method. *Adv. Eng. Forum* **2013**, *8–9*, 301–308. [CrossRef]
89. Das, P.; Bathula, S.; Gollapudi, S. Evaluating the effect of grain size distribution on thermal conductivity of thermoelectric materials. *Nano Express* **2020**, *1*, 020036. [CrossRef]
90. Morais, E.A.; Scalvi, L.V.A.; Ribeiro, S.J.L.; Geraldo, V. Poole-Frenkel effect in Er doped SnO_2thin films deposited by sol-gel-dip-coating. *Phys. Status Solidi A* **2005**, *202*, 301–308. [CrossRef]

91. Kityk, I.V.; Halyan, V.V.; Yukhymchuk, V.O.; Strelchuk, V.V.; Ivashchenko, I.A.; Zhydachevskyy, Y.; Suchocki, A.; Olekseyuk, I.D.; Kevshyn, A.H.; Piasecki, M. NIR and visible luminescence features of erbium doped Ga2S3–La2S3 glasses. *J. Non-Crystalline Solids* **2018**, *498*, 380–385. [CrossRef]
92. Yang, J.-H.; Gao, J.-F.; Yong, S.-L.; Ma, X.-L.; Liu, L.-J. Synthesis, characterization and optical studies on Sm^{2+}-doped CdSe nanocrystals: A blueshift and fixed emission with high quantum yields. *Rare Met.* **2019**, *38*, 1097–1104. [CrossRef]
93. Han, J.; Wang, L.; Wong, S.S. Observation of Photoinduced Charge Transfer in Novel Luminescent CdSe Quantum Dot-$CePO_4$:Tb Metal Oxide Nanowire Composite Heterostructures. *J. Phys. Chem. C* **2014**, *118*, 5671–5682. [CrossRef]
94. Wang, M.; Zhang, Y.; Yao, Q.; Ng, M.; Lin, M.; Li, X.; Bhakoo, K.K.; Chang, A.Y.; Rosei, F.; Vetrone, F. Morphology Control of Lanthanide Doped $NaGdF_4$ Nanocrystals via One-Step Thermolysis. *Chem. Mater.* **2019**, *31*, 5160–5171. [CrossRef]

magnetochemistry

MDPI

Article

New Manufacturing Process for Granular Texture Management in Polycrystalline BaM Hexaferrites through the Goethite Crystallite Laths Aspect Ratio, and a Specialized Law of Approach to the Magnetic Saturation for Partly Polarized Uniaxial Materials

Antoine Hoëz, Jean-Luc Mattei * and Alexis Chevalier

Lab-STICC, UMR CNRS 6285, 6 av. Le Gorgeu, 29238 Brest, France
* Correspondence: jean-luc.mattei@univ-brest.fr

Abstract: This study is aimed at the manufacture and the magnetic properties of polycrystalline M-type hexaferrites $BaFe_{12}O_{19}$ (barium ferrite, or BaM) materials of different magnetic texturing grades, going from a random distribution of the BaM crystallites to their almost complete stacking. Our target is to optimize the value of reduced-remanence magnetization M_R/M_S, which is among the most significant features of the self-polarized materials. In this study, we focus on the role played by the precursors hematite (isotropic spherical shape) and goethite (anisotropic lath shape). Therefore, 11 samples with a flat cylinder shape are fabricated, with an increasing hematite to goethite ratio. We demonstrate that this ratio drives the texturization of the samples by producing self-polarized materials with different M_R/M_S from the simple green compaction of the precursors, followed by a heat treatment. Most importantly, our study reveals the orientation of BaM particles after compaction; therefore, M_R/M_S, is strongly influenced by the aspect ratio of the lath-shaped goethite crystallites. Additionally, we show that finer goethite crystallites yield higher-value M_R/M_S. We optimize the aspect ratio of the goethite crystallites for an improved BaM texture. The optimization of the morphology of the goethite crystallites leads to an increase in the BaM particles' orientation and stacking. The salient outcome of this work, which distinguishes it significantly from recent works, is that the particles stacking increases with the value of the shape factor η (defined as the ratio of the diameter of the laths to their length) of the goethite, evidenced by XRD results. The Rietveld refinements of powder diffractograms and the measured magnetic properties reveal a particle-stacking enhancement caused by not only the ratio of hematite: goethite but mainly by an optimal aspect ratio of the goethite crystallites. Based on this study, the BaM materials are further manufactured with a controlled magnetic texture; thus, they are partly self-polarized. They show reduced-remanence magnetization M_R/M_S varying from 0.5 and 0.81, while the angular dispersion of the BaM particles' easy axis of magnetization varies from 60° to 10°. The magnetic properties of the samples are further studied in microwave experiments, from which the value of the magnetocrystalline anisotropy field H_K = 16.6 kOe is deduced. The first magnetization curves of each sample are obtained using a VSM. A law of approach to the saturation suitable for the case of the uniaxial polycrystalline materials, and for which the particle stacking is only partial, is proposed for the fitting of the magnetization process. It is suggested that by using the proposed law with a known magnetocrystalline anisotropy constant K_1, the angular grain-dispersion can be found.

Keywords: barium hexaferrites; self-polarized hexaferrites; cold compaction; controlled magnetic texturization; magnetocrystalline anisotropy; law of approach to saturation

Citation: Hoëz, A.; Mattei, J.-L.; Chevalier, A. New Manufacturing Process for Granular Texture Management in Polycrystalline BaM Hexaferrites through the Goethite Crystallite Laths Aspect Ratio, and a Specialized Law of Approach to the Magnetic Saturation for Partly Polarized Uniaxial Materials. *Magnetochemistry* **2023**, *9*, 30. https://doi.org/10.3390/magnetochemistry9010030

Academic Editor: Carlos J. Gómez García

Received: 16 November 2022
Revised: 23 December 2022
Accepted: 30 December 2022
Published: 12 January 2023

1. Introduction

In recent years, a revival in the interest in rare-earth-free permanent magnets, such as M-type hexaferrites [1–3], has been observed. Among this class of materials, strontium hexaferrites and barium hexaferrites, $BaFe_{12}O_{19}$ (BaM), present intermediate performances between standard ferrite and rare-earth magnets. In addition to a strong magnetocrystalline anisotropy, the BaM hexaferrites engineered to have a sufficient magnetic texture can present the great advantage of possessing high-remanence magnetization. For these reasons, they have attracted attention for use in microwave and millimeter wave (MMW) applications [4–6]. The interest of such properties is obvious in the design of the integrated and miniature MMW devices, in which the external permanent magnet would be eliminated [7–9].

The crystals of BaM grow in a flat hexagonal platelet shape, with an aspect ratio of about 1/3. Their easy axis of magnetization is perpendicular to the basal plane. The critical single-domain size is between 0.5 and 1 μm [10]. The achievement of stacking these single-domain particles is desired in order to produce a self-polarized material.

Among the manufacturing techniques used to effectively fabricate these magnets, those operating with standard ceramic technology and mechanical processes [10] must be distinguished from those implementing soft chemistry methods [6,10–13]. The latter fabrication methods involve topotactical reactions. In addition to being inexpensive, unlike other techniques, soft chemistry methods present the notable advantage of leading to the spontaneous stacking of particles [6]. It should be made clear, however, that even if the polycrystalline material has a high-quality magnetic texture, there is always some misorientation of the easy axes of magnetization, which could hardly be limited only by applying a strong magnetic field during the compaction process. Recent works [14,15], showed that the combined use of goethite and hematite crystallites to fabricate strontium hexaferrites ($SrFe_{12}O_{19}$), allowed some degree of alignment of the crystallites after a cold compaction and a heat treatment with M_R/M_S = 0.71. The magnitude of this value, indicative of the degree of particle packing, is insufficient for many applications. The required M_R/M_S value for a self-polarized material used in a circulator, one of the major applications at present [4,16–19], should be higher than 0.83 [6,20]. In our previous work, using a coprecipitation technique, we created highly oriented bulk compacts made of BaM particles with reduced-remanence magnetization M_R/M_S = 0.88, which are suitable for self-biased applications [6]. In order to increase this value, we optimized several stages of the fabrication process. In the present work, we focus on the role played by the morphology of the goethite crystallites (which are used as a precursor) and on the quality of BaM particle packing. We show that the shape and aspect ratio of the goethite crystallites are critical parameters for obtaining high M_R/M_S values, and that they were not taken into account before [14,15].

In the case where the dispersion of the axes of easy magnetization of the material is random, i.e., when the material has no magnetic texture, the law of approach to saturation (LAS) is commonly used to describe the first magnetization curve [21]. However, when used under its most usual formulation, the LAS applies only to the specific case where the material is constituted of particles dispersed at random, and they consequently show a random distribution of their easy axes of magnetization. This state defines the lower degree of the magnetic texture. The LAS is not adapted to the case of a partly or a fully oriented material and should not be used as is. Consequently, it is desirable to have a LAS that is applicable to textured materials. Our purpose is to determine the magnetization curve of a material where the magnetic texture lies between random and perfectly oriented.

The aim of the present work is twofold. Firstly, we present a new method for manufacturing $BaFe_{12}O_{19}$ materials with different magnetic texturing grades. The novelty and salient point of our method lie in the use of goethite crystallites with an optimized shape and aspect ratio. The value obtained for the remanent magnetization M_R/M_S of the BaM material increases from 0.54 to 0.81 when the angular dispersion of the easy axis of magnetization varies from 60° to 10° (XRD measurements). The magnetocrystalline field H_K is measured from microwave experiments. Secondly, we propose a new formulation of the LAS, partly based on structural characterizations, where the spatial dispersion of the easy axis of magnetization in grain-oriented polycrystals of hexagonal materials is considered, and from which the angular dispersion value of the BaM particles could be inferred.

2. Experimental Details

X-ray diffraction measurements are carried out on a Panalytical Empyrean with Chi-Phi-Z configuration using Cu Kα radiation over a 2θ range from 15° to 60°. The parameters Z, Chi, and Omega are adjusted for each sample.

The microstructures of the samples are imaged by a scanning electron microscope Hitachi-S-3200N (Japan). The hysteresis loops of the sintered samples are measured at room temperature using a Vibrating Sample Magnetometer EZ9 from MicroSense (Lowell, MA, USA), with maximum applied field intensity equal to 1500 kA/m. Each sample was maintained on a Pyrex sample holder, with its c-axis (this axis being defined by the direction of the applied pressure during compaction, as explained below) aligned firstly parallel, and then perpendicular, to the applied field, in order to measure the hysteresis loops in both directions. The sample oscillated at a frequency of 70 Hz and averaged for 3 s. The demagnetizing correction is further carried out by applying the relationship $H_i = H\text{-}NM$, where H_i is the internal filed and H the applied field. The values of the demagnetizing factor N are tabulated in [22]. The frequency evolution of the imaginary part of the magnetic permeability is extracted from S parameters measured by a ZVA67 vector network analyzer (Rohde & Schwarz, Munich, Germany) in the frequency band [10 GHz–60 GHz]. The frequency evolution of the imaginary part of the magnetic permeability is extracted in using a Rohde and Schwarz ZVA67 vector network analyzer in the frequency band [10 GHz–60 GHz]. The anisotropy field H_K is deduced from the resonant frequency f_R determined by the maximum of the imaginary part μ'' of the intrinsic permeability of the sample from the relationship between H_K and f_R, which writes $f_R = \gamma H_K$, where γ is the gyromagnetic ratio ($\gamma = 35.12 \times 10^{-3}$ MHz/A.m^{-1}).

3. Results and Discussion

3.1. Manufacture of a Self-Polarized BaM Sample

The polycrystalline $BaFe_{12}O_{19}$-oriented samples were made using barium carbonate $BaCO_3$ (Acros-organics, Nidderau, Germany, 99%) and goethite FeO(OH), which was synthetized in the laboratory from iron III hexahydrate (Sigma-Aldrich, Saint Louis, MS, USA, 98%). For the sake of confidentiality, the synthesis process is not detailed. These components were carefully mixed with a manual mortar. The mixtures obtained were pressed into cylindrical pellets (7 mm diameter and 5 mm thickness) at a pressure of 320 MPa using a uniaxial hydraulic press. The direction of the applied pressure defined the c-axis, which was normal to the upper side of the pellets. The pellets were further heated to 1040 °C with zero dwell time and heating and cooling rates of 3 K/min.

During compaction, the applied pressure orients the goethite crystallites along the compaction axis (Figure 1a). We first considered the synthesized goethite crystallites: they are lath-shaped, with a crystallographic axis [001] that is oriented perpendicular to the largest face of the crystallite (Figure 1d). The pressure applied during the molding step leads to a preferential orientation of the goethite crystallites, the [001] axes of which being predominantly parallel to the direction of the applied pressure. The X-ray diffraction pattern of the goethite before and after the pressing process confirm a preferential orientation of the goethite close to the direction [001]. The direction [101], which is closest to [001] among these that contribute to the diffraction pattern, is clearly the one that remains, with a strong intensity (Figure 1a,b), whereas all the others Bragg diffraction peaks are extinguished. The Rietveld analyses performed (Table 1) using Jana2020 (Institute of Physics, Prague, Czech Republic) [23] give a preferred orientation axis [001] with an r-factor of 0.39. This preferential orientation of the [001] axes is maintained during the various topotactic reactions that occur during the subsequent heat treatment leading to the formation of $BaFe_{12}O_{19}$ [24–26]. The chemical processes that lead to $BaFe_{12}O_{19}$ from oriented goethite crystallites are as follows:

$$2\ \alpha FeO(OH)[001] \rightarrow \alpha Fe_2O_3[001] + H_2O$$
$$\alpha Fe_2O_3[001] + BaCO_3 \rightarrow BaFe_2O_4[001]$$
$$BaFe_2O_4[001] + \alpha Fe_2O_3[001] \rightarrow BaFe_{12}O_{19}[001]$$

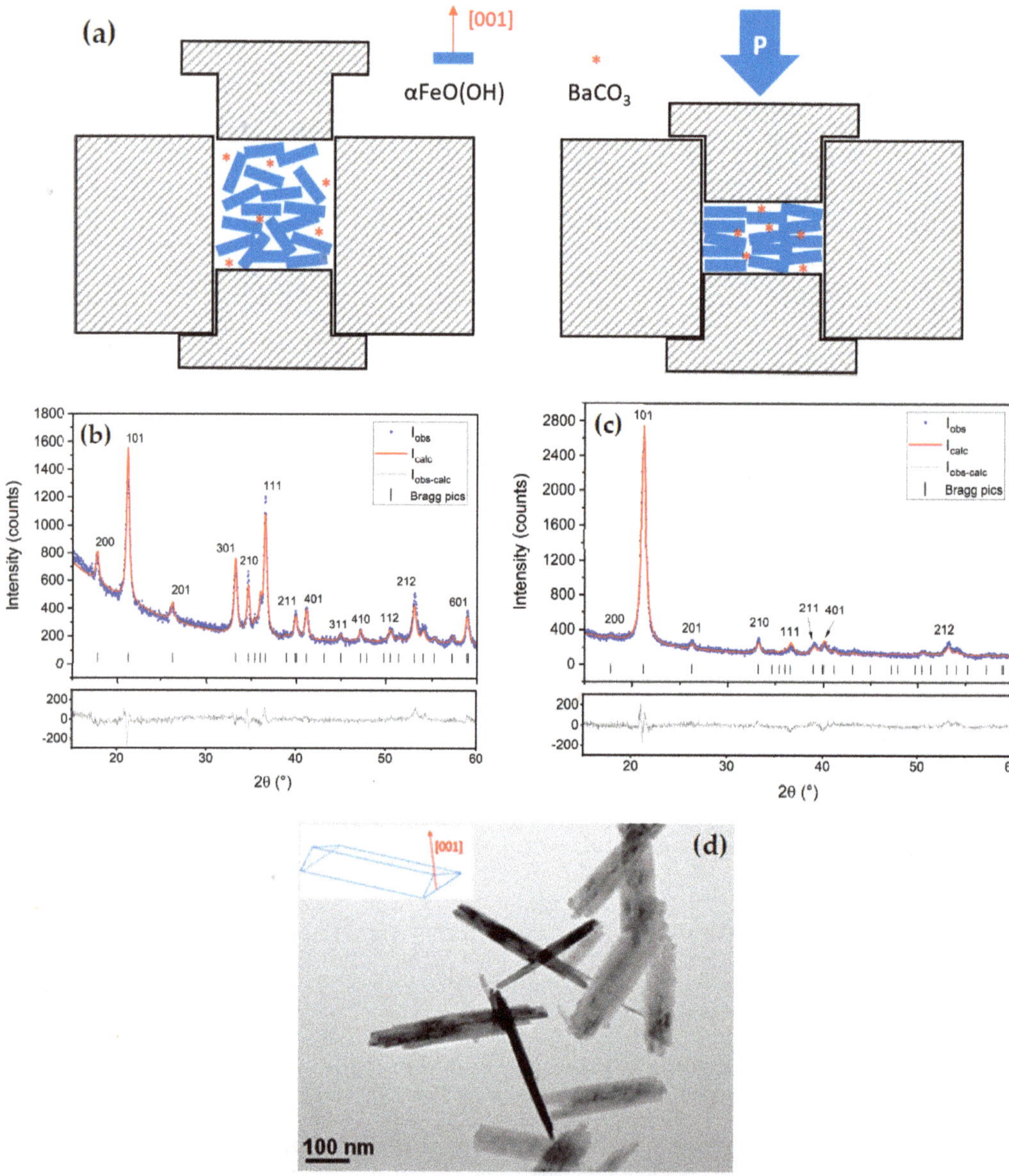

Figure 1. (**a**) Schematic representation of the alignment in the basal plane of goethite crystallites, observed after uniaxial pressure. (**b**,**c**) Rietveld analysis of goethite before and after pressing process, respectively. (**d**) TEM images of lath-shaped goethite crystallites.

Table 1. Rietveld refinement values (self-polarized material).

	Goethite before Compaction	Goethite after Compaction
a, *b*, *c* (Å)	9.9734(12), 3.0288(4), 4.6257(7)	9.9740(6), 3.0255(15), 4.6151(8)
Space Group	*Pnma*	*Pnma*
March–Dollase r	1	0.3997
R_{obs}, $_wR_{obs}$, R_{all}, $_wR_{all}$, GOF	6.15, 6.53, 6.57, 6.54, 1.39	13.98, 13.15, 14.36, 13.16, 1.39

3.2. The Aspect Ratio of the Goethite Particles: Its Major Role on the Quality of the BaM Particles Stacking

It was found that the aspect ratio of width to length η of the goethite impacts the stacking quality of the resulting BaM crystallites. The X-ray patterns of the three samples obtained with G1, G2, and G3 clearly show an improvement of the stacking quality with increasing η (Figure 2d), and the intensity ratio I(008)/I(107) increases with η (Table 2). It is expected that these results might be significantly improved by optimizing the heat treatment after compaction. Further studies are ongoing.

Table 2. Influence of the aspect ratio of the goethite on BaM particles stacking.

Goethite	G1	G2	G3
η	0.06	0.15	0.2
I(008)/I(107)	0.5	0.8	1.2

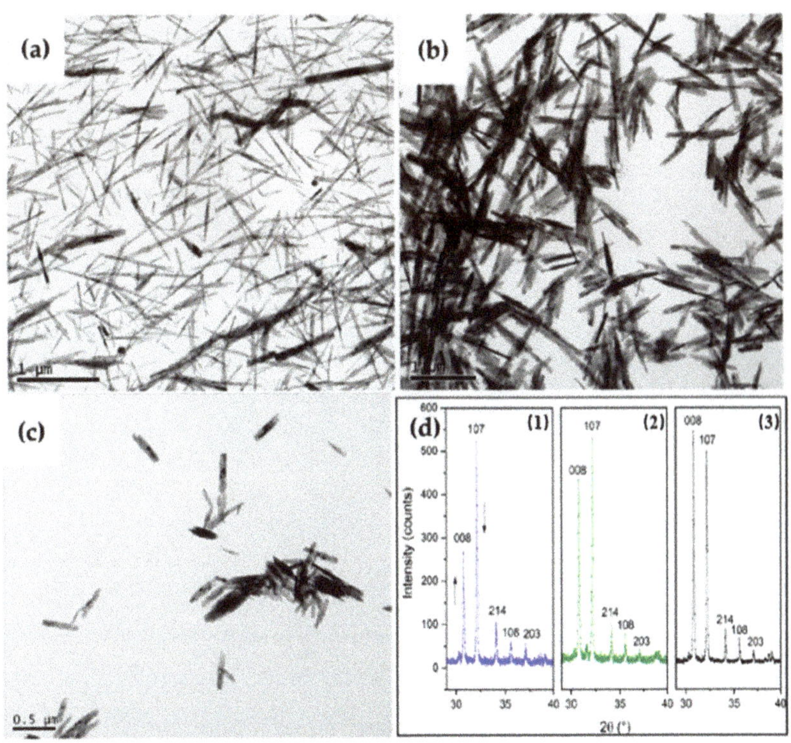

Figure 2. (**a**–**c**) Goethites G1, G2, G3, respectively. (**d**) X-Ray pattern of oriented BaM samples (**1**–**3**) obtained with G1, G2, G3, respectively.

3.3. Manufacturing of BaM Samples with Differents Grades of Orientation

The polycrystalline $BaFe_{12}O_{19}$ samples were synthesized using barium carbonate $BaCO_3$ (Acros-organics, 99%), ferric oxide Fe_2O_3 (Sigma-Aldrich, 99%), and goethite FeO(OH) synthetized in the laboratory from iron III hexahydrate (Sigma-Aldrich, 98%). Keeping the 1:12 molar ratio for Ba:Fe and with different molar proportions of Fe_2O_3 and FeO(OH), a total of eleven samples were prepared. These components were carefully mixed with a mortar, with molar ratios $\varrho = Fe_2O_3$:FeO(OH) varying from 0 to 1 (Table 3). The so-obtained mixtures were pressed into cylindrical pellets (with dimensions of 7 mm diameter and 5 mm thickness) at a pressure of 320 MPa using a uniaxial hydraulic press. The direction of the applied pressure defined the c-axis, perpendicular to the upper side of the pellets. The pellets were further heated to 1040 °C, with zero dwell time and heating and cooling rates set to 3 °C/min.

Table 3. Details on the prepared samples.

Goethite (% molar) Hematite (% molar)	0 100	5 95	10 90	20 80
Notation in this paper	BaM-0	BaM-5	BaM-10	BaM-20
Goethite (% molar) Hematite (% molar)	30 70	40 60	50 50	60 40
Notation in this paper	BaM-30	BaM-40	BaM-50	BaM-60
Goethite (% molar) Hematite (% molar)	70 30	80 20	90 10	100 0
Notation in this paper	BaM-70	BaM-80	BaM-90	BaM-100

The effect of the applied pressure was different on hematite particles than on goethite crystallites. This is a key point of the manufacturing process, which we will emphasize. Now we consider the hematite particles, which have a spherical shape (Figure 3): because the applied pressure does not have any effect on the orientation of these particles, their crystallographic axes remain randomly oriented.

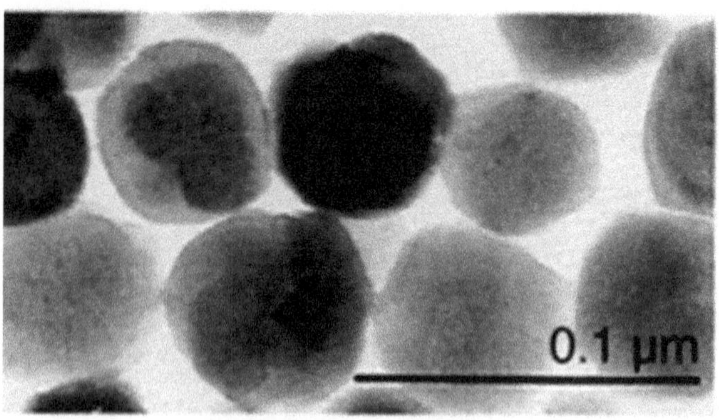

Figure 3. TEM images of spherical crystallites of hematite.

The initial ratio of Fe_2O_3:FeO(OH) drives the amount of randomly oriented or oriented $BaFe_{12}O_{19}$ crystals. Figure 4a,b are SEM images evidencing the differences in texture between randomly oriented crystallites in sample BaM-0 (prepared with hematite only) and stacked crystallites in sample BaM-100 (prepared with goethite only).

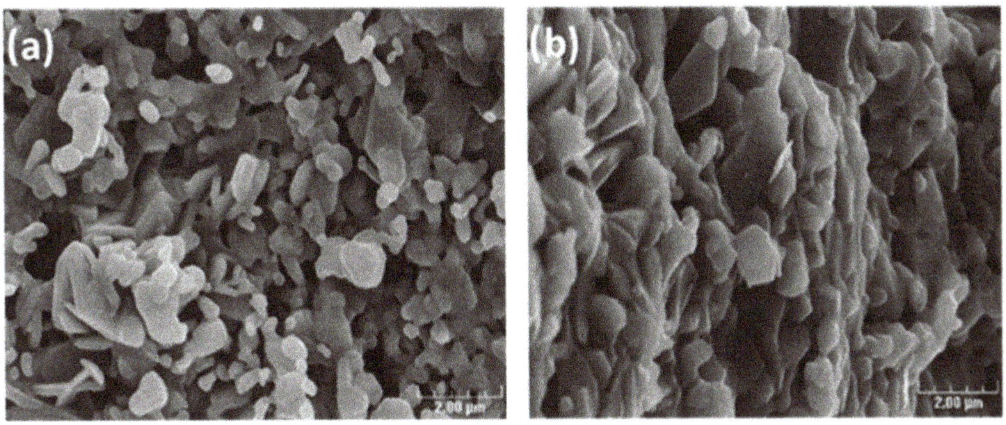

Figure 4. SEM images show the evolution of the texturization between BaM-0 (**a**) and BaM-100 (**b**).

3.4. *Crystallographic Study of BaM Samples Manufactured with Different Texturations*

X-ray diffraction data were processed with HighScore Plus software (Panalytical, UK) [27] for phase analysis and Jana2020 for Rietveld analysis. The latter made it possible to examine the purity of the samples and to characterize the stacking of the crystallites. The crystallographic model used comes from the ICSD 259873 CIF file with cell parameters $a = b = 5.8909$ Å, $c = 23.1882$ Å and space group symmetry $P6_3/mmc$. The Rietveld analysis shows that all samples are single-phase (Figure 5). After refinement, the cell parameters and R factors for BaM-0 are a = b = 5.8928 Å, c = 23.2295 Å and Robs = 3.67, wRobs = 4.47, Rall = 4.78, wRall = 4.68, GOF = 1.19, whereas and BaM-100, they are a = b = 5.8921 Å, c = 23.2161 Å and Robs = 5.41, wRobs = 5.60, Rall = 6.42, wRall = 5.79, GOF = 1.26.

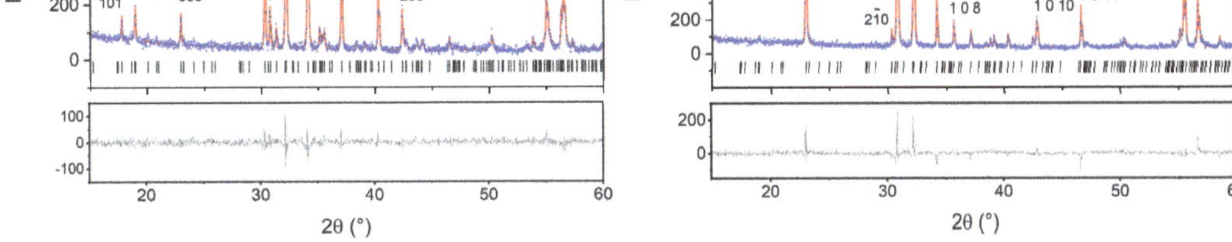

Figure 5. Diffraction pattern and Rietveld analysis of BaM-0 (**left**) and of BaM-100 (**right**).

When comparing the X-Ray diffraction pattern of all the samples from BaM-0 to BaM-100, it appears that (00l) basal reflections become stronger (Figure 5). This is explained by an increasing orientation of the crystallites [28]. More specifically the (008) reflection becomes stronger with the increasing goethite to hematite molar ratio, meaning that the rate of oriented BaM particles increases as well (Figure 6).

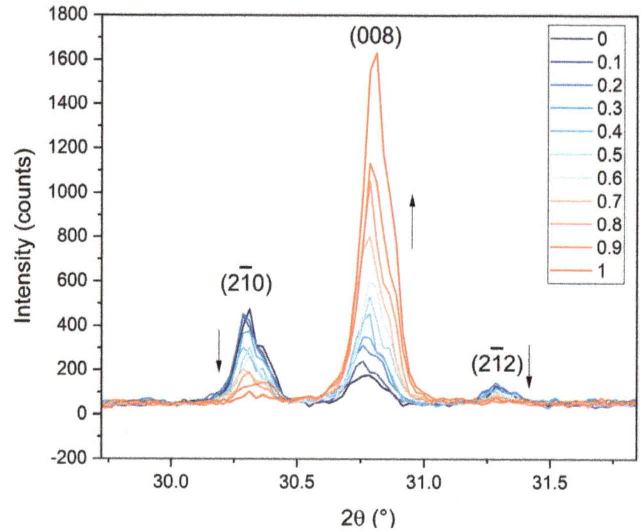

Figure 6. The intensity of XRD (008) reflection increases with the molar ratio between goethite and hematite.

XRD data make it also possible to obtain information on the distribution of the orientation of the crystallites. Among the several mathematical models available (e.g., Sasa-Uda [29], Capkova, and Valvoda [30]), the March–Dollase approach [31] is used for the determination of the degree of preferred orientation. The March–Dollase model, developed to describe the compaction of platelike grains under uniaxial stress [32], states that the fraction P(θ) of crystallites having the inclination angle θ between the normal to the diffraction plane and the hkl plane is defined:

$$P(\theta) = \left(r^2 \cos^2(\theta) + \frac{\sin^2(\theta)}{r}\right)^{-\frac{3}{2}} \quad (1)$$

The March–Dollase factor, denoted r, defines the spread of angular distribution of the crystallite inclinations. It is extracted from XRD data using Rietveld refinement with Jana2020. If the maximum of the orientation distribution does not occur for the [001] direction, which is most often the case, the distribution is shifted so that $\theta = 0$ represents the easy axis of magnetization [001]. Thus, for the determination of the factor r with a direction [hkl] different than [001], the relation P(θ) with ω the angle between [001] and [hkl] becomes:

$$P(\theta) = \left(r^2 \cos^2(\theta - \omega) + \frac{\sin^2(\theta - \omega)}{r}\right)^{-\frac{3}{2}} \quad (2)$$

Rietveld analyses are used to extract the r factor for each sample, with the vector [1 0 16] to define the preferred orientation. The Rocking curve (Figure 7a) shows a difference of 3.6° between the maximum intensity of (008) reflection and the diffraction plane. This result confirms that the preferred orientation vector is around [1 0 16] because the angle between [1 0 16] and [0 0 1] is 3.56°. The r-factor decreases linearly as a function of the molar percentage of goethite (Figure 7b). This result shows that the preferential orientation according to [001] is more important, as the initial proportion of goethite is greater. The variation of the orientation distributions as a function of the molar proportion of goethite is shown in Figure 8. Obviously, these conclusions might be directly applied to the spatial distribution of the easy axis of the hexagonal BaM platelets in each sample.

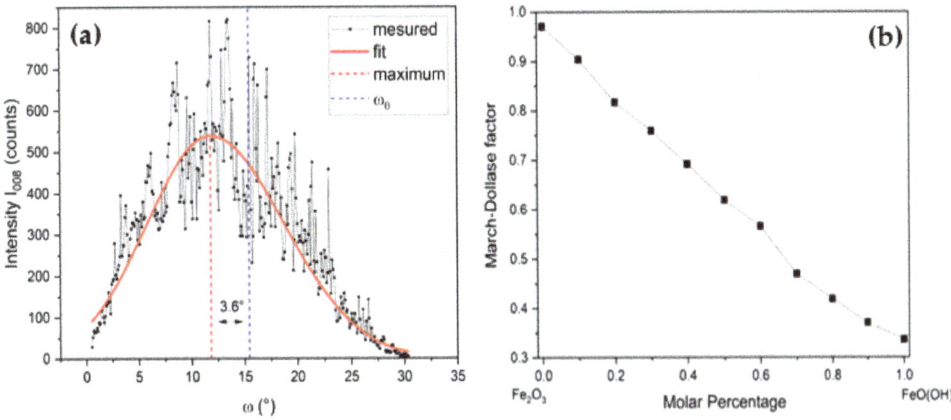

Figure 7. (**a**) Rocking curve for the intensity of the (008) Bragg diffraction pic. (**b**) Evolution of the March–Dollase factor as a function of mole percentage of goethite.

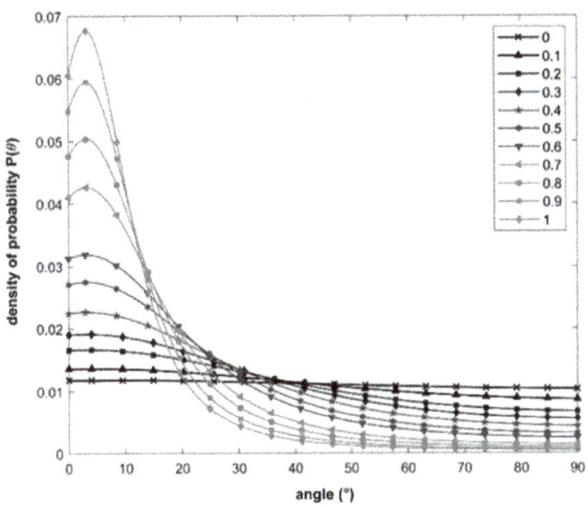

Figure 8. Probability density of crystallite orientations, from March–Dollase model.

3.5. Experimental Investigations of the Magnetocrystalline Anisotropy

In the case of a hexagonal crystal, and omitting higher terms than the second order, the magnetocrystalline anisotropy energy is expressed as [2]:

$$E_K(\theta_m) = K_1 \sin^2(\theta_m) + K_2 \sin^4(\theta_m) \tag{3}$$

where K_1 and K_2 are magnetic anisotropy constants and θ_m is the angle between the magnetization and the c-axis (easy axis of magnetization.)

Then, the anisotropy field H_K is given by [33]:

$$H_K = \frac{2K_1 + 4K_2}{\mu_0 M_S} \tag{4}$$

For non-substituted BaM, $K_2 = 0$. The value of K_1 is positive, equal to 5.4 10^5 erg/g [34], meaning that the easy axis of magnetization is parallel to the hexagonal c-axis. The uniaxial anisotropy field H_K of BaM is typically in the 1320–1360 kA/m or 16.5–17 kOe range [10,35].

3.6. Microwave Measurement Method for Determining Anisotropy

A successful way of determining the magnetocrystalline anisotropy field of a magnetic material involves the measurement of its dynamic permeability in the high-frequency range. Thus, it is established that the maximum of the imaginary part μ″ of the magnetic permeability μ (μ = μ′-jμ″) occurs at a frequency f_R which, in the absence of demagnetizing effects, is written $f_R = \gamma H_K$, with γ and H_K being the gyromagnetic ratio (γ = 2.8 MHz/kOe = 1.7608 × 10^{11} $s^{-1}T^{-1}$) and the magnetocrystalline anisotropy field, respectively.

The broadband complex permeability of the BAM-0 sample was measured in the frequency band [10 GHz–60 GHz] by the transmission/reflection method using a Rohde and Schwarz ZVA67 vector network analyzer. The demagnetized sample was placed on the top of a high-frequency microstrip line to measure the transmission S21 and reflection S11 parameters. First, the effective permeability of the structure (microstrip line and sample) was calculated from the measured S-parameters using a specially written Matlab code based on the NRW equations [36]. Then, conformal mapping applied to permeability [37,38] was used to extract the intrinsic permeability of the sample from the effective multilayers structure. Finally, the anisotropy field H_K of the hexaferrite was deduced from the resonant frequency determined by the maximum of the imaginary part μ″ of the intrinsic permeability. The frequency evolution of μ″(*f*) normalized to its maximum is shown in Figure 9. The maximum of μ″(f) occurs for f_R = 46.6 GHz, to which corresponds H_K = 1330 kA/m (i.e., 16.6 kOe). This result agrees with the value of the anisotropy field H_K of BaM ferrites, which is close to 1360 kA/m (i.e., 17 kOe). However, a distribution of switching fields inevitably appears in a given assembly of particles with shapes, sizes, morphological, and structural defects that vary from one particle to another, and then the anisotropy field thus determined should correspond to the higher value of the anisotropy field distribution $f(H_K)$.

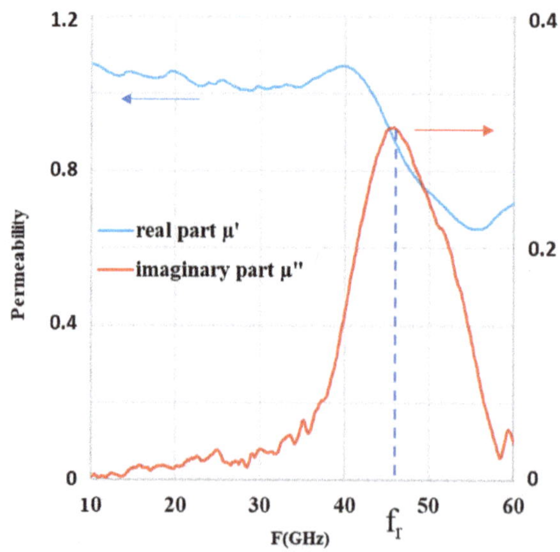

Figure 9. Frequency evolution of the normalized intrinsic permeability of sample BAM-0.

3.7. Isothermal Remanence Measurements

In order to clarify this point, we discriminated the anisotropy field distribution $f(H_K)$ from the switching field distribution. It was established that by measuring the initial remanence curve $M_R(H)$ a switching field distribution could be calculated [39], from which the H_K-distribution function $f(H_K)$ could be obtained [39]. The switching field distribution was similar to the anisotropy field distribution in the situation where the particles diameter was below the critical single-domain size D_C. For BaM, D_C is about 0.5–1 μm [10], which is much larger than the typical diameter observed in this study. Therefore, it is legitimate to admit that SFD and $f(H_K)$ are similar. In the present study, the distribution $f(H_K)$ was obtained by the isothermal remanence measurement (IRM). The IRM measures the remanent magnetization as a function of an increasing magnetizing field starting from a demagnetized state. After demagnetizing the sample, a small field was applied and subsequently removed, after which the remanent magnetization was measured. This magnetization is plotted against the previously applied field. Next, a somewhat larger field was applied and subsequently removed, after which the next remanent magnetization was measured. The resulting curve $M_R(H)$ looks comparable to the virgin curve. For a random assembly of particles with uniaxial anisotropy, the anisotropy field distribution can be obtained by differentiating the reduced IRM curve ($m_R(H) = M_R(H)/M_\infty$, where M_∞ is for the maximum remanent magnetization value) after considering the effects of demagnetizing fields. The distribution function $f(H_K)$ is then obtained by readjusting the field scale, and where H_i is the internal magnetic field:

$$f(H_K) = \left[\frac{dm_R(H_i)}{dH_i}\right]_{H_i=\frac{H_K}{2}} \tag{5}$$

Figure 10 shows $f(H_K)$ for the less-textured sample (BaM-0). In the higher fields, the distribution decreases slowly until the value $(H_K)_{Max} = 1370$ kA/m (i.e. 16.6 kOe) is reached (triangle symbol in Figure 10). The behavior in the high fields is still unclear; however, it could be ascribed to dipolar interactions between grains.

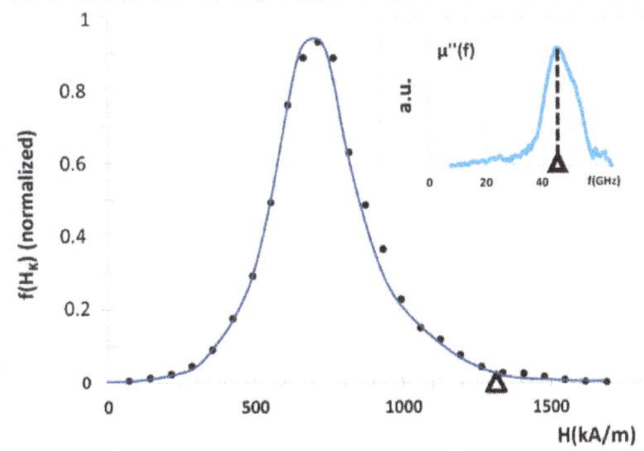

Figure 10. Anisotropy field distribution *f*(*HK*) of the sample BaM-0. The triangle shows the value of *HK* as measured by using microwave experiment (see main text). Inset: the frequency variation of the imaginary part of permeability (μ″) shows a maximum at frequency *fR* = 46.6 GHz. To this value of *fR* corresponds the maximum value H_K of the anisotropy field distribution (both are marked by a triangle), the relationship beween *fR* and *HK* being *fR* = γH_K.

The anisotropy field value obtained from microstrip experiments, $(H_K)_{Max} = 16.6$ kOe, is very close to the one for which $f(H_K)$ cancels. While the IRM measurements give the anisotropy field distribution, the microstrip line measurement method provides the maximum value of anisotropy field, which corresponds to the field to be applied to complete the coherent spin rotation in the Stoner and Wolfarth model [40].

3.8. The First Magnetization Curve and the Law of Approach to Saturation

The first magnetization curve of an isotropic polycrystalline material, as it approaches magnetic saturation, is commonly described using the law of approach to saturation, which expresses [2,41];

$$M(H) = M_S\left(1 - \frac{a}{H_i} - \frac{b}{H_i^2}\right) + x_P H_i \tag{6}$$

M_S is the saturation magnetization and H_i is the internal field (i.e., the applied field H_0 corrected from the demagnetizing field: H_i = H_0-H_d). The term a/H, which results from the presence of inclusions and defects, must vanish at high enough magnetic fields. The last term χ_P is a small high-field susceptibility, called paraprocess. It is due to high-field band splitting, and can be neglected below moderately applied fields intensities [2]. The b/H^2 term arises from the magnetic moments reorientation when the anisotropy axis is misaligned with the applied field. It is directly related to the magnetocrystalline anisotropy (even H_K or the first anisotropy constant K_1). Then, the following relation (1) is commonly used to derived M_S and the anisotropy field for an isotropic distribution of magnetic moments which can change direction only by rotating against the magnetic anisotropy [42,43]:

$$M(H) = M_S\left(1 - \frac{b}{H^2}\right) \tag{7}$$

If such a polycrystalline ferromagnet consists of randomly oriented, single-domain crystallites having uniaxial anisotropy, the coefficient b writes:

$$b = \frac{H_k^2}{15} \tag{8}$$

The variation of M as a function of $1/H^2$ being linear for field values close to saturation therefore allows to determine both M_S, and H_K, as well as K_1 (if the constant K_2 is negligible compared to K_1.)

The LAS can then be used to determine the magnetocrystalline anisotropy constant K_1 and the saturation magnetization M_S, for a random distribution of uniaxial crystals, from (7):

$$M(H) = M_S\left(1 - \frac{4}{15}\frac{K_1^2}{\mu_0 M_S^2}\frac{1}{H^2}\right) \tag{9}$$

However, the LAS as such does not apply when the spatial distribution of the particles is no longer random. This situation arises in the case of polycrystalline materials intended for the production of permanent magnets. In order to quantify more precisely the expected modifications to be applied to the LAS regarding the amount of granular texturization in the partly oriented hexaferrites used in this study, we propose to introduce in relation (9) an additional factor, which is intended to take into account the degree of disorientation of the particles, itself being a function of the initial ratio Fe_2O_3:FeO(OH).

To the best of our knowledge, the only work published in this area comes from Celasco et al. [44]. In [44], the saturation approach of polycrystalline magnetic materials made of randomly oriented, single-domain crystallites with cubic anisotropy and a preferential orientation was studied theoretically, and an adapted saturation-approach law was proposed. The orientation of the particles was described by an angular distribution of the axes of easy magnetization carried by the grains. The authors showed that the LAS writes:

$$M = M_S\left(1 - \frac{A_{cub}(K_1, K_2, \tau)}{\mu_0 \times M_S^2} \times \frac{1}{H^2}\right) \tag{10}$$

where τ quantifies the orientation of the axes of easy magnetization. Following [45], the influence of the grain dispersion on the saturation-approach law is fully contained in the $A_{cub}(K_1, K_2, \tau)$ factor. In the case of a random dispersion of grains the function $A_{cub}(K_1, K_2, \tau)$ simply reduces to the usual expression for a cubic anisotropy, and to the first order [45]:

$$A_{cub}(K_1, K_2, \tau) = \frac{8}{105}K_1^2 \tag{11}$$

However, there is no equivalent expression in the published literature for a LAS that could be used in the case where the crystallites exhibit uniaxial anisotropy. Therefore, we have adopted an experimental approach, leading to the determination of a function $A_{uni}(K_1, \tau)$ to the case of polycrystalline magnetic materials consisting of BaM grains of uniaxial anisotropy and with preferential spatial orientation (K_2 = 0 for unsubstituted BaM). Hence, we assumed that the LAS can be written in a similar way as in [44]:

$$M = M_S\left(1 - \frac{A_{uni}(K_1,\rho)}{\mu_0 M_S^2}\frac{1}{H^2}\right) \tag{12}$$

Our purpose is to find out the experimental variation of the factor $A_{uni}(K_1,\rho)$, where ϱ is a parameter intended to take the grade of texturization into account. We made the reasonable assumption that the anisotropy field H_K, as well as the anisotropy constant K_1, are independent of the spatial dispersion of the platelets: the value of H_K was supposed to be the one given in section IV: H_K = 16.6 kOe. The value of K_1 will be derived further from Equation (4), with K_2 = 0.

The direction of the normal $\vec{n}$ to the basal surface of a sample defines its C axis. A magnetic field H_0 is applied parallel to the C axis. The magnetization M is measured along this direction. The demagnetizing effects were taken into account, the internal field H_i being obtained by the relation H_i = H_0-N_ZM, where N_Z is the macroscopic demagnetizing field coefficient along the C axis. The value of N_Z is fixed by the aspect ratio of the sample (thickness: diameter), as given in [22]. As the goethite content increases, the obtained hysteresis cycles evolve steadily from that characteristic of a random dispersion of uniaxial and monodomain particles for sample BaM-0 to that characteristic of a strongly self-polarized material for sample BaM-100 (Figure 11). For BaM-0, the hysteresis loops measured along the basal plane of the sample ($\vec{n}$ direction) and perpendicular to $\vec{n}$ are completely identical due to the isotropy of the sample, whereas for BaM-100, these hysteresis loops are characteristic of an anisotropic material. The LAS was applied to the first magnetization curves using the relation (12) in the range of internal fields between 1384 kA/m and 1000 kA/m, typically.

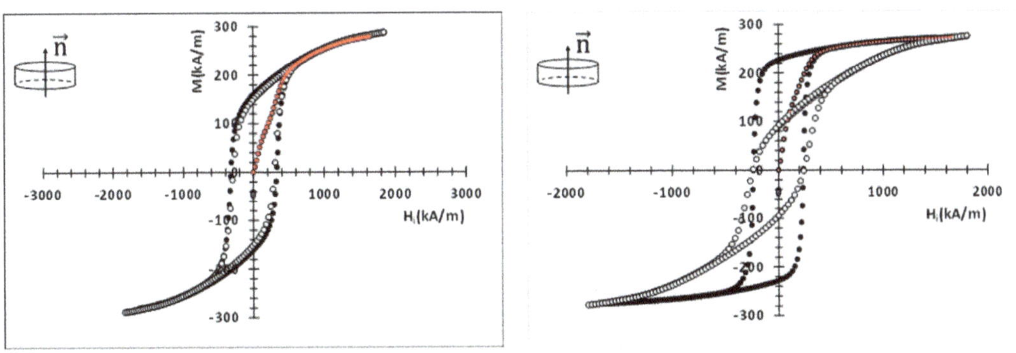

Figure 11. Hysteresis cycles measured for samples BaM-0 (**left**) and BaM-100 (**right**). Black symbols: $\vec{H}//\vec{n}$, open symbols: $\vec{H}\perp\vec{n}$. The dashed lines in red are the first magnetization curves.

The behavior of the measured magnetization M as a function of $1/H_i^2$ is strongly linear in this range (Figure 12), which confirms that both the constant a and the susceptibility χ_p of Equation (6) can be neglected in the high-field domain. The data extracted from M(H) loops and the LAS are reported in Table 4.

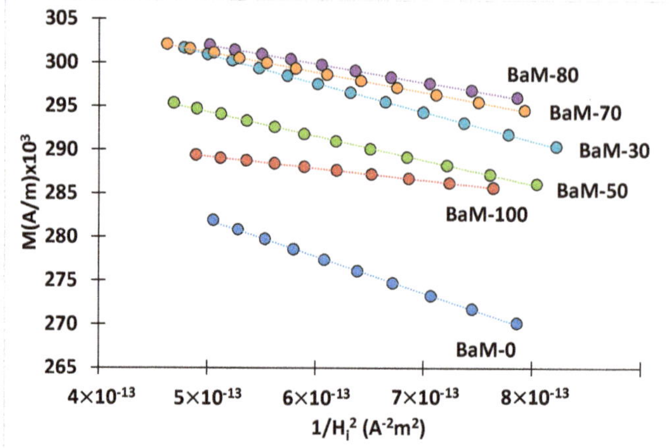

Figure 12. Law of approach to saturation of some BaM samples. The determination coefficients R2 in M vs. 1/Hi2 linear relations are better than 0.999 in any case.

Table 4. Data obtained from hysteresis loops and the law of approach to saturation. M_S is the saturation magnetization, $M_{R//}$ and $M_{R\perp}$ the remanent magnetizations measured along the easy axis and perpendicular to the easy axis, respectively. α_{50} is the full width at half remanence measured in ARM. For $A_{uni}(K_1,\varrho)$, see Equation (12) and main text.

	M_S (kA/m)	$M_{R//}$ (kA/m)	$M_{R//}/M_S$ (kA/m)	$M_{R//}/M_{R\perp}$ (kA/m)	α_{50} (°)	$A_{uni}(K_1,\varrho)/K_1^2$
BaM-0	304	166	0.55	1.06	154	0.33
BaM-5	319	174	0.55	1	150	0.29
BaM-10	308	173	0.56	0.98	128	0.29
BaM-20	319	185	0.58	0.88	125	0.28
BaM-30	317	200	0.63	0.83	123	0.24
BaM-40	308	201	0.65	0.78	118	0.23
BaM-50	308	203	0.66	0.69	105	0.21
BaM-60	311	229	0.74	0.64	101	0.19
BaM-70	313	230	0.73	0.57	93	0.17
BaM-80	312	237	0.76	0.54	89	0.16
BaM-90	303	242	0.8	0.46	87	0.12
BaM-100	299	238	0.81	0.40	70	0.1

3.9. Discussion

The variation of the remanent magnetization $M_{R//}/M_S$ with the percentage of molar mass of goethite (Figure 13) attests the evolution of the texture of the samples from a random structure towards a state of ordered alignment of the easy axes. These results suggest that the actual angular dispersion of easy axis of magnetization (as described by $M_{R//}/M_S$ in Figure 13) is significantly larger than the angular dispersion of the crystal grains (Figure 8). This difference could be due to dipolar interactions between grains, which cannot be detected by XRD measurements.

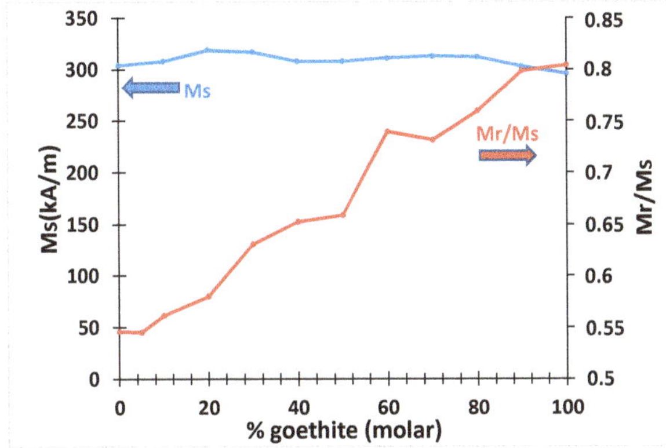

Figure 13. Variations of the saturation magnetization (MS) and reduced-remanence magnetization ($M_{R//}/M_S$) with the molar percentage of goethite.

From these fits, the value of $A_{uni}(K_1,\varrho)/K_1{}^2$ was determined for each sample (Table 4). The choice of the parameter for quantifying the spatial dispersion of the axes can be made among the following quantities: (a) the angular dispersion of the axes of easy magnetization, as determined in X-ray diffraction; (b) the ratio $M_{R\perp}/M_{R/\!/}$, where $M_{R/\!/}$ and $M_{R\perp}$ are the remanent magnetization measured perpendicular and parallel to the sample plane (i.e., presumably along the easy direction and in the magnetic hard plane, respectively) [4]; (c) the width at half height of the angular remanence measurement (ARM). The ARM measures the remanent magnetization as a function of the angle. This is achieved by turning the sample to a certain angle, applying the maximum field, reducing the field to zero, and measuring the resulting moment. The zero crossing marks the hard axis angle. The narrower the transition, the better defined the anisotropy direction is. By taking the derivative of this curve, a spread in the easy axis directions can be obtained (Figure 14). It is convenient to characterize the data by the full width at half-remanence α_{50} (see Table 4).

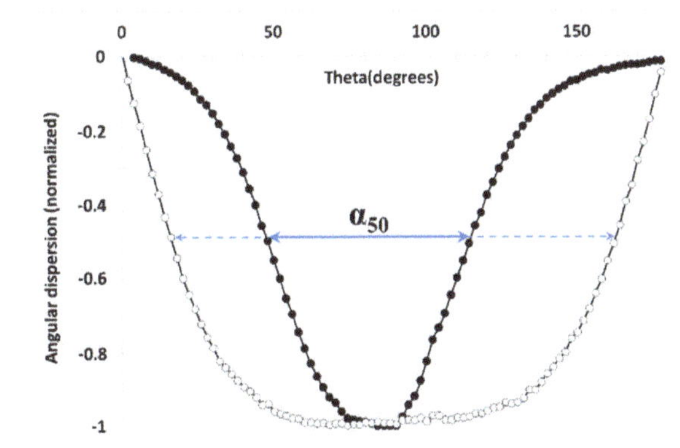

Figure 14. Measured angular dispersion and full width at half-remanence α_{50} for BaM-0 (open circles) and BaM-100 (full circles).

In [44], the angle between the cubic axis that forms the smallest angle with the direction of the applied field is chosen as to represent the grain-dispersion parameter τ (Equation (10)). There is a strong correlation between the angular dispersion of the axes of easy magnetization and both the ratio $M_{R\perp}/M_{R/\!/}$ on one hand, and with the width at half height α_{50} of the ARM, on the other hand (Figure 15).

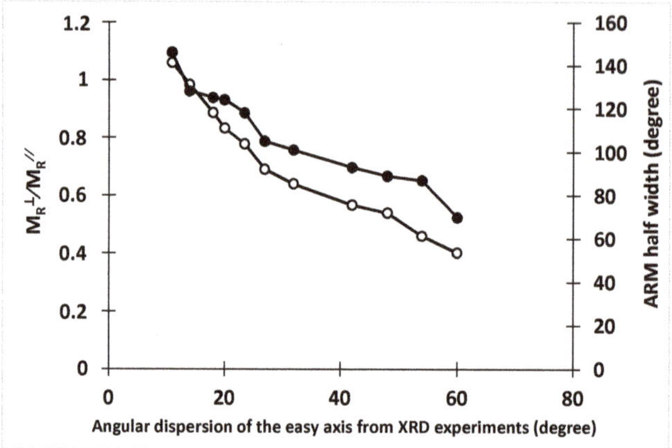

Figure 15. Correlation between 2θ, $M_R{}^{\perp}/M_R{}^{//}$ (open symbols) and ARM (full symbols).

Therefore, we admitted that these two parameters have a common significance, and we chose, as a parameter for quantifying the orientation of the easy axes, the half-height width of the ARM.

The variation of $A_{uni}(K_1,\varrho)/K_1{}^2$ is shown in Figure 16. Two points are added to the experimental data: one corresponding to a monocrystal, for which $A_{uni} = 0$, and the other corresponding to a polycrystal with an ideal random distribution of easy axes, for which $A_{uni} = 4/15$. Interestingly, these variations are not linear and have a maximum. This behavior is also reported on in [44] with regard to the calculated $A_{cub}(K_1, K_2, \tau)/K_1^2$. This was presented as a real effect and not due to the particular distribution function used to describe the dispersion of the grains in the polycrystal. As seen in Figure 16, it is obvious that the law of approach to saturation for isotropic crystals given by relation (9) is no longer suitable for textured crystals, even if the texture is weak. Instead, it is necessary to use a LAS of the same type as in relation (12), where the influence of the grain dispersion on the law of approach to saturation is fully contained in the $A_{uni}(K_1,\rho)$ factor. In this present study, we found that once the constant K_1 is determined, the angular grain-dispersion can be deduced from saturation-approach measurements.

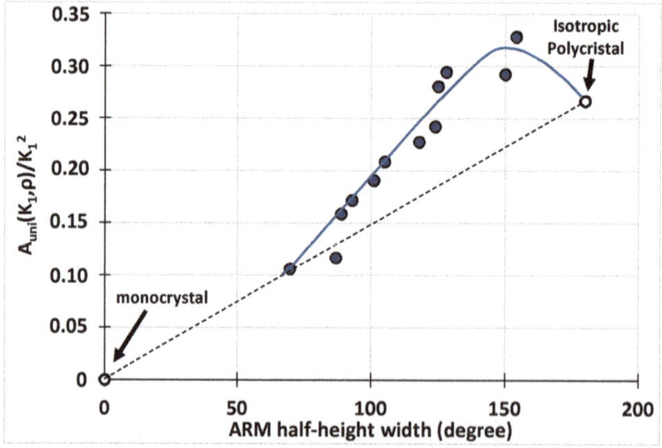

Figure 16. Variation of $A_{uni}(K_1, \rho)/K_1^2$ as a function of the ARM half-height width (the full line is only a guide for the eyes).

4. Conclusions

BaM samples with different magnetic texturing grades are manufactured by controlled topotactical reaction by using hematite and goethite particles. It is demonstrated that the hematite: goethite ratio drives the texturization of the samples. The novelty of this study lies in the optimization of the aspect ratio of the goethite crystallites in view of an improved BaM texturization. We show that the optimization of the morphology of goethite crystallites improves the BaM particles' orientation and stacking. The salient result of this study is the demonstration that by using a single cold compression process and a heat treatment, we can organize BaM particles stacking, which increases with the value of the aspect ratio of the goethite crystallites. The Rietveld refinements of powder diffractograms clearly revealed a particles-stacking enhancement, which is dependent not only of the hematite: goethite ratio but also of the optimal aspect ratio of goethite crystallites. This optimization resulted in a significant improvement of the remanent magnetization value, increasing it to 0.82 compared with the most recent literature. Additionally, we expect in the near future to further improve this value by optimizing the heat treatment after compaction. Based on this study, BaM materials are further manufactured with a controlled magnetic texture; therefore, they are partly self-polarized. They show a reduced-remanence magnetization M_R/M_S varying from 0.5 to 0.81, while the angular dispersion of the BaM particles' easy axis of magnetization varies from 60° to 10°. The magnetocrystalline anisotropy field was measured: microwave measurements provided its maximum value, while its distribution function is obtained from IRM experiments. A law of approach to saturation was proposed and adapted to the case of uniaxial polycrystalline materials for which the particles stacking is only partial. In this law, the influence of the grain dispersion on the saturation approach is fully contained in an additional factor $A_{uni}(K_1,\varrho)$. The experimental values of $A_{uni}(K_1,\varrho)$ compare well with numerical data from the single study on this topic. It is suggested that by using the proposed law once the constant K_1 is determined, the angular grain-dispersion can be deduced from saturation-approach measurements.

Author Contributions: A.H. has carried out the synthesis and the structural characterizations of the samples, he has contributed to the writing of the paragraphs dealing with the chemical parts and more generally to the review and editing of this research article. He has also contributed to the conceptualization and analysis of the materials. A.C. has performed the microwave measurements, and he has contributed to the writing of the paragraphs dealing with microwave measurements and the law of approach to saturation, as well as to the review of the article. J.-L.M. is a supervisor of the research project, supporting the work reported in this article. He has contributed to the conceptualization, methodology, and review of the work, he performed the magnetic measurements, and contributed to the writing of the associated paragraphs. All authors have read and agreed to the published version of the manuscript.

Funding: This work is partially supported by a public grant overseen by the French National Research Agency (ANR-20-ASTR-0010, CONTACT project).

Institutional Review Board Statement: No ethical approval required.

Informed Consent Statement: Not applicable.

Data Availability Statement: Data sharing is not applicable to this article.

Conflicts of Interest: The authors declare no conflict of interest.

References

1. Mazaleyrat, F.; Zehani, K.; Pasko, A.; Loyau, V.; LoBue, M. XXIst century ferrites. *J. Phys. Conf. Ser.* **2012**, *365*, 12001. [CrossRef]
2. Coey, J.M.D. *Magnetism and Magnetic Materials*; Cambridge University Press: Cambridge, UK, 2010.
3. Coey, J.M.D. Permanent magnets: Plugging the gap. *Scr. Mater.* **2012**, *67*, 524–529. [CrossRef]
4. Harris, V.G.; Geiler, A.; Chen, Y.; Yoon, S.D.; Wu, M.; Yang, A.; Chen, Z.; He, P.; Parimi, P.V.; Zuo, X.; et al. Recent advances in processing and applications of microwave ferrites. *J. Magn. Magn. Mater.* **2009**, *321*, 2035–2047. [CrossRef]
5. Geiler, A.; Daigle, A.; Wang, J.; Chen, Y.; Vittoria, C.; Harris, V. Consequences of magnetic anisotropy in realizing practical microwave hexaferrite devices. *J. Magn. Magn. Mater.* **2012**, *324*, 3393–3397. [CrossRef]

6. Mattei, J.L.; Le, C.N.; Chevalier, A.; Maalouf, A.; Noutehou, N.; Queffelec, P.; Laur, V. A simple process to obtain anisotropic self-biased magnets constituted of stacked barium ferrite single domain particles. *J. Magn. Magn. Mater.* **2018**, *451*, 208–213. [CrossRef]
7. Turki, H.; Huitema, L.; Monediere, T.; Lenoir, B.; Breuil, C. Complete Methodology of Low-loss Ultra-wideband Junction Circulator. In Proceedings of the IEEE MTT-S International Microwave Symposium Digest, Philadelphia, PA, USA, 10–15 June 2018; Volume 2018, pp. 746–749.
8. Olivier, V.; Huitema, L.; Lenoir, B.; Turki, H.; Breuil, C.; Pouliguen, P.; Monediere, T. Dual-Band Ferrite Circulators Operating on Weak Field Conditions: Design Methodology and Bandwidths' Improvement. *IEEE Trans. Microw. Theory Tech.* **2020**, *68*, 2521–2530. [CrossRef]
9. Turki, H.; Huitema, L.; Monediere, T.; Lenoir, B.; Breuil, C. New concept validation of low-loss dual-band stripline circulator. *IEEE Trans. Microw. Theory Tech.* **2019**, *67*, 845–850. [CrossRef]
10. Pullar, R.C. Hexagonal ferrites: A review of the synthesis, properties and applications of hexaferrite ceramics. *Prog. Mater. Sci.* **2012**, *57*, 1191–1334. [CrossRef]
11. El Shater, R.; El-Ghazzawy, E.; El-Nimr, M.K. Study of the sintering temperature and the sintering time period effects on the structural and magnetic properties of M-type hexaferrite BaFe12O19. *J. Alloys Compd.* **2018**, *739*, 327–334. [CrossRef]
12. Barrera, V.; Betancourt, I. Hard magnetic properties of nanosized Sr(Fe,Al)12O19 hexaferrites obtained by Pechini method. *J. Phys. Chem. Solids* **2016**, *93*, 1–6. [CrossRef]
13. Lotgering, F.K. Topotactical reactions with ferrimagnetic oxides having hexagonal crystal structures—I. *J. Inorg. Nucl. Chem.* **1959**, *9*, 113–123. [CrossRef]
14. Vijayan, H.; Povlsen, A.; Thomas-Hunt, J.; Mørch, M.I.; Christensen, M. Exploiting different morphologies of non-ferromagnetic interacting precursor's for preparation of hexaferrite magnets. *J. Alloys Compd.* **2022**, *915*, 165333. [CrossRef]
15. Vijiayan, H. Ultrathin nanoplatelets of six-line ferrihydrite enhances the magnetic properties of hexaferrite. *Mater. Chem. Front.* **2021**, *5*, 3699–3709. [CrossRef]
16. Sakai, T.; Chen, Y.; Chinnasamy, C.N.; Vittoria, C.; Harris, V.G. Textured Sc-Doped Barium—Ferrite Compacts for Microwave Applications Below 20 GHz. *IEEE Trans. Magn.* **2006**, *42*, 3353–3355. [CrossRef]
17. Chen, Y.; Fitchorov, T.; Gao, J.; Koblischka-Veneva, A.; Koblischka, M.R.; Vittoria, C.; Harris, V.G. Topochemical growth of textured polycrystalline barium hexaferrite from oriented antiferromagnetic α-FeOOH nanorods. *Nanotechnology* **2009**, *20*, 445606. [CrossRef]
18. Wang, J.; Yang, A.; Chen, Y.; Chen, Z.; Geiler, A.; Gillette, S.M.; Harris, V.G.; Vittoria, C. Self biased Y-junction circulator at Ku band. *IEEE Microw. Wirel. Compon. Lett.* **2011**, *21*, 292–294. [CrossRef]
19. Harris, V.G.; Sokolov, A.S. The Self-Biased Circulator: Ferrite Materials Design and Process Considerations. *J. Supercond. Nov. Magn.* **2019**, *32*, 97–108. [CrossRef]
20. Chen, Y.; Geiler, A.L.; Sakai, T.; Yoon, S.D.; Vittoria, C.; Harris, V.G. Microwave and magnetic properties of self-biesed barium ferrite sreen printed thick films. *J. Appl. Pysics* **2006**, *99*, 08M904. [CrossRef]
21. Andreev, S.; Bartashevich, M.; Pushkarskya, V.; Maltsev, V.; Pamyatnykh, L.; Tarasov, E.; Kudrevatykh, N.; Goto, T. Law of approach to saturation in highly anisotropic ferromagnets Application to Nd-Fe-B melt-spun ribbons. *J. Alloys Compd.* **1997**, *260*, 196–200. [CrossRef]
22. Chen, D.X.; Brug, J.A.; Goldfarb, R.B. Demagnetizing factors for cylinders. *IEEE Trans. Magn.* **1991**, *27*, 3601–3619. [CrossRef]
23. Petrícek, V.; Dušek, M.; Palatinus, L. Crystallographic computing system JANA2006: General features. *Z. Krist.* **2014**, *229*, 345–352. [CrossRef]
24. Cudennec, Y.; Lecerf, A. Topotactic Transformations of Goethite and Lepidocrocite into Hematite and Maghemite. *Solid State Sci.* **2005**, *7*, 520–529. [CrossRef]
25. Pankov, V.; Bartholdson, Å.; Stukalov, O.; Smolenchuk, S.; Babushkin, O.; Gremenok, V. Growth of BaFe12O19 Thin Films Formed by Reactive Diffusion. *J. Cryst. Growth* **2003**, *252*, 382–390. [CrossRef]
26. Timofeev, A.V.; Kostishin, V.G.; Chitanov, D.N. The influence of power ferritization technology on the degree of magnetic structure in plates of BaFe12O19 and SrFe12O19 hexaferrites. *Technical Physics Letters* **2019**, *45*, 401–403. [CrossRef]
27. Degen, T.; Sadki, M.; Bron, E.; König, U.; Nénert, G. The HighScore Suite. *Powder Diffr.* **2014**, *29*, S13–S18. [CrossRef]
28. Cullity, B.D.; Stock, S.R. *Elements of X-ray Diffraction*; Prentice-Hall: Hoboken, NJ, USA, 2001; ISBN 0201610914.
29. Uda, M. The Structure of synthetic Fe3S4 and the Nature of Transition to FeS *Z. Anorg. Allg. Chem.* **1967**, *350*, 105. [CrossRef]
30. Capkova, P.; Valvoda, V. Preferred orientation in powder samples of magnesium and magnesium-cadmium alloys. *Czech. J. Phys.* **1974**, *24*, 891–900.
31. Dollase, W.A. Correction of intensities for preferred orientation in powder diffractometry: Application of the March model. *J. Appl. Crystallogr.* **1986**, *19*, 267. [CrossRef]
32. Zevin, L.; Kimmel, G. *Quantitative X-ray Diffractometry*; Springer: New York, NY, USA, 1995.
33. Durst, K.D.; Kronmüller, H. Determination of intrinsic magnetic material parameters of Nd2Fe14B from magnetic measurements of sintered Nd15Fe77B8 magnets. *J. Magn. Magn. Mater.* **1986**, *59*, 86–94. [CrossRef]
34. Kreisel, J.; Vincent, H.; Tasset, F.; Paté, M.; Ganne, J.P. An investigation of the magnetic anisotropy change in BaFe12-2xTixCoxO19 single crystals. *J. Magn. Magn. Mater.* **2001**, *24*, 17–29. [CrossRef]

35. Smit, J.W. *Ferrites: Physical Properties of Ferrimagnetic Oxydes in Relation to Their Technical Applications*; Wiley: New York, NY, USA, 1959.
36. Baker-Jarvis, J.; Janezic, M.D.; Riddle, B.F.; Johnk, R.T.; Kabos, P.; Holloway, C.L.; Geyer, R.G.; Grosvenor, C.A. *Measuring the Permittivity and permeability of Lossy Materials: Solid, Metals, Building Materials, and Negative-Index Materials*; U.S. Government Printing Office: Washington, DC, USA, 2004.
37. Pucel, R.; Masse, D. Microstrip propagation on magnetic substrates—Part I: Design theory. *IEEE Microw. Theory Tech.* **1972**, *20*, 304–308. [CrossRef]
38. Wan, C.; Hoorfar, A. Improved Design Equations for Multilayer Microstrip Lines. *IEEE Microw. Guided Wave Lett.* **2000**, *10*, 223–224.
39. Pfeiffer, H. Determination of Anisotropy Field Distribution. *Phys. Satus Soliidi A* **1990**, *118*, 295–306. [CrossRef]
40. Stoner, E.C.; Wolhfarth, E.P. A mecanism of hysteresis in heterogeneous alloys. *Philos. Trans. R. Soc. A Math. Phys. Eng. Sci.* **1948**, *840*, 599–642.
41. Grossinger, R. A Critical Examination of the Law of Approach to Saturation. *Phys. Stat. Sol.* **1981**, *66*, 665–674. [CrossRef]
42. Néel, L.; Pautenet, R.; Rimet, G.; Giron, V. On the laws of magnetization of ferromagnetic single crystals and polycrystals. Application to uniaxial compounds. *J. Appl. Phys.* **1960**, *31*, S27–S29. [CrossRef]
43. Herbst, J.; Pinkerton, F. Law of approach to saturation for polycrystalline ferromagnets: Remanent initial state. *Phys. Rev. B Condens. Matter Mater. Phys.* **1998**, *57*, 10733–10739. [CrossRef]
44. Celasco, M.; Mazzeti, P. Saturation Approach Law for Grain-Oriented Polycrystalline Magnetic Materials. *IEEE Trans. Magn.* **1969**, *5*, 372–378. [CrossRef]
45. Grössinger, R.; Sassik, H.; Holzer, D.; Pillmayr, N. Magnetic characterization of soft magnetic materials—Experiments and analysis. *J. Magn. Magn. Mater.* **2003**, *254–255*, 7–13. [CrossRef]

magnetochemistry

Article

Shielding Effectiveness Measurement Method for Planar Nanomaterial Samples Based on CNT Materials up to 18 GHz

Andrea Amaro [1,*], Adrian Suarez [1], Jose Torres [1], Pedro A. Martinez [1], Roberto Herraiz [1], Antonio Alcarria [2], Adolfo Benedito [3], Rocio Ruiz [3], Pedro Galvez [3] and Antonio Penades [3]

[1] Department of Electronic Engineering, University of Valencia, 46010 Burjassot, Spain; adrian.suarez@uv.es (A.S.); jose.torres@uv.es (J.T.); pedro.a.martinez@uv.es (P.A.M.); roberto.herraiz@uv.es (R.H.)
[2] Würth Elektronik eiSos GmbH & Co. KG, 74638 Waldenburg, Germany; antonio.alcarria@we-online.de
[3] AIMPLAS Technological Institute of Polymers, 46980 Valencia, Spain; abenedito@aimplas.es (A.B.)
* Correspondence: andrea.amaro@uv.es

Abstract: The study and measurement of the shielding effectiveness (SE) of planar materials is required to predict the suitability of a certain material to form an enclosed electromagnetic shield. One of the most widely used standards for measuring the SE of planar materials is ASMT D4935-18. It is based on a coaxial sample holder (CSH) that operates up to 1.5 GHz. Due to this standard's frequency limitations, new variants with higher frequency limits have been developed by decreasing the size of the CSH conductors and the samples. However, this method and its high-frequency variants require two types of samples with very specific geometries and sizes. This method is unsuitable for certain types of nanomaterials due to their complex mechanization at such undersized scales. This contribution proposes an alternative SE measurement method based on an absorber box that mitigates the problems presented by the ASTM D4935-18 standard. The SE of rigid nanomaterial samples based on several concentrations of multi-walled carbon nanotubes (MWCNT) and two different fiber reinforcements have been obtained.

Keywords: shielding effectiveness (SE); nanomaterials; absorber box; electromagnetic compatibility (EMC)

Citation: Amaro, A.; Suarez, A.; Torres, J.; Martinez, P.A.; Herraiz, R.; Alcarria, A.; Benedito, A.; Ruiz, R.; Galvez, P.; Penades, A. Shielding Effectiveness Measurement Method for Planar Nanomaterial Samples Based on CNT Materials up to 18 GHz. *Magnetochemistry* **2023**, *9*, 114. https://doi.org/10.3390/magnetochemistry9050114

Academic Editor: Cătălin-Daniel Constantinescu

Received: 24 March 2023
Revised: 21 April 2023
Accepted: 23 April 2023
Published: 25 April 2023

1. Introduction

The fast-paced advancements in electronic devices, information technology, wearable devices, and 5G technology have significantly increased electromagnetic interference (EMI) and radiation pollution [1]. This has required developing new materials with advanced shielding capabilities to reduce the effects of EMI. Developing this kind of material aims to increase the security of sensitive devices or systems that can be susceptible to EMI. Moreover, new advanced materials must ensure the protection of human health to reduce the risk of problems derived from exposure to electromagnetic radiation. Consequently, there is a high demand for advanced materials that can significantly address the challenges posed by EMI.

The investigation of lightweight EMI shielding materials will allow the possibility of increasing safety in 5G communications. When it comes to shielding materials, one of the most determining parameters for the application of the material is shielding effectiveness (SE). The SE indicates the attenuation intensity experimented by an electromagnetic wave traveling through a medium, A, after interacting with a medium, B (shield).

Magnetic materials, which are the main ones responsible for magnetic losses, achieve electromagnetic wave (EMW) absorption through magnetic hysteresis loss, eddy current effects, and ferromagnetic resonance [2–4]. Ferrites are widely used as EMW absorbers due to their high magnetic permeability, saturation magnetization (Ms), and resistivity (Ω), as well as a significant flexibility that allows the modification of their chemical composition to adapt their magnetic properties to specific applications [5–7]. A recent study evaluated the efficiency of magnetostatic protection using nanostructured permalloy shielding coatings, demonstrating their potential for enhancing the shielding efficiency of electronic devices achieving a maximum SE value of 29 dB [8]. On the other hand, carbonaceous materials (e.g., carbon nanotubes (CNT), MXenes, or graphene foams) are excellent candidates for enhancing the absorption of the incident EMW due to their interesting electromagnetic and molecular properties, such as a unique combination of high conductivity and low density [9,10]. Specifically, multi-walled carbon nanotubes (MWCNTs) possess a shielding effect against EMI owing to their conductivity and unique internal porous structure and morphology. Consequently, they are an excellent and cost-effective choice for the primary material in composites. Moreover, the availability of specific heteroatomic groups in MWCNTs makes them suitable for convenient modification in subsequent applications. The combination of high conductivity and the presence of numerous internal interfaces because of either their porous structure or molecular arrangement enhances the dielectric loss through interfacial polarization. This also introduces an additional absorption mechanism based on multiple reflections, consisting of the continuous reflection of the incident wave in the different interfaces of the particle; thus, enhancing the attenuation of the EMW reflections [9,11]. The dispersion of carbonaceous particles in a polymer matrix, either thermoplastic or thermosetting, entails the improvement of the SE capabilities of the matrix [12] This allows the obtaining of specific compounds for lightweight applications overcoming the limitations of metals in terms of high density and corrosion susceptibility. Increasing the content of carbonaceous particles was found to have a positive effect on the SE of the resulting compound, achieving a maximum absorption of −38 dB in the X-band with 5% wt MWCNTs [13] and −66 dB for a polystyrene (PS) compound containing 20% wt MWCNT obtained via compression molding [14].

However, further increasing the content of carbon fillers also entails the increment of the compound viscosity and, therefore, hinders its processability, as noted in [15]. Fiber-reinforced polymer (FRP) composite materials are characterized by their heterogeneity and anisotropy, which imparts to them the property of not exhibiting plastic deformation. FRP composites have found widespread use in a diverse range of contemporary applications, such as space, aviation, and automotive. Carbon-fiber-reinforced polymer (CFRP) and glass-fiber-reinforced polymer (GFRP) composite materials, among other fiber-reinforced materials, have gained increasing popularity due to their outstanding strength and low specific weight properties, leading them to replace conventional materials in various applications [16]. The use of CFRP is a prominent alternative to address the manufacturing problems derived from the use of particle-based composites [17]. Although recent studies have demonstrated the feasibility of using woven prepreg laminates to produce shielding effectiveness of more than 100 dB at low frequencies (<1 GHz) [18], the specific contribution of each type of shielding mechanism and the effect of the typology of the fibers at higher frequencies remains an open question. Suitable compositions and orientation of fibers made desired properties and functional characteristics of some GFRP composites equal to steel, had higher stiffness than aluminum, and the specific gravity was one-quarter of the steel [19]. Martinez et al. performed SE measurements in the frequency range of 300 kHz–8 GHz on GF composites in combination with different conductive materials, such as MWCNT and copper mesh, reporting an attenuation of approximately −40 dB up to 1.5 GHz for the copper mesh case [20]. Another study investigated the EMI shielding performance of carbon-nanomaterial-embedded fiber-reinforced polymer composites, revealing that the EMI shielding effectiveness of the composites was significantly improved with the addition of carbon nanomaterials. A sample with 3% CNT-GNP CFRP composition demonstrated an EMI shielding effectiveness higher than 15 dB [21].

To evaluate the suitability of these novel shielding materials for integration into a 5G system or other high-frequency applications, it is necessary to perform a characterization of their SE. Due to the wide variety of applications and shapes that a shielding material can adopt, this is generally characterized as a planar material. Depending on the frequency range where the material will operate, there are various measurement methods to determine their shielding effectiveness. Currently, the most widely used standard for measuring the effectiveness of shielding of planar materials is the ASTM D4935-18 standard [22]. However, this standard is limited in frequency to 1.5 GHz, making it significantly restricted when considering 5G technology that operates at much higher frequencies. Alternative measurement methods based on free-space measurements, such as the IEEE 299 standard, are available but are hindered by the complexity of measurement due to the dimensions of the material sample required and the infrastructure required to perform the measurements [23]. Considering these limitations, alternative measurement methods derived from existing standards are currently being developed to address these issues.

In this work, the proposed measurement method is based on an absorbing box that overcomes the limitations that standard methods present. This method eliminates the sample size and mechanization issue, as it does not require a sample that is either too large or too small. Additionally, it makes it possible to measure in a frequency range that goes further than the region defined by the method presented in the ASTM D4935-18 standard. Furthermore, this method mitigates the problems of surrounding influences as measurements are taken within a controlled, absorbing environment. The results of EMI shielding effectiveness of the developed samples are reported in the frequency range of 700 MHz–18 GHz, covering the sub-6 GHz band of the 5G spectrum and part of the mmWave band.

This manuscript is organized as follows. First, Section 2 describes the manufacturing process of the five samples and their most relevant characteristics. Section 3 illustrates the main planar material measurement methods. This section also describes the current standard methods, their limitations, and the non-standardized measurement techniques. In this section, it is described the measurement setup to determine the SE of the different samples. Subsequently, the obtained results are presented in Section 4. This section also discusses the performance of the samples in the entire frequency range in terms of the attenuation that they provide. Finally, the main conclusions obtained in this research are summarized in Section 5.

2. Material Characterization

2.1. Material Selection

Before thermoplastics were extended, thermosets were widely used for various applications. However, as the industry evolves, thermosets have some serious limitations. When heating a thermoset once, it becomes irreversibly hardened when cured due to heating. Once cured, a thermoset plastic cannot be modified in shape by applying heat or pressure because the curing process has set a permanent chemical bond. The only way to break that chemical bond is by exposing it to a high-temperature source where the thermoset plastic is burned off. Hence the recyclability is zero compared to thermoplastic polymers, which can be repeatedly heated and remolded into desired shapes or forms [24].

Acrylonitrile butadiene styrene (ABS) is an amorphous thermoplastic copolymer built by polymerizing styrene and acrylonitrile in the presence of polybutadiene. The combination of the three confers to ABS a wide range of characteristics, such as impact resistance, toughness, heat resistance, or weather and chemical resistance [24]. ABS is widely used in the plastic industry for modern processes such as plastic injection molding for end products such as protective housings, stiff packaging, and structurally robust parts, as well as in the production of polymer blends, such as polycarbonate + ABS or polyamide + ABS, and can be regularly found in the automotive sector [25–27]. Additionally, the use of ABS has also been reported for EMI shielding applications throughout the manufacturing of ABS/MWCNT compounds, due to its processability and dimensional stability [28,29]. Raw ABS 118HF pellets supplied by Elix Polymers are used in this study in the production of seven samples containing different MWCNT concentrations and reinforcements. These samples are summarized in Table 1. A CNT Masterbatch from Nanocyl, the Plasticyl ABS1501, is used to manufacture Sample 7.

Table 1. List of samples.

Sample ID	Description	Particle Type	Particle Percentage
1	ABS	NA [1]	NA
2	ABS + 5% CNT	MWCNT	5%
3	ABS + 10% CNT	MWCNT	10%
4	CF LFP ABS + 3% CNT	MWCNT	3%
5	GF LFP ABS + 3% CNT	MWCNT	3%
6	2 × CF LFP ABS + 3% CNT	MWCNT	3%
7	MB [2] ABS + 15% CNT	MWCNT	15%

[1] NA: Not applicable. [2] MB: Masterbatch

2.2. *Samples Manufacturing*

The samples studied to determine their SE to prevent EMI has been manufactured following a two-step process encompassing the production of the raw thermoplastic compounds and the subsequent obtention of the testing samples. The compounding stage included using an extruder setup with ABS/CNT pellets and a thermoplastic pultrusion line for LFP production. Finally, the samples underwent compression molding to produce the final rectangular specimens.

2.2.1. Compounding

The initial stage consists of obtaining the raw compounds used in manufacturing the testing samples. These initial compounds carry diverse concentrations of multi-walled carbon nanotube (MWCNT) particles. Furthermore, a sample consisting of 100% raw ABS (Sample 1) allowed the definition of a minimum reference value of EMI shielding. The compounds are obtained following three different and independent techniques.

The compounds used in the manufacturing of Samples 2 to 5 are produced by means of a PRISM 16 L/D 25 twin screw extruder and located at the compounding facilities at AIMPLAS (Figure 1a). In this process, ABS pellets and particles are fed together into the extruder via a specific hopper. During the extrusion process, the raw materials are mixed due to the effect of temperature and the shearing forces exerted by the twin screws. Taking into consideration 240 °C as the processing temperature of the ABS matrix used, a flat temperature profile of 260 °C is settled to process the compounds to ensure good processability of the materials (Table 2). Then, at the end of the extruder, a continuous filament with a pre-defined diameter is obtained and cooled down. Finally, a cutting unit located at the end of the setup generates the ABS/CNT pellets with the desired length.

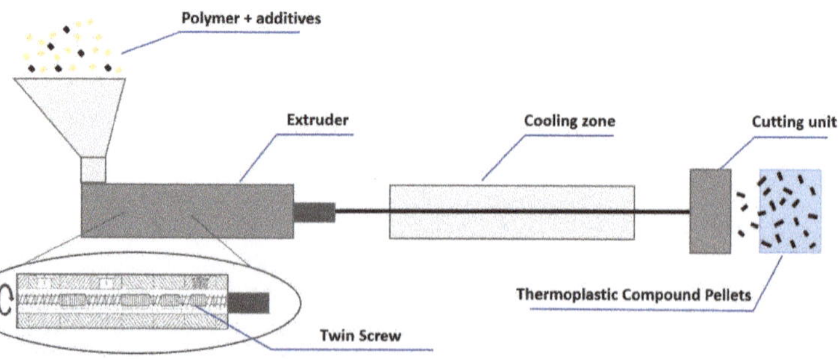

(a)

Figure 1. *Cont.*

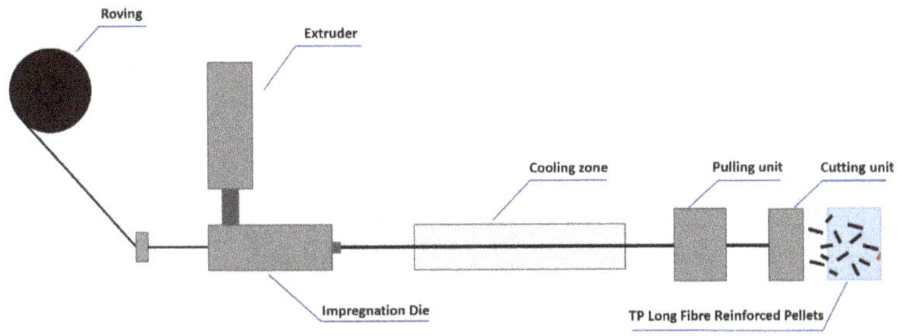

(b)

Figure 1. Compounding process. (**a**) Schematic of the extruder setup. (**b**) Schematic of the LFP production line.

Table 2. Extrusion parameters.

EXTRUSION PARAMETERS										
FEEDER	EXTRUSION ZONE		TEMPERATURE PROFILE							
SET POINT (Kg/h)	SCREW (RPM)	PRESSURE (bar)	ZONE 1 (°C)	ZONE 2 (°C)	ZONE 3 (°C)	ZONE 4 (°C)	ZONE 5 (°C)	ZONE 6 (°C)	ZONE 7 (°C)	ZONE 8 (°C)
1	200	8.2	260	260	260	260	260	260	260	260

Long fiber pellets (LFP) are produced in a thermoplastic pultrusion line developed by AIMPLAS, allowing Ø4 mm unidirectional threads of glass or carbon fiber impregnated by the ABS/CNT compounds previously developed (Figure 1b). In this process, the pellets produced in the previous step are fed to the same extruder described previously in order to melt the polymer and facilitate the subsequent impregnation of the fiber. A temperature profile of 270 °C is used to process the ABS/CNT compound in the extruder to ensure good processability and further fiber impregnation. The resulting melted polymer matrix compound is then transferred to the impregnation die to impregnate the fiber thread effectively. The continuous thread of thermoplastic impregnated fiber leaving the impregnation die at a rate of 600 g/h is cooled down and cut into 12 mm pellets reinforced with oriented and continuous either carbon or glass fiber. The resulting carbon and glass LFP are used to manufacture Samples 4 and 5, respectively.

2.2.2. Compression Molding

Following the production of the pellets, rectangular specimens of dimensions 210 mm × 297 mm × 2 mm are obtained (Figure 2) via compression molding using a CUYMA PH1000 hot press. In this process, the pellets obtained in Section 2.2.1 are placed in a mold of pre-defined dimensions and located between two rigid and hot plates. The effect of the temperature and pressure upon the closure of the mold induced the melting of the pellets and obtaining testing samples with the required dimensions (Figure 3). For all specimens, the temperature is kept constant at 250 °C through the entire compression cycle, which accounted for an initial force of 60 kN for 5 min, followed by an increase in the force to 100 kN and holding for 10 min (Table 3).

Seven testing samples are produced, one per each type of compound. An additional specimen is produced with the same formulation as used for the manufacturing of Sample 6. Coupling these two specimens together (Sample 7) would shed light on the relationship between the increase in the thickness by a factor of 2 and the EMI shielding.

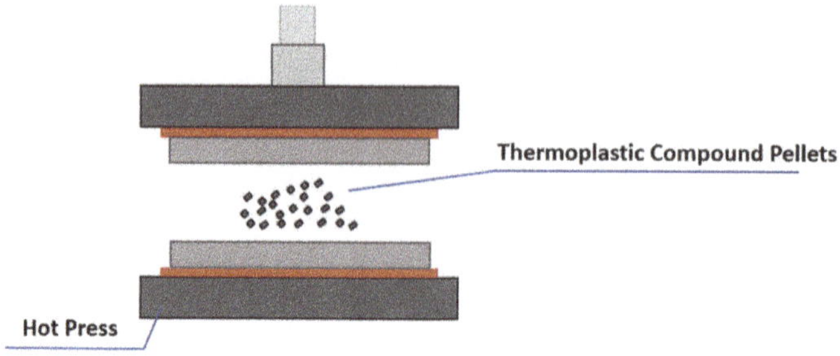

Figure 2. Schematic of the hot press molding process.

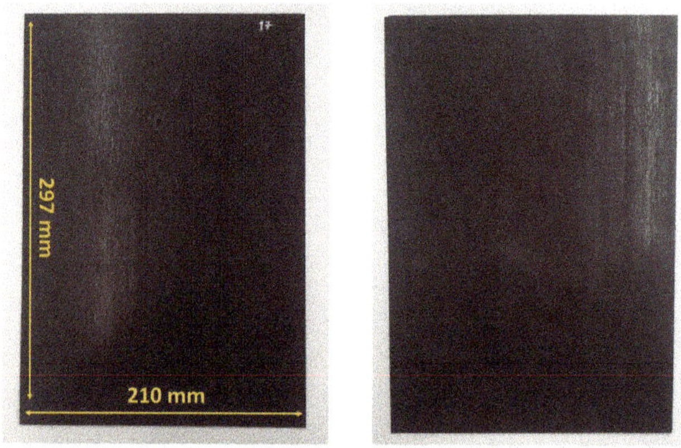

(a) **(b)**

Figure 3. CF LFP ABS + 3% CNT sample. (**a**) Front side. (**b**) Back side.

Table 3. Hot press molding parameters.

Time (min)	Pressure (kN)	Temperature (°C)
5	60	250
10	100	250

2.3. Dispersion Analysis

Following the preparation of the samples, an analysis of the dispersion of the EMI shielding particles in the polymeric matrix is deemed necessary to assess the quality of the manufacturing process. Scanning electron microscopy (SEM) imaging is performed using a Hitachi S-4800 Scanning Electron Microscope which allowed the characterization of the surface of Samples 2, 4, and 7, generated in Section 2.2 (Figure 4). The selected samples are representative of each of the three manufacturing processes and materials described previously.

CNTs appear as thin and elongated structures and are homogeneously distributed across the three samples, indicating a high-quality manufacturing process. Sample 7 contains the highest concentration of CNTs after visual assessment, confirming the nominal specifications of this sample (Figure 4a,b). It is worth noting that the combined effect of the low apparent density of the CNTs (~0.23 g/cm^3) and the high weight percentage of CNTs (15%) contained by this sample, results in a dense network of CNT that hinders the visualization of the polymeric matrix.

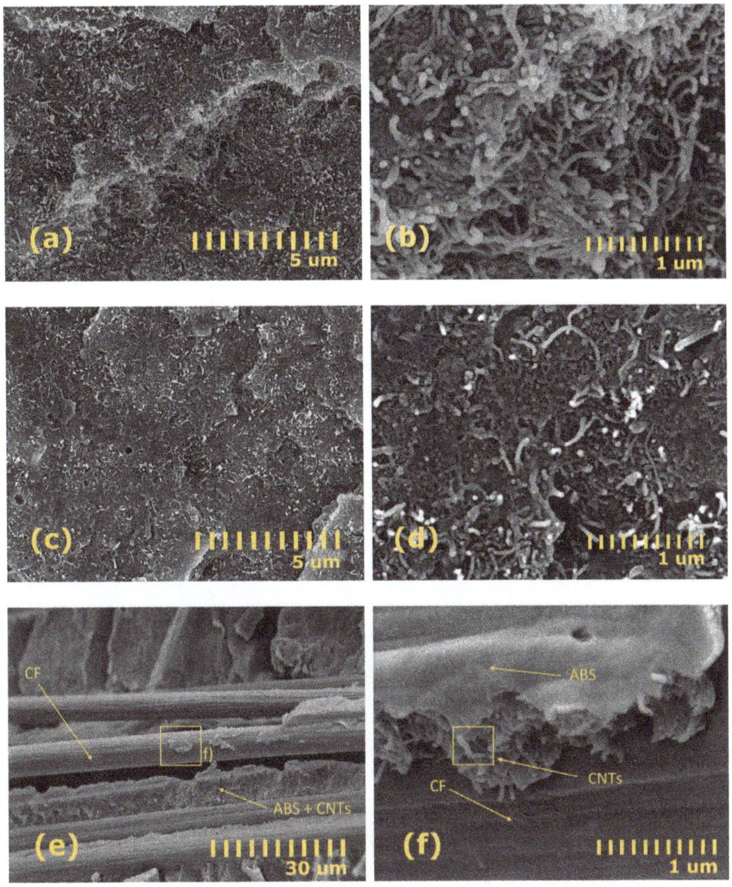

Figure 4. SEM images of three of the developed samples. General view (**left**). Detailed view (**right**). (**a,b**) MB ABS + 15% CNT (**c,d**) ABS + 5%CNT (**e,f**) CF LFP ABS + 3% CNT.

As the concentration of CNT decreases (Samples 2 and 4), the polymeric matrix is rendered visible, and the network of CNT becomes less dense (Figure 4c,d). Additionally, the correct impregnation of the carbon fibers by the polymer matrix can be observed in Figure 4e, ruling out the delamination of the fibers produced by the sample preparation.

Regarding the CNT distribution, it can be observed that a similar distribution is obtained across the samples included in Figure 4. This effect indicates a high level of reproducibility of the manufacturing process. Following a visual examination, empty spaces and aggregates are not appreciated, and CNTs are homogeneously dispersed in the polymer matrix.

3. Planar Material Measurement Methods

The current standard method that defines the measurement procedure to determine the shielding effectiveness of planar materials is the ASTM 4935-18. Nevertheless, the frequency region of this measurement method is limited in frequency (up to 1.5 GHz). Other measurement techniques used to analyze the performance of planar materials are based on the IEEE 299 standard. Due to the wide variety of applications that need to be shielded by using housing with a specific size and shape, planar materials are generally characterized, considering different field conditions. Thereby, depending on the frequency range and the sample features, different measurement techniques may be used to cover the entire frequency range of interest. The 5G operates in a wide range of frequencies. Currently, two different frequency ranges are available for the 5G technology, FR1 and FR2. The bands in the FR1 spectrum are envisaged for the operation of traditional cellular communication, whereas FR2 bands aim to provide short-range very high data rate capability. The 5G FR1 range covers frequencies up to 7.125 GHz, and FR2 encompasses frequencies above 24.5 GHz.

3.1. Standard Measurement Methods

The main techniques for the measurement of the SE of planar materials are based on using mono-mode coaxial TEM cells, according to the standard ASTM D4935-18, or the use of emitting and receiving antennas as in the IEEE 299 standard. The first method has a frequency limit of about 1.5 GHz; the last one applies for higher frequencies, but it requires large sheet samples, which is a disadvantage when dealing with novel materials that rely on rare raw materials. The cost of the shielding particles can be very high, making unfeasible the manufacturing of large-scale samples. Consequently, measuring these materials using a method that requires covering the entire door of an anechoic chamber becomes impractical.

The ASTM D4935-18 is the standard test method for measuring the electromagnetic shielding effectiveness of planar materials. This method allows measuring planar samples in a narrow frequency range from 30 MHz to 1.5 GHz. The technique measures the insertion loss (IL) that results when introducing test samples in a coaxial two-conductor transmission line holder, supporting transverse electromagnetic (TEM) propagation mode. The procedure requires two types of specimens with the same thickness to make SE measurements: the reference and the load specimens (Figure 5). The difference between the measurements of the load and the reference specimen provides the measurement of the SE, caused by the reflection and absorption of the material between the two flanks of the coaxial probe. The upper-frequency limit that can be measured with this method depends on the cut-off frequency for the transverse electric propagation mode of the coaxial cell holder. At frequencies higher than the cut-off, higher-order modes other than TEM can propagate, changing the field distribution inside the cell and causing resonances in the measured results, which have an adverse effect on the accuracy of the measured results. Therefore, the main disadvantage of the fixture is the narrow frequency band of operation, which is limited, considering the operating frequencies of current electronic devices and systems.

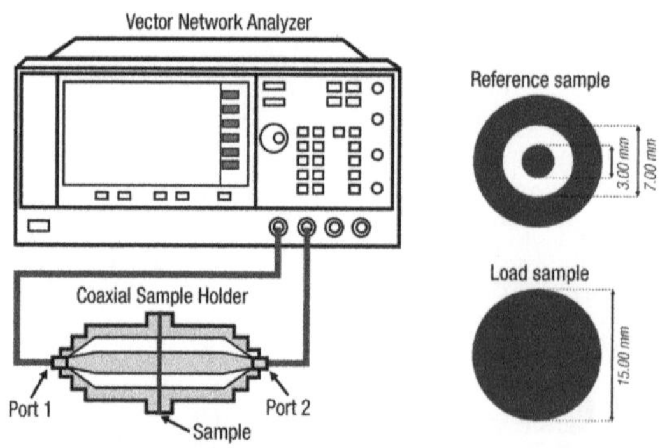

Figure 5. Measurement test setup based on the ASTM D4935-18 procedure and sample geometries and dimensions.

The IEEE 299 standard defines how to measure the effectiveness of electromagnetic shielding enclosures. This method is carried out by placing a sample of the material under test between two antennas connected to a vector network analyzer equipment (VNA) that provides the emitting signal to one of the antennas and receives the field measured by the receiving antenna (Figure 6). Thereby, it is possible to obtain the shielding effectiveness of the material under test by analyzing the S-parameters obtained through a reference measurement (without the material sample) and a load measurement (by placing the material between the two antennas). The reference can be taken in free space or through an open aperture in an anechoic chamber wall. The SE is obtained by taking the difference between the received field strength (in dB units) with the sample absent and with the sample present. The nature of the illuminating field varies with frequency and the type of antenna used.

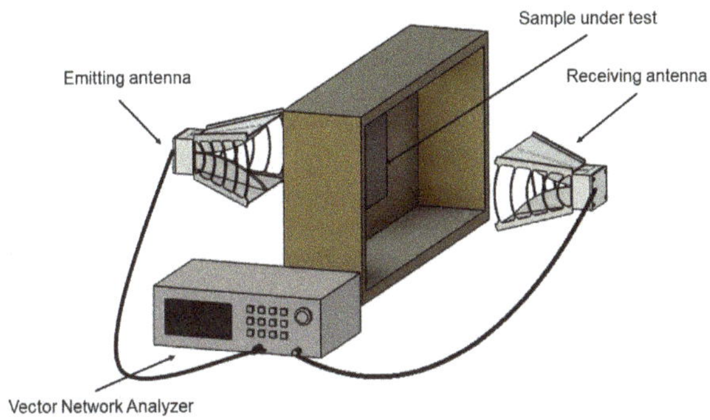

Figure 6. Measurement test setup based on the IEEE 299 standard procedure.

To these limitations is added the high influence between the characteristics of the anechoic chamber as well as the proper location/orientation of the antennas and the sample in the space.

3.2. *Non-Standardized Measurement Methods*

Due to the increase in operating frequencies and the evolution of 5G towards FR2, it is important to develop non-standardized measurement methods through setups, fixtures, and techniques compatible with the operating frequencies of 5G technologies and the samples manufactured.

From the standards, some derivative methods can be highlighted: nested reverberation chambers [30], vibrating intrinsic reverberation chambers [31], TEM cell methods [32], ASTM D4935-18 high-frequency variants [33–35], or absorber box methods [36,37]. The last one proposes an alternative to free space measurements in an anechoic chamber, where the sample size is significantly reduced, and no complex sample preparation is required. Moreover, the equipment and the sample size determine the cutoff frequency, so the method is considerably adaptable to the type of material to be measured. This alternative also eliminates the main problems that the other techniques present, making this method suitable for this study.

As some bibliographic sources indicate, extending the upper-frequency limit of the ASTM D4935-18 standard could be possible. Some institutions have modified this standard coaxial cell holder to perform SE measurements at higher frequencies and on smaller-size materials under test [34,35]. Basically, as the cut-off frequency, and consequently the upper-frequency limit, depends on cell dimensions, new versions of coaxial sample holders can be designed and fabricated, reducing the inner radius of the outer conductor and the radius of the center one.

On the other hand, the measurements based on the IEEE 299 standard are another interesting line of research since they can be performed with RF antennas (inside an anechoic chamber), extending the frequency region defined by the ASTM D4935-18 standard. The upper frequency for IEEE 299 methods is limited by the chamber leakage and the need for the sample to be large enough to minimize edge diffraction effects. Nevertheless, test-to-test variations arise from normal differences between instruments, from discrepancies between transmitting and receiving antennas, including their positions, and primarily from differences between test techniques. Most of the methods based on free space require sample dimensions too large, in addition to being strongly conditioned by the environmental conditions and the directivity of the antennas. Consequently, the proposed alternative method is based on an EMI absorber box lined with absorbent material and two antennas: one transmitter and one receiver.

3.3. Proposed Shielding Effectiveness Measurement System

The proposed measurement method is an adaptation of one of the methods included in the P2715 standard, a guide for the characterization of the shielding effectiveness of planar materials. This method provides the SE of planar materials, adapting to the specific requirements of the study. The main advantage is that the sample machining is simple, which represents a significant breakthrough when dealing with rigid and delicate materials that cannot be machined with very specific geometries or tiny dimensions. Furthermore, no electrical connection to the sample is required, which allows the measurement of samples with low conductivity, contrary to the measurement method proposed in the ASTM D4935 standard. These facts mean that a wide range of measurements can be made with reasonable speed due to the easiness of the measurement procedure. Another notable advantage is the elimination of frequency limitation, making it a suitable method for measurements in the frequency range where 5G technology operates.

The scheme of the proposed prototype is shown in Figure 7. The receiving antenna embedded in the absorber and the receiving antenna are commercial 700 MHz to 18 GHz A-Info LB-7180 ridged waveguide horns. The dimensions of the cavity are 300 × 500 mm, thus, adapting to the rectangular shape of the horn antennas [38]. In one of the sides of the box, a hole has been drilled where an SMA-type connector is placed and connected to the transmitting antenna inside the cavity. The laminated absorber is arranged inside the cavity of the absorber box cut to fit the geometry and dimensions of the antenna. The absorber material is a commercial series made from polyurethane foam that is treated with carbon and assembled in a laminate construction to generate a controlled conductivity gradient. The upper layers above the emitting antenna have a square opening of 100 × 100 mm where the sample is placed to measure the SE of the sample. Assuming no contact exists between the equipment and the sample, two more sheets of absorbent material are placed on top of the sample to mitigate diffraction losses due to the edge effect and to flatten the sample in case it has a concave or convex shape due to the manufacturing process. These top layers of absorbers have an opening of the same dimensions as the sample layer to illuminate the sample as uniformly as possible.

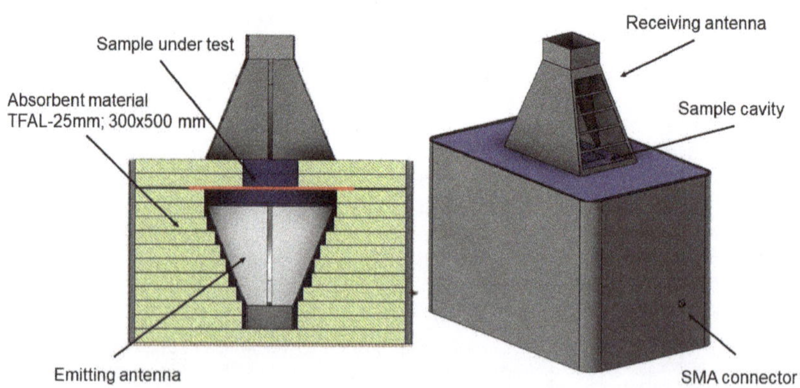

Figure 7. Cross-sectional view of the prototype system.

The selected size of the developed samples is 210 × 297 mm. These dimensions correspond to a fairly standard size for sample prototyping, although it is possible to perform measurements with samples up to 300 × 500 mm, which corresponds to the maximum dimensions of the cavity. Furthermore, with this size, the samples aim to be large enough to reduce the edge effect previously mentioned, but at the same time, small enough to ensure that production is simple and cost-effective.

The antennas are connected to port 1 (emitting channel) and port 2 (receiving channel) of a VNA measuring equipment through an SMA-type connector and two cables Megaphase KB18-S1S1-48 SMA. The frequency range of the final system is 700 MHz–18 GHz, limited by the maximum and minimum operating frequencies of the antennas.

The procedure to obtain the shielding effectiveness of the different samples is based on measuring the scattering parameter S_{21} by taking the transmission ratio through the system without any sample ($S_{21,ref}$), and with the sample present in the cavity ($S_{21,sample}$). Subsequently, the SE is calculated according to Equation 1:

$$SE\ (dB) = S_{21,sample} - S_{21,ref} \tag{1}$$

The prototype and the final measurement setup are shown in Figure 8, where the receiving horn antenna is situated above a material under test and connected to the VNA equipment.

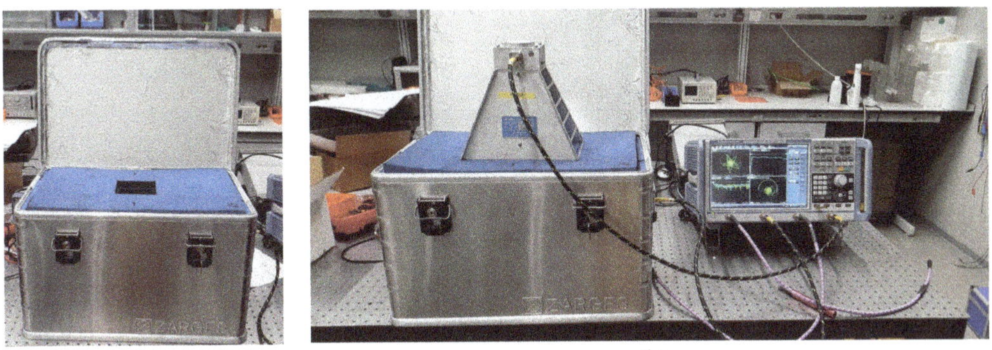

(a) **(b)**

Figure 8. Experimental measurement setup. (**a**) Sample window layer. (**b**) Experimental setup performing a sample measurement.

To characterize the system, the measurement of the dynamic range of the proposed method has been carried out. The SE of the cavity without a sample has been compared with the SE resulting from placing a perfect electric conductor (PEC) with the maximum size of the cavity (300 × 500 mm) to avoid the effect of diffraction at the edges. In this case, the sample is an aluminum sheet with a thickness of t = 2 mm. Figure 9 shows the resulting dynamic range of the system, which is approximately -100 dB and is expected to be suitable to measure the SE of the developed materials according to the values obtained in previous studies of composites with similar characteristics [20,39].

The most notable features of the measurement method used to evaluate the developed samples are summarized in Table 4.

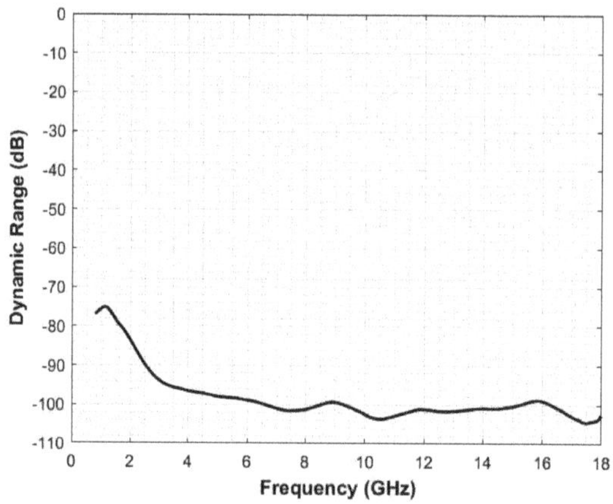

Figure 9. Measured dynamic range of the system with a 2 mm thick aluminum sample.

Table 4. Hot press molding parameters.

Measured Parameter	Unit of Measure	Cavity Dimensions	Sample Size	Frequency Range	Measured Dynamic Range
Shielding effectiveness	dB	300 × 500 mm	210 × 297 mm	700 MHz–18 GHz	−100 dB

4. Results and Discussion

This section is focused on showing the results corresponding to the measurement of seven sample composites under test. Firstly, it is compared the different composite samples based on ABS (samples 1, 2, 3, and 7) to analyze how the increase in the concentration of CNT is turned into an improvement of the SE parameter. Subsequently, the influence of the reinforcement material used to manufacture the composite is studied by comparing the samples based on glass fiber and carbon fiber (samples 4 and 5). Finally, the effect of introducing a thicker carbon fiber reinforcement in the composite is analyzed (samples 4 and 6).

Figure 10 shows the results obtained in terms of SE of different composites with an ABS matrix without fiber reinforcement and three different concentrations of %w CNT. The ABS trace represents the outcome of the SE measurement conducted on the ABS matrix without any filler material. This particular measurement is used as a reference to compare the SE values obtained from the other samples. It can be observed how this trace does not provide considerable attenuation but the increase in CNT filler leads to a rise in the SE provided by the material. If we take the value of 7.125 GHz as a reference, which corresponds to the upper limit of the FR1 band in the 5G spectrum, a value of −24.75 dB is obtained for the trace with a concentration of 5w%CNT. For the sample with twice the weight concentration of CNT, the SE value increases to −39.65 dB. In the last case, for the 15w%CNT sample at the reference frequency, the SE value obtained is −81.30 dB.

Below the reference frequency, the behavior of the traces is slightly different. The red trace corresponding to sample 2 shows a linear behavior, whereas the traces corresponding to the samples with the highest CNT concentration (samples 3 and 7) show an increase in SE as the frequency increases. On the other hand, starting at approximately 14 GHz, it can be observed that the red trace (sample 2) continues to exhibit a flat behavior, whereas the blue trace (sample 3) continues to decrease. However, the black trace (sample 7) shows a change in slope, taking an ascending trend. This fact leaves the door open for further study at higher frequencies to determine if, at a given frequency, the sample with the highest concentration of CNT may not necessarily present the greatest attenuation.

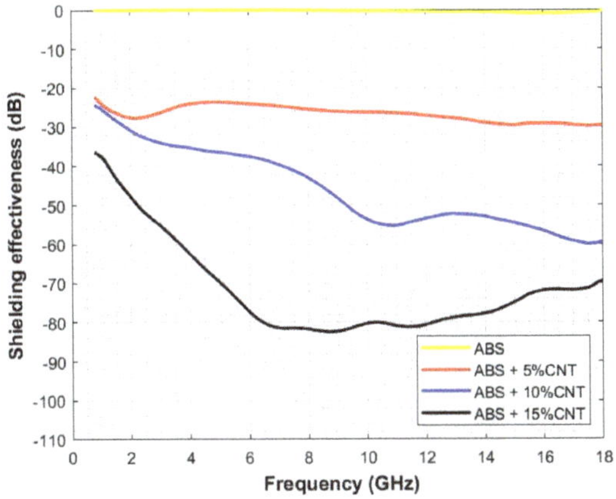

Figure 10. Comparison of ABS polymer samples with three different w%CNT concentrations.

It has been observed that to achieve higher levels of attenuation, it is necessary to increase the concentration of CNT. However, this presents a challenge during the machining process as the addition of CNT increases the viscosity of the composite. To address this issue, two compounds have been developed with the addition of different reinforcements, one based on CF and the other based on GF. These reinforcements provide rigidity to the material, thereby improving its mechanical properties.

Figure 11 shows the shielding effectiveness provided by the CF reinforcement sample comparing it with the GF reinforcement sample (samples 4 and 5, respectively) with a 3w%CNT. It can be observed that the CF sample exhibits considerably higher attenuation compared to the GF sample when the same %w filler is introduced. Quantitatively comparing the results, for the selected reference frequency of 7.125 GHz, it can be observed that the SE value of the CF sample is −60.03 dB. In contrast, the SE value obtained for the GF sample is −14.24 dB, which significantly differs from the CF sample, even though the weight percentage of CNT is the same for both samples. This is attributed to the nature of the fiber reinforcements, as carbon fiber alone exhibits a certain level of shielding depending on the fiber density due to the highly conductive nature of carbon fibers, whereas fiberglass does not cause significant attenuation.

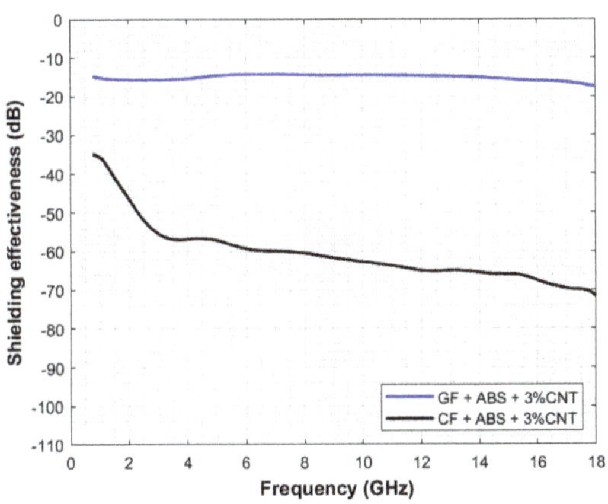

Figure 11. Carbon-fiber- and glass-fiber-reinforced polymers.

Although GF does not provide a significant attenuation by itself, these two compounds have been compared due to the fact that GF provides some advantages in terms of cost, production, and machining. GF is generally more affordable because the materials used to produce it are widely available. Additionally, the manufacturing of GF is simpler and requires fewer processing steps. On the other hand, this type of reinforcement is easier to manipulate and has a longer lifespan than CF reinforcement. Therefore, the type of reinforcement chosen will depend on the final application of the composite and the required level of attenuation.

Figure 12 shows the influence of the sample thickness in the measurement of the SE. To achieve this, a CF reinforcement sample (sample 4) with a thickness of t = 2 mm and 3w%CNT filling is compared with two stacked sheets of the sample (sample 6). On the other hand, the red trace shows the measurement result of the two overlapping sheets. These observations suggest that the sample thickness significantly affects the shielding capability of the material. Below the reference frequency, the behavior of the traces is similar up to approximately 3.5 GHz, where the two traces diverge. It can be observed how the CF sample (black trace) presents considerable attenuation, particularly from 4 GHz. From this point up to 7.125 GHz, the black trace (sample 2) shows a linear behavior with a slightly decreasing slope, whereas the red trace (sample 6) shows a more abrupt decrease until the reference frequency, where it flattens out. Comparing the results for the selected frequency of 7.125 GHz, it can be observed that the SE value of the single-layer CF sample is −60.03 dB, whereas the SE value of the double-layer CF sample shows an increase of 21.29 dB, reaching a SE value of −81.32 dB. Moreover, it is noted that the red trace shows an increase in slope starting at approximately 16 GHz. This phenomenon may be attributed to multiple reflections that occur between the two sheets of material.

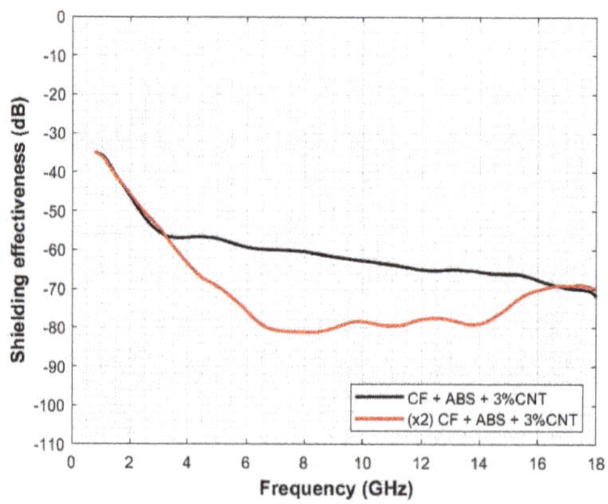

Figure 12. Carbon-fiber-reinforced polymers. Comparison of one sample (black trace) and two superposed samples (red trace).

The following Table 5 provides a summary of the most representative results for each of the seven samples in three different frequency ranges. This table shows the maximum and minimum SE values in decibels.

The ability to detect these variations is essential for optimizing the performance of shielding materials in a given application. By identifying the frequency ranges in which a material provides the most significant attenuation in terms of attenuation, it is possible to design more effective shielding systems. These results demonstrate how this method is capable of detecting variations in the shielding performance of the material across different frequency ranges.

Table 5. Summary of the most representative results in three different frequency ranges.

Sample ID	700 MHz to 6 GHz Min SE (dB)	700 MHz to 6 GHz Max SE (dB)	Frequency Ranges 6 GHz to 12 GHz Min SE (dB)	6 GHz to 12 GHz Max SE (dB)	12 GHz to 18 GHz Min SE (dB)	12 GHz to 18 GHz Max SE (dB)
1	−0.12	−0.25	−0.21	−0.54	−0.43	−1.30
2	−21.87	−28.62	−24.28	−27.36	−27.17	−30.11
3	−24.16	−37.41	−37.41	−56.35	−51.81	−60.44
4	−34.92	−59.89	−59.78	−65.40	−63.56	−71.76
5	−14.25	−15.89	−14.19	−14.67	−14.60	−17.62
6	−34.78	−77.53	−77.23	−82.61	−68.19	−80.49
7	−36.24	−77.99	−77.99	−83.84	−69.58	−80.95

5. Conclusions

The proposed measurement methodology shows significant advantages, including the simplicity of the sample machining, which means that very specific geometries or tiny dimensions are not required. This method allows the measurement of samples with low conductivity, contrary to the measurement method proposed in the ASTM D4935 standard. The sample insertion and removal process can be completed within a matter of seconds, which supposes that a wide range of measurements can be made with reasonable speed due to the easiness of the measurement procedure. The measured dynamic range is approximately −100 dB, allowing us to analyze the samples developed in this study in the frequency range where 5G technology operates.

On the other hand, EMI shielding effectiveness in the 700 MHz–18 GHz frequency range has been studied. Two types of materials have been compared, depending on the presence of fiber reinforcement. It has been observed the performance of the materials studied in a wider frequency spectrum than specified by ASTM4935-18 standard and controlling the surrounding effects by using the absorber box method.

Some of the samples analyzed have demonstrated to provide a significant attenuation. For those composites based on a polymer matrix with different concentrations of CNT, a value of −81.30 dB has been obtained for the frequency of 7.125 GHz for the 15w%CNT composite, which is a very significant SE value considering the nature of these materials. Whilst it is true that an increased volume fraction of filler may lead to a decrease in the mechanical performance of the host matrix by means of deterioration of its inherent morphology, it is necessary to incorporate a higher filler content in order to achieve higher SE. It is desirable to employ CNT/polymer composites at low filler loadings to produce cost-effective and versatile conductive composites.

The inclusion of a carbon fiber reinforcement has proven to be an effective strategy for achieving significant attenuation in composite materials, providing higher attenuation than the GF-reinforced composite with the same CNT concentration. This approach offers an advantage over using a high percentage of filler, which may lead to undesirable mechanical properties. This is due to the highly conductive nature of carbon fibers, which can effectively attenuate electromagnetic radiation. Furthermore, using CF reinforcement can also provide additional benefits such as increased stiffness and strength, as well as reduced weight. This, in turn, makes these materials an excellent alternative to replace traditional shielding materials.

It has to be highlighted that these types of characterizations are very relevant from a technological and industrial point of view. Specifically, for those sectors related to 5G technology, since the use of EMI shielding based on plastic materials has many advantages, such as manufacturing cost reduction.

Author Contributions: Conceptualization, A.S., J.T., A.B. and R.R.; Methodology, A.A. (Andrea Amaro), P.A.M. and A.A. (Antonio Alcarria); Validation, A.S., J.T. and P.A.M.; Formal analysis, A.A. (Andrea Amaro), R.H. and A.A. (Antonio Alcarria); Investigation, A.A. (Andrea Amaro), A.S. and R.H.; Resources, A.B., R.R., P.G. and A.P.; Writing—original draft, A.A. (Andrea Amaro), A.S., R.H.,

P.G. and A.P.; Writing—review & editing, J.T., P.A.M, A.B. and R.R.; Project administration, J.T.; Funding acquisition, J.T. All authors have read and agreed to the published version of the manuscript.

Funding: The project with reference PLEC2021-007994 has been funded by MCIN/AEI/10.13039/501100011033 and by the European Union NextGenerationEU/PRTR.

Institutional Review Board Statement: Not applicable.

Informed Consent Statement: Not applicable.

Data Availability Statement: Not applicable.

Conflicts of Interest: The authors declare no conflict of interest.

References

1. Martinez, P.A.; Navarro, E.A.; Victoria, J.; Suarez, A.; Torres, J.; Alcarria, A.; Perez, J.; Amaro, A.; Menendez, A.; Soret, J. Design and Study of a Wide-Band Printed Circuit Board Near-Field Probe. *Electronics* **2021**, *10*, 2201. [CrossRef]
2. Ibrahim, I.R.; Matori, K.A.; Ismail, I.; Awang, Z.; Rusly, S.N.A.; Nazlan, R.; Mohd Idris, F.; Muhammad Zulkimi, M.M.; Abdullah, N.H.; Mustaffa, M.S.; et al. A Study on Microwave Absorption Properties of Carbon Black and Ni0.6Zn0.4Fe2O4 Nanocomposites by Tuning the Matching-Absorbing Layer Structures. *Sci. Rep.* **2020**, *10*, 3135. [CrossRef] [PubMed]
3. Zhao, B.; Deng, J.; Zhang, R.; Liang, L.; Fan, B.; Bai, Z.; Shao, G.; Park, C.B. Recent Advances on the Electromagnetic Wave Absorption Properties of Ni Based Materials. *Eng. Sci.* **2018**, *3*, 5–40. [CrossRef]
4. Victoria, J.; Suarez, A.; Torres, J.; Martinez, P.; Alcarria, A.; Martos, J.; Garcia-Olcina, R.; Soret, J.; Muetsch, S.; Gerfer, A. Transmission Attenuation Power Ratio Analysis of Flexible Electromagnetic Absorber Sheets Combined with a Metal Layer. *Materials* **2018**, *11*, 1612. [CrossRef]
5. Houbi, A.; Aldashevich, Z.A.; Atassi, Y.; Bagasharova Telmanovna, Z.; Saule, M.; Kubanych, K. Microwave Absorbing Properties of Ferrites and Their Composites: A Review. *J. Magn. Magn. Mater.* **2021**, *529*, 167839. [CrossRef]
6. Suarez, A.; Victoria, J.; Torres, J.; Martinez, P.A.; Alcarria, A.; Perez, J.; Garcia-Olcina, R.; Soret, J.; Muetsch, S.; Gerfer, A. Performance Study of Split Ferrite Cores Designed for EMI Suppression on Cables. *Electronics* **2020**, *9*, 1992. [CrossRef]
7. Suarez, A.; Victoria, J.; Alcarria, A.; Torres, J.; Martinez, P.; Martos, J.; Soret, J.; Garcia-Olcina, R.; Muetsch, S. Characterization of Different Cable Ferrite Materials to Reduce the Electromagnetic Noise in the 2–150 KHz Frequency Range. *Materials* **2018**, *11*, 174. [CrossRef]
8. Zubar, T.; Grabchikov, S.; Kotelnikova, A.; Kaniukov, E.; Kutuzau, M.; Leistner, K.; Nielsch, K.; Vershinina, T.; Tishkevich, D.; Kanafyev, O.; et al. Efficiency of Magnetostatic Protection Using Nanostructured Permalloy Shielding Coatings Depending on Their Microstructure. *Nanomaterials* **2021**, *11*, 634. [CrossRef]
9. Iqbal, A.; Kwon, J.; Kim, M.-K.; Koo, C.M. MXenes for Electromagnetic Interference Shielding: Experimental and Theoretical Perspectives. *Mater. Today Adv.* **2021**, *9*, 100124. [CrossRef]
10. Qin, M.; Zhang, L.; Wu, H. Dielectric Loss Mechanism in Electromagnetic Wave Absorbing Materials. *Adv. Sci.* **2022**, *9*, 2105553. [CrossRef]
11. Wu, N.; Hu, Q.; Wei, R.; Mai, X.; Naik, N.; Pan, D.; Guo, Z.; Shi, Z. Review on the Electromagnetic Interference Shielding Properties of Carbon Based Materials and Their Novel Composites: Recent Progress, Challenges and Prospects. *Carbon* **2021**, *176*, 88–105. [CrossRef]
12. Wang, L.; Qiu, H.; Song, P.; Zhang, Y.; Lu, Y.; Liang, C.; Kong, J.; Chen, L.; Gu, J. 3D Ti3C2Tx MXene/C Hybrid Foam/Epoxy Nanocomposites with Superior Electromagnetic Interference Shielding Performances and Robust Mechanical Properties. *Compos. Part A Appl. Sci. Manuf.* **2019**, *123*, 293–300. [CrossRef]
13. Sachdev, V.K.; Sharma, S.K.; Tomar, M.; Gupta, V.; Tandon, R.P. EMI Shielding of MWCNT/ABS Nanocomposites in Contrast to Graphite/ABS Composites and MWCNT/PS Nanocomposites. *RSC Adv.* **2016**, *6*, 45049–45058. [CrossRef]
14. Arjmand, M.; Apperley, T.; Okoniewski, M.; Sundararaj, U. Comparative Study of Electromagnetic Interference Shielding Properties of Injection Molded versus Compression Molded Multi-Walled Carbon Nanotube/Polystyrene Composites. *Carbon* **2012**, *50*, 5126–5134. [CrossRef]
15. Chung, D.D.L. Electromagnetic Interference Shielding Effectiveness of Carbon Materials. *Carbon* **2001**, *39*, 279–285. [CrossRef]
16. Altin Karataş, M.; Gökkaya, H. A Review on Machinability of Carbon Fiber Reinforced Polymer (CFRP) and Glass Fiber Reinforced Polymer (GFRP) Composite Materials. *Def. Technol.* **2018**, *14*, 318–326. [CrossRef]
17. Martins, L.C.; Pontes, A.J. Fiber Reinforced Thermoplastics Compounds for Electromagnetic Interference Shielding Applications. *J. Reinf. Plast. Compos.* **2022**, *41*, 206–214. [CrossRef]
18. Munalli, D.; Dimitrakis, G.; Chronopoulos, D.; Greedy, S.; Long, A. Electromagnetic Shielding Effectiveness of Carbon Fibre Reinforced Composites. *Compos. Part B Eng.* **2019**, *173*, 106906. [CrossRef]
19. Awan, G.H.; Ali, L.; Ghauri, K.M.; Ramzan, E.; Ehsan, E. Effect of various forms of glass fiber reinforcements on tensile properties of polyester matrix composite. *J. Fac. Eng. Technol.* **2010**, *16*, 33–39.

20. Martinez, P.A.; Victoria, J.; Torres, J.; Suarez, A.; Alcarria, A.; Amaro, A.; Galindo-Galiana, B.; Losada-Fernandez, C.; Ramirez-Monsell, V.; Lopez-Rius, B. Analysis of EMI Shielding Effectiveness for Plastic Fiber Composites in the 5G Sub-6 GHz Band. In Proceedings of the 2021 Joint IEEE International Symposium on Electromagnetic Compatibility Signal and Power Integrity, and EMC Europe, EMC/SI/PI/EMC Europe 2021, Raleigh, NC, USA, 26 July–13 August 2021.
21. *ASTM D4935-18*; Standard Test Method for Measuring the Electromagnetic Shielding Effectiveness of Planar Materials. ASTM International: West Conshohocken, PA, USA, 2018. [CrossRef]
22. *IEEE Std 299-2006 (Revision of IEEE Std 299-1997)*; IEEE Standard Method for Measuring the Effectiveness of Electromagnetic Shielding Enclosures. IEEE: Piscataway, NJ, USA, 2007; pp. 1–52. [CrossRef]
23. Chanda, M.; Roy, S.K. *Plastics Technology Handbook*; CRC Press: Boca Raton, FL, USA, 2006; ISBN 9780429124020.
24. Harper, C.A. (Ed.) *Handbook of Plastic Processes*; Wiley: New York, NY, USA, 2006; ISBN 9780471662556.
25. Moore, J.D. Acrylonitrile-Butadiene-Styrene (ABS)—A Review. *Composites* **1973**, *4*, 118–130. [CrossRef]
26. Pradeep, S.A.; Iyer, R.K.; Kazan, H.; Pilla, S. Automotive Applications of Plastics: Past, Present, and Future. In *Applied Plastics Engineering Handbook*; Elsevier: Amsterdam, The Netherlands, 2017; pp. 651–673.
27. Kapoor, S.; Goyal, M.; Jindal, P. Effect of Functionalized Multi-Walled Carbon Nanotubes on Thermal and Mechanical Properties of Acrylonitrile Butadiene Styrene Nanocomposite. *J. Polym. Res.* **2020**, *27*, 40. [CrossRef]
28. Shen, H.Y.; Jiao, Q.Z.; Zhao, Y.; Li, H.S.; Sun, Z. Electrical Conductivity and Electromagnetic Interference Shielding Effectiveness of Multiwalled Carbon Nanotubes Filled ABS Composites. *Adv. Mater. Res.* **2011**, *194–196*, 1554–1557. [CrossRef]
29. Holloway, C.L.; Hill, D.A.; Ladbury, J.; Koepke, G.; Garzia, R. Shielding Effectiveness Measurements of Materials Using Nested Reverberation Chambers. *IEEE Trans. Electromagn. Compat.* **2003**, *45*, 350–356. [CrossRef]
30. Hara, M.; Yoshikai, T.; Takahashi, Y.; Vogt-Ardatjew, R.; Leferink, F. Numerical Analysis of Vibrating Intrinsic Reverberation Chamber between Various Shielding Effectiveness Measurement Techniques. In Proceedings of the 2020 International Symposium on Electromagnetic Compatibility—EMC EUROPE, Rome, Italy, 23–25 September 2020; pp. 1–6.
31. Pocai, M.; Dotto, I.; Festa, D. Fabrics SE Measurements in a Reverberation Chamber by an Apertured TEM Cell and by a Small Nested Reverberation Chamber. In Proceedings of the 2011 IEEE International Symposium on Electromagnetic Compatibility, Long Beach, CA, USA, 14–19 August 2011; pp. 693–698.
32. Tamburrano, A.; Desideri, D.; Maschio, A.; Sabrina Sarto, M. Coaxial Waveguide Methods for Shielding Effectiveness Measurement of Planar Materials Up to 18 GHz. *IEEE Trans. Electromagn. Compat.* **2014**, *56*, 1386–1395. [CrossRef]
33. Sarto, M.S.; Tamburrano, A. Innovative Test Method for the Shielding Effectiveness Measurement of Conductive Thin Films in a Wide Frequency Range. *IEEE Trans. Electromagn. Compat.* **2006**, *48*, 331–341. [CrossRef]
34. Tamburrano, A.; Sarto, M.S. Electromagnetic Characterization of Innovative Shielding Materials in the Frequency Range up to 8 Gigahertz. In Proceedings of the 2004 International Symposium on Electromagnetic Compatibility (IEEE Cat. No.04CH37559), Silicon Valley, CA, USA, 9–13 August 2004; pp. 551–556.
35. Marvin, A.C.; Dawson, L.; Flintoft, I.D.; Dawson, J.F. A Method for the Measurement of Shielding Effectiveness of Planar Samples Requiring No Sample Edge Preparation or Contact. *IEEE Trans. Electromagn. Compat.* **2009**, *51*, 255–262. [CrossRef]
36. Austin, A.N.; Dawson, J.F.; Flintoft, I.D.; Marvin, A.C. Analysis of the Shielding Properties of Metalised Nonwoven Materials. In Proceedings of the 2013 International Symposium on Electromagnetic Compatibility, Brugge, Belgium, 2–6 September 2013; pp. 526–531.
37. LB-7180 0.7-18GHz Broadband Horn Antenna Technical Specification. Available online: http://www.ainfoinc.com.cn/en/p_ant_h_brd.asp (accessed on 10 April 2023).
38. Amaro, A.; Suarez, A.; Tamburrano, A.; Torres, J.; Marra, F.; Martinez, P.A.; Galindo, B.; Soriano, N.; Victoria, J.; Alcarria, A. EMI Shielding Effectiveness Study for Innovative Carbon Nanotube Materials in the 5G Frequency Region. *IEEE Trans. Electromagn. Compat.* **2023**, *65*, 177–185. [CrossRef]
39. Boroujeni, A.Y.; Tehrani, M.; Manteghi, M.; Zhou, Z.; Al-Haik, M. Electromagnetic Shielding Effectiveness of a Hybrid Carbon Nanotube/Glass Fiber Reinforced Polymer Composite. *J. Eng. Mater. Technol.* **2016**, *138*, 041001. [CrossRef]

magnetochemistry MDPI

Article

AC Susceptibility Studies of Magnetic Relaxation in Mn_{12}-Stearate SMMs on the Spherical Silica Surface

Oleksandr Pastukh, Piotr Konieczny, Magdalena Laskowska * and Łukasz Laskowski

Institute of Nuclear Physics Polish Academy of Sciences, PL-31342 Krakow, Poland; oleksandr.pastukh@ifj.edu.pl (O.P.); piotr.konieczny@ifj.edu.pl (P.K.); lukasz.laskowski@ifj.edu.pl (Ł.L.)
* Correspondence: magdalena.laskowska@ifj.edu.pl

Abstract: The study of magnetic relaxations in Mn_{12}-stearate single-molecule magnets deposited on the surface of spherical silica nanoparticles was performed. For such a purpose, the investigation of AC magnetic susceptibility dependence on the frequency and temperature was performed. Based on the Argand plots obtained for different temperatures and temperature dependencies of susceptibility, obtained for different frequencies of AC field, the corresponding relaxation times were derived. Fitting to the Arrhenius law revealed the values of an effective energy barrier and a mean relaxation time, which were consistent for both measuring techniques ($U_{eff}/k_B \sim 50$ K and $\tau_0 \sim 10^{-7}$ s) and similar to the corresponding values for the analogous bulk compounds. Additionally, the obtained relaxation parameters for the Mn_{12}-stearate molecules on the spherical silica surface were compared with corresponding values for the Mn_{12}-based single-molecule magnets deposited upon other types of nanostructured silica surface.

Keywords: nanostructured silica; single-molecule magnets; nanocomposites; magnetic relaxation

Citation: Pastukh, O.; Konieczny, P.; Laskowska, M.; Laskowski, Ł. AC Susceptibility Studies of Magnetic Relaxation in Mn_{12}-Stearate SMMs on the Spherical Silica Surface. *Magnetochemistry* **2021**, *7*, 122. https://doi.org/10.3390/magnetochemistry7090122

Academic Editor: Andrea Caneschi

Received: 26 July 2021
Accepted: 31 August 2021
Published: 8 September 2021

1. Introduction

Nowadays, significant effort is being dedicated to the design and manufacture of new nanosized materials with specific magnetic properties. Very attractive from such a point of view seems to be the single-molecule magnets (SMMs)—low dimensional molecules consisting of coupled paramagnetic metal ions, possessing unique magnetic properties (magnetic hysteresis, slow magnetic relaxations, quantum tunneling effects) available at the molecular level [1,2]. Due to such magnetic characteristics, SMMs have high applicative potential in the fields of molecular spintronics, quantum computers and dense magnetic memory devices [3]. However, to have the possibility of manipulating individual molecules for the realization of nanoelectronic devices, it is necessary to deposit and separate SMMs on the solid surface [4]. In this aim, different techniques and methods, as well as different types of SMMs and solid substrate materials were applied and described in the literature. A gold surface was the substrate of choice in the earliest investigations of SMM surface deposition [5]. A variety of complexes of the most commonly used families of SMMs, Mn_{12} and Fe_4, have been specifically designed for direct deposition on gold substrate [6–9]. Furthermore, in numerous studies, the described approaches were extended to more technologically relevant substrates (carbon, metal oxides, silicon) as well as other SMM complexes [4,10]. However, despite the successful efforts for the organization of such SMMs on the surface and using sensitive measurement techniques, the fact that SMMs retain magnetic hysteresis and slow relaxation behavior when anchored to surfaces is not always evident.

In our previous studies, the possibility of the successful deposition of Mn_{12}-stearate (Mn_{12}-st) SMMs on the surface of silica nanoparticles was shown [11,12]. The nanostructured silica substrate seems to be the ideal material for the deposition of such molecules providing the preservation of their inner structure and magnetic properties [13,14]. Furthermore, the application of the substrate with a developed surface allows for a very detailed analysis of the separated SMMs. In our case, considering the weight of the measured sample—0.1 g—the corresponding surface area was close to 1.3 m^2 [11]. Such a property gives the possibility for a very detailed analysis of the behavior of the SMMs distributed on this surface. The detailed study of magnetic relaxation in surface-deposited SMMs could reveal new interesting features of such structures, important for the practical applications of similar magnetic systems. The sensitive instrument for the characterization of magnetic materials that allows for obtaining additional information on the mechanism of slow relaxation in SMMs is the dynamic magnetic susceptibility technique, also called AC susceptibility, χ_{AC} [15]. Therefore, in the current study, the detailed analysis of magnetic relaxation in Mn_{12}-st SMMs deposited on the surface of spherical silica was performed with the use of AC susceptibility measurements. Furthermore, obtained data were compared to the results of magnetic relaxations in similar magnetic materials.

2. Materials and Methods

The experimental material was a nanocomposite developed by our team: silica spheres with a diameter of approximately 300 nm containing Mn_{12}-st single-molecule magnets anchored at the surface with assumed concentration, as depicted in Figure 1. The control of the distribution was realized by means of the spacer units and anchoring groups, as we reported earlier [12]. In this case, the material containing six spacer molecules per single anchored SMM was chosen. Such a material presented the most interesting magnetic properties [14], and in our opinion, is worthy of further analysis. The synthesis procedure was described in our previous work [12]. The investigated sample was denoted as SilS–Mn_{12}.

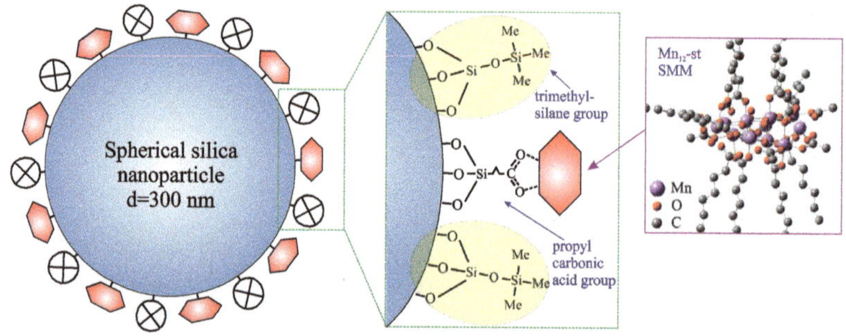

Figure 1. Schematic representation of the investigated material—spherical silica nanoparticles containing deposited Mn_{12}-st SMMs with controlled distance between molecules. Trimethyl silane units play the role of spacers/separators between SMMs, maintaining the assumed distances between them. The number of spacers per single magnetic molecules determines the distribution of Mn_{12}-st at the silica's surface.

Magnetic measurements were carried out with the Quantum Design Magnetic Property Measurement System (MPMS) magnetometer on the sample in a powder form. The mass of the sample was 0.1 g, which includes the mass of the silica nanoparticles with deposited Mn_{12}-st magnetic molecules. The measurements of the phase-sensitive AC susceptibility were based on the measurement of in-phase χ' and out-of-phase χ'' components of χ_{AC} dependence on the temperature. The oscillating magnetic fields of T f = 10 Hz and f = 120 Hz were used in the temperature range from 2.0 K to 8.0 K. The frequency dependence of χ_{AC} susceptibility was measured in the temperature range 3.0 K–5.0 K with the amplitude of the oscillating field of 3 Oe and frequencies changing from 0.1 Hz to 1 kHz.

3. Results and Discussion

The frequency dependence of the χ' and χ'' signals are a consequence of one or more relaxation processes with characteristic relaxation time constants. A common way to determine the relaxation time is the so-called Cole–Cole plot [16], known in magnetism as the Argand plot [17], where χ'' versus χ' is plotted for a given temperature and different angular frequencies ω. If the relaxation process is characterized by a distribution of relaxation times, then the Cole–Cole expression is modified to the so-called extended Debye model [16]:

$$\begin{aligned} \chi(\omega) &= \chi_S + \frac{\chi_T - \chi_S}{1 + (i\omega\tau)^{1-\alpha}}, \\ \chi'(\omega) &= \chi_S + (\chi_T - \chi_S)\frac{1 + (\omega\tau)^{1-\alpha}\sin(\pi\alpha/2)}{1 + 2(\omega\tau)^{1-\alpha}\sin(\pi\alpha/2) + (\omega\tau)^{2-2\alpha}}, \\ \chi''(\omega) &= (\chi_T - \chi_S)\frac{(\omega\tau)^{1-\alpha}\cos(\pi\alpha/2)}{1 + 2(\omega\tau)^{1-\alpha}\sin(\pi\alpha/2) + (\omega\tau)^{2-2\alpha}}, \end{aligned} \tag{1}$$

where χ_T is isothermal susceptibility, χ_S is adiabatic susceptibility and α is a measure of the distribution of the relaxation times. The latter parameter can be derived by the experimental AC data in the Argand plot because the semicircle becomes an arc of a circle with a central angle equal to $(1-\alpha)\pi$. The angular frequency ω at which the magnetic absorption $\chi''(\omega)$ reaches its maximum determines the relaxation time τ of the relevant relaxation process as $\tau = \omega^{-1}$. As a result, by performing such an analysis for the sample at different temperatures, it is possible to obtain the temperature dependence of the relaxation time $\tau(T)$. In order to obtain information about the relaxation process for the case of surface-deposited SMMs at the sample SilS-Mn_{12}, the AC susceptibility studies were performed. Firstly, dynamic magnetic properties were studied with frequency-dependent AC susceptibilities in the zero-applied DC field for the frequency range 0.1–1000 Hz. The corresponding Cole–Cole plots for the temperatures from 3.0 K to 5.0 K with a driving AC field amplitude of 3 Oe are presented in Figure 2.

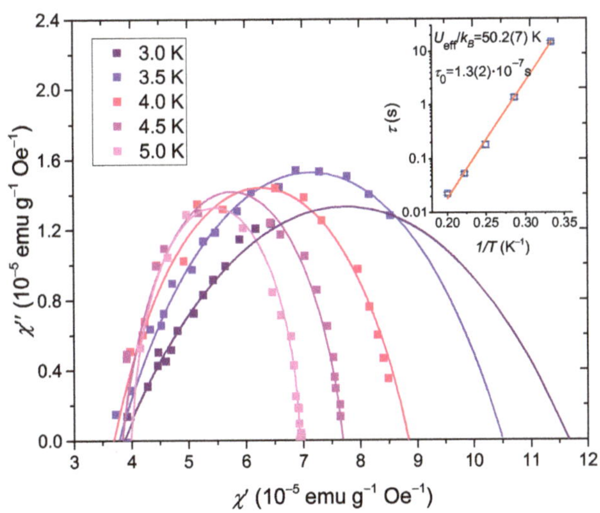

Figure 2. Argand plots for the sample SilS-Mn_{12} measured in the frequency range of 0.1 Hz–1000 Hz with an oscillating magnetic field amplitude of H_{AC} = 3 Oe for different temperatures (solid lines represent the best fit to Equation (1)) (inset: corresponding data of temperature dependence of the relaxation time on a log scale (solid line stands for the best fit to the Arrhenius function (2))).

As can be seen, the obtained experimental data for each temperature differ considerably from perfect semicircles and form arcs of size $(1-\alpha)\pi$. Moreover, for each temperature, most of the data lie in little flattened semi-circles. Such a behavior is typical of a wide distribution of relaxation times [18]. The obtained results can be analyzed using the extended Debye model by fitting AC susceptibility data to Equation (1). All of the parameters corresponding to the best fits are summarized in Table 1. The fit reveals that parameter α, which gives information about the distribution of the relaxation, is temperature dependent and continuously decreases up to the highest temperature of 5.0 K. The fitting determines that for a temperature of 5.0 K, the parameter $\alpha \to 0$, which indicates that the system reverts to the Debye form of a single relaxation time, while with the decreasing temperature, a wide distribution of relaxation times is observed. The average distribution of relaxation times is equal to approximately 0.3. This result is consistent with similar studies performed on other Mn_{12} complexes [19–21] and probably results from a distribution of effective energy barriers, as a consequence of the distribution of environments of the molecules in a sample in powder form [19]. Furthermore, the relaxation time increases with the decreasing temperature.

Table 1. Fit parameters obtained from AC susceptibility data for sample SilS-Mn_{12} using the extended Debye model.

T (K)	χ_S (10^{-5} emu g^{-1} Oe^{-1})	χ_T (10^{-5} emu g^{-1} Oe^{-1})	τ (s)	α
3.0	3.84 (9)	11.65 (8)	14.55672 (4)	0.582 (4)
3.5	3.78 (3)	10.51 (6)	1.36302 (1)	0.457 (7)
4	3.72 (1)	8.84 (2)	0.18202 (2)	0.352 (3)
4.5	3.84 (1)	7.69 (1)	0.05274 (4)	0.192 (4)
5	4.003 (7)	6.94 (2)	0.0226 (1)	0.065 (5)

In addition to the α parameters, the effective time constants can be obtained for each temperature from the reciprocal angular frequency at maximum absorption. The zero-field relaxation times are the function of inverse temperature collected in the inset of Figure 2. In the temperature range 3.0 K–5.0 K, the relaxation times can be approximated by an Arrhenius law:

$$\tau(T) = \tau_0 e^{\frac{U_{eff}}{k_B T}}, \tag{2}$$

where τ_0 is the relaxation time at the high temperature limit, the U_{eff} is effective energy barrier and k_B is Boltzmann constant. The fit of Equation (2) yielded $U_{eff}/k_B = 50.2(7)$ K and $\tau_0 = 1.27(26) \times 10^{-7}$ s. These results slightly differ from the values obtained from the AC susceptibility measurements of bulk Mn_{12} complexes ($U_{eff}/k_B \sim 60$ K, $\tau_0 \sim 10^{-7}$ s) [1,22], but the order of magnitude remains the same.

The next step was the measurement of temperature dependence of the in-phase and out-of-phase components of the magnetic susceptibility for the sample SilS-Mn_{12}. The study was performed for three different values of frequencies f = 1 Hz, f = 10 Hz and f = 120 Hz in the temperature range from 2.0 K to 8.0 K with the amplitude of an oscillating magnetic field 3 Oe. As it can be seen from Figure 3, χ_{AC} of such a sample shows strong dependence on frequency. Upon decreasing temperature, χ' reaches the frequency-dependent maximum, and when the moments cannot follow the oscillating field, start to decrease. From the other side, the out-of-phase signal shows the shifting of peaks to a higher temperature on the increasing frequency. This indicates that the relaxation rate increases with temperature. Such a frequency-dependent shift in the temperature of the maxima of $\chi(\omega, T)$ can be analyzed by the Mydosh parameter [23]:

$$\Phi = \frac{\Delta T_m / T_m}{\Delta(\log(f))}, \tag{3}$$

where T_m is the temperature of the maximum of χ, ΔT_m is the difference between the peaks corresponding to the extremes of χ and f is the frequency. The obtained value for the SilS-Mn_{12} sample is equal to 0.21(1), which is close to the reported value $\Phi = 0.19$ for bulk Mn_{12}-ac [24]. This confirms the superparamagnetic character of the compound. From the frequency dependence of χ'' peaks, it is also possible to estimate the temperature dependence of the relaxation time at the zero applied field. By taking the temperature positions of the χ'' maxima for corresponding frequency, the relaxation time can be plotted as the function of inverse temperature (see the inset of Figure 3). Such a dependence can then be fitted to the Arrhenius law (2). As a result of the fitting, we obtained U_{eff}/k_B = 51(1) K and $\tau_0 = 4.7(4) \times 10^{-7}$ s. These results are consistent with those obtained earlier from the Argand plots and are close to the data reported in the literature [1,22].

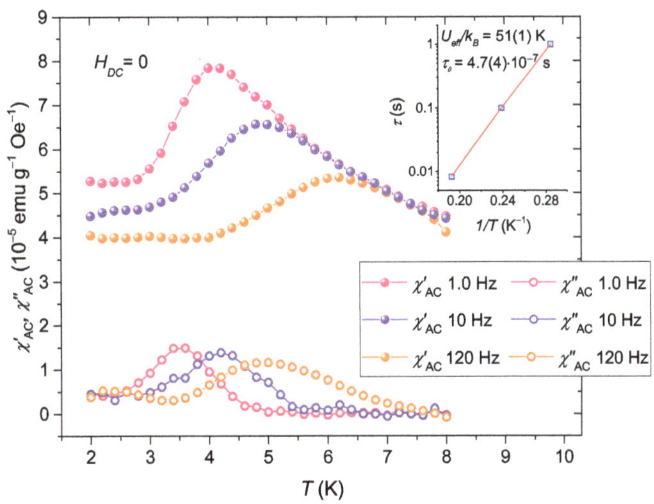

Figure 3. Temperature dependence of in-phase and out-of-phase susceptibility at various frequencies in the zero static field for the sample SilS-Mn_{12} (lines are a guide for eyes) (inset: corresponding relaxation times in function of inverse temperature (solid line stands for the best fit to the Arrhenius function (2))).

We can also compare the results of relaxation time dependence on the reciprocal temperature for both used approaches, as shown in Figure 4. Additionally, the relaxation dynamics of Mn_{12}-st SMMs deposited into spherical nanoparticles can be compared with the corresponding results for analogous molecules incorporated into mesoporous silica matrix SBA-15 (corresponding sample is denoted as SBA-Mn_{12} in Figure 4). As it was reported [25], Mn_{12}-st molecules incorporated into mesoporous silica show strong frequency-dependent χ_{AC} susceptibility with a wide distribution of relaxation times. The relaxation vs. reciprocal temperature dependence of such a sample shows quite good consistency with the results of the data for Mn_{12}-st molecules on the spherical silica substrate. The values of an effective energy barrier and mean relaxation time, derived from AC susceptibility measurements, are similar for both silica substrates.

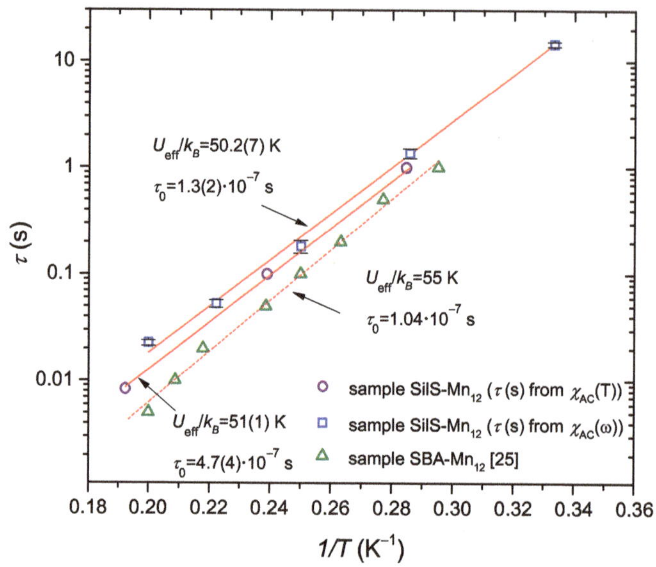

Figure 4. Relaxation times in function of inverse temperature for sample SilS-Mn_{12} in comparison to the SBA-Mn_{12} sample (solid lines stand for the best fit to the Arrhenius function (2) for each measurement).

It should be noted that using nanostructured silica as a substrate for the deposition of Mn_{12} derivatives was described in the literature and mostly based on applying incorporation inside various mesoporous silica structures using different chemical strategies (see Table 2). The silica matrices SBA-15 and MCM-41 belong to the most common type of host material for such a purpose [26,27]. Numerous studies show the preservation of the slow magnetic relaxation of deposited SMMs; however, very often, the simple impregnation of a mesoporous matrix cannot guarantee the SMMs' separation and regular distribution of functionalities inside silica pores. Any practical applications of such nanocomposites require the controlled organization and separation of SMMs in different dimensions to allow for easy read-and-write processes. Furthermore, a nanostructuration process should provide the preservation of chemical integrity and unique magnetic properties of deposited molecules. Unfortunately, the control of the inter-molecular distances between individual SMMs inside the channels of mesoporous silicas was not yet achieved [28]. From such a point of view, spherical silica as a substrate provides not only the successful deposition of molecules, but also the possibility of the horizontal distribution of deposited SMMs. As described in our previous studies [13], the use of spacer units creates the possibility of the regular organization of SMMs on the spherical silica surface. Such observation could be of great importance for the nanotechnological applications of surface-deposited SMMs.

Table 2. Relaxation parameters derived from AC susceptibility measurements for Mn_{12}-based SMMs organized on different nanostructured silica substrates.

Type of Silica Substrate	Applied SMMs	U_{eff}/k_B (K)	τ_0 (s)	Reference
Spherical nanoparticles	Mn_{12}-stearate	51(1)	$4.7(4) \times 10^{-7}$	—
SBA-15	Mn_{12}-stearate	55	1.04×10^{-7}	[25]
SBA-15	Mn_{12}-acetate	72	1.03×10^{-8}	[26]
SBA-type	Mn_{12}-benzonate	63	4.6×10^{-9}	[29]
MCM-41	Mn_{12}-acetate	58.4	1.3×10^{-8}	[27]

The important thing that should also be taken into consideration is the analysis of magnetic relaxation by an AC susceptibility study along with magnetic relaxation obtained from DC studies. As shown in previous studies for the analogous sample [14], the values of an effective energy barrier and mean relaxation time, obtained from time-dependent magnetization measurements, were slightly different to those obtained by AC susceptibility measurements (for DC measurements: U_{eff}/k_B = 33.6(9) K and τ_0 = 4.6(3) $\times$ 10^{-4} s; for AC measurements: $U_{eff}/k_B \sim 50$ K and $\tau_0 \sim 10^{-7}$ s), which can be seen in Figure 5. It can be explained by considering the different temperature range of each measurement. At relatively low temperatures, below the blocking temperature, magnetic relaxation in the sample is enhanced by the quantum tunneling of magnetization (QTM), as can be concluded from the typical shape of the $M(H)$ curve at zero field (see Figure 3a at [14]). Moreover, the time decay of the magnetization study revealed that the relaxation rate significantly changes with the temperature. Such a behavior indicates that quantum tunneling is thermally assisted and becomes more important as the temperature is lowered. This relaxation mechanism is well-known for Mn_{12} complexes [24,30] and results in the reduced value of an effective energy barrier and accelerates the relaxations. In contrast to AC measurements, at a higher measuring frequency (and temperature), the shortcuts due to tunneling in the middle of the energy barrier lose their importance, the apparent energy barrier is larger and τ_0 approaches the value of 10^{-7} s. Nevertheless, there is a possibility of correlating the relaxation data obtained from the AC and DC measurements in order to evaluate the average relaxation parameters for the sample in the whole temperature region (see Figure 5). This resulted in the fit to the Arrhenius law revealing the following values: U_{eff}/k_B = 45.2(7) K and τ_0 = 3.4(9) $\times$ 10^{-6} s.

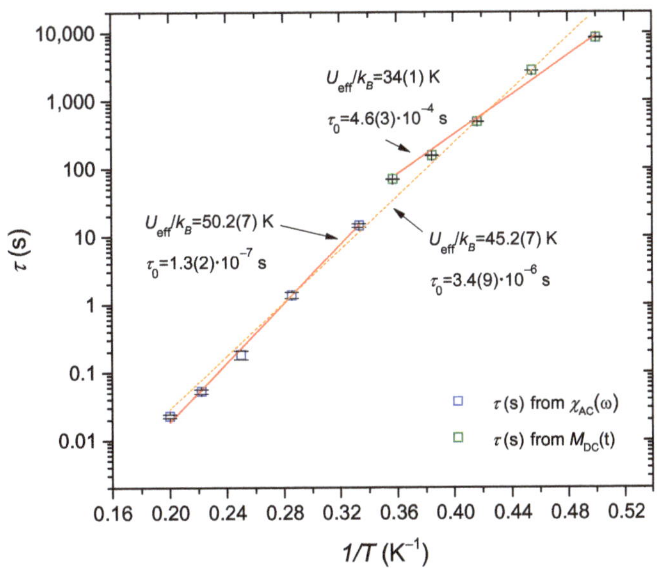

Figure 5. Relaxation times in function of the inverse temperature for sample SilS-Mn_{12}, derived from the AC susceptibility measurements and DC magnetization measurements (solid lines stand for the best fit to the Arrhenius function (2) for each measurement and the dashed line represents the best fit to such a function for both the DC and AC relaxation data).

4. Conclusions

In this work, we analyzed magnetic relaxation in Mn_{12}-stearate single-molecule magnets deposited on the surface of spherical silica with assumed distribution. As we can conclude, the obtained results of the AC susceptibility study of magnetic relaxations confirm that molecular magnets in sample SilS-Mn_{12} exhibit considerable magnetic anisotropy and slow magnetic relaxations, which remain almost intact after the surface deposition. Obtained values of the effective energy barrier and relaxation time show full consistency for both applied techniques (frequency dependence of AC susceptibility and temperature dependence of χ_{AC}) and are similar to the corresponding values for analogous bulk compounds. Obtained data could be useful for further analysis of SMMs' surface behavior and the investigation of similar nanocomposite structures.

Author Contributions: Conceptualization: O.P., M.L. and Ł.L.; methodology, O.P. and P.K.; investigation, O.P. and P.K.; resources, M.L. and Ł.L.; data curation, O.P.; writing—original draft preparation, O.P. and Ł.L.; writing—review and editing, M.L., Ł.L. and P.K.; visualization, O.P.; project administration, O.P., M.L. and Ł.L. All authors have read and agreed to the published version of the manuscript.

Funding: This work was supported by the resources of the National Centre of Science (Grant-No: 2017/26/E/ST5/00162 (Ł.L.)).

Conflicts of Interest: The authors declare no conflict of interest.

References

1. Sessoli, R.; Gatteschi, D.; Caneschi, A.; Novak, M. Magnetic bistability in a metal-ion cluster. *Nature* **1993**, *365*, 141–143. [CrossRef]
2. Christou, G.; Gatteschi, D.; Hendrickson, D.N.; Sessoli, R. Single-molecule magnets. *MRS Bull.* **2000**, *25*, 66–71. [CrossRef]
3. Bartolome, S.J.; Luis, F.; Fernández, J.F. *Molecular Magnets: Physics and Applications*; Springer: Berlin/Heidelberg, Germany, 2014; p. 395.
4. Cornia, A.; Mannini, M.; Sainctavit, P.; Sessoli, R. Chemical strategies and characterization tools for the organization of single molecule magnets on surfaces. *Chem. Soc. Rev.* **2011**, *40*, 3076–3091. [CrossRef] [PubMed]
5. Cornia, A.; Fabretti, A.C.; Pacchioni, M.; Zobbi, L.; Bonacchi, D.; Caneschi, A.; Gatteschi, D.; Biagi, R.; Del Pennino, U.; De Renzi, V.; et al. Direct observation of single-molecule magnets organized on gold surfaces. *Angew. Chem. Int. Ed.* **2003**, *42*, 1645–1648. [CrossRef]
6. Mannini, M.; Pineider, F.; Sainctavit, P.; Danieli, C.; Otero, E.; Sciancalepore, C.; Talarico, A.M.; Arrio, M.A.; Cornia, A.; Gatteschi, D.; et al. Magnetic memory of a single-molecule quantum magnet wired to a gold surface. *Nat. Mater.* **2009**, *8*, 194–197. [CrossRef] [PubMed]
7. Voss, S.; Burgert, M.; Fonin, M.; Groth, U.; Rüdiger, U. A comparative study on the deposition of Mn 12 single molecule magnets on the Au (111) surface. *Dalton Trans.* **2008**, 499–505. [CrossRef]
8. Zobbi, L.; Mannini, M.; Pacchioni, M.; Chastanet, G.; Bonacchi, D.; Zanardi, C.; Biagi, R.; Del Pennino, U.; Gatteschi, D.; Cornia, A.; et al. Isolated single-molecule magnets on native gold. *Chem. Commun.* **2005**, *12*, 1640–1642. [CrossRef]
9. Mannini, M.; Pineider, F.; Danieli, C.; Totti, F.; Sorace, L.; Sainctavit, P.; Arrio, M.A.; Otero, E.; Joly, L.; Cezar, J.C.; et al. Quantum tunnelling of the magnetization in a monolayer of oriented single-molecule magnets. *Nature* **2010**, *468*, 417–421. [CrossRef] [PubMed]
10. Cornia, A.; Mannini, M. Single-Molecule Magnets on Surfaces. In *Molecular Nanomagnets and Related Phenomena*; Springer: Berlin/Heidelberg, Germany, 2014; pp. 293–331.
11. Laskowski, L.; Kityk, I.; Konieczny, P.; Pastukh, O.; Schabikowski, M.; Laskowska, M. The Separation of the Mn_{12} Single-Molecule Magnets onto Spherical Silica Nanoparticles. *Nanomaterials* **2019**, *9*, 764. [CrossRef]
12. Laskowska, M.; Pastukh, O.; Kuźma, D.; Laskowski, Ł. How to Control the Distribution of Anchored, Mn_{12}–Stearate, Single-Molecule Magnets. *Nanomaterials* **2019**, *9*, 1730. [CrossRef]
13. Laskowska, M.; Pastukh, O.; Konieczny, P.; Dulski, M.; Zalsiński, M.; Laskowski, L. Magnetic behaviour of Mn_{12}-stearate single-molecule magnets immobilized on the surface of 300 nm spherical silica nanoparticles. *Materials* **2020**, *13*, 2624. [CrossRef]
14. Pastukh, O.; Konieczny, P.; Czernia, D.; Laskowska, M.; Dulski, M.; Laskowski, Ł. Aging effect on the magnetic properties of Mn_{12}-stearate single-molecule magnets anchored onto the surface of spherical silica nanoparticles. *Mater. Sci. Eng. B* **2020**, *261*, 114670. [CrossRef]
15. Balanda, M. AC susceptibility studies of phase transitions and magnetic relaxation: Conventional, molecular and low-dimensional magnets. *Acta Phys. Pol. A* **2013**, *124*, 964–976. [CrossRef]
16. Cole, K.S.; Cole, R.H. Dispersion and absorption in dielectrics I. Alternating current characteristics. *J. Chem. Phys.* **1941**, *9*, 341–351. [CrossRef]
17. Dekker, C.; Arts, A.; De Wijn, H.; Van Duyneveldt, A.; Mydosh, J. Activated dynamics in a two-dimensional Ising spin glass: Rb 2 $Cu_{1-x}Co_x F_4$. *Phys. Rev. B* **1989**, *40*, 11243. [CrossRef]
18. Evangelisti, M.; Bartolomé, J. Secondary magnetic relaxations in Mn_{12} complexes. *J. Magn. Magn. Mater.* **2000**, *221*, 99–102. [CrossRef]

19. Chakov, N.E.; Soler, M.; Wernsdorfer, W.; Abboud, K.A.; Christou, G. Single-molecule magnets: Structural characterization, magnetic properties, and 19F NMR spectroscopy of a Mn_{12} family spanning three oxidation levels. *Inorg. Chem.* **2005**, *44*, 5304–5321. [CrossRef] [PubMed]
20. Boskovic, C.; Brechin, E.K.; Streib, W.E.; Folting, K.; Bollinger, J.C.; Hendrickson, D.N.; Christou, G. Single-molecule magnets: A new family of Mn_{12} clusters of formula $[Mn_{12}O_8X_4 (O_2CPh)_8L_6]$. *J. Am. Chem. Soc.* **2002**, *124*, 3725–3736. [CrossRef] [PubMed]
21. Bałanda, M.; Pełka, R.; Fitta, M.; Laskowski, Ł.; Laskowska, M. Relaxation and magnetocaloric effect in the Mn 12 molecular nanomagnet incorporated into mesoporous silica: A comparative study. *RSC Adv.* **2016**, *6*, 49179–49186. [CrossRef]
22. Bagai, R.; Christou, G. The drosophila of single-molecule magnetism:$[Mn_{12}O_{12}(O_2CR)_{16}(H_2O)_4]$. *Chem. Soc. Rev.* **2009**, *38*, 1011–1026. [CrossRef]
23. Mydosh, J.A. *Spin Glasses: An Experimental Introduction*; Taylor and Francis: Abingdon, UK, 1993.
24. Novak, M.; Folly, W.; Sinnecker, J.; Soriano, S. Relaxation in magnetic nanostructures. *J. Magn. Magn. Mater.* **2005**, *294*, 133–140. [CrossRef]
25. Laskowska, M.; Bałanda, M.; Fitta, M.; Dulski, M.; Zubko, M.; Pawlik, P.; Laskowski, Ł. Magnetic behaviour of Mn_{12}-stearate single-molecule magnets immobilized inside SBA-15 mesoporous silica matrix. *J. Magn. Magn. Mater.* **2019**, *478*, 20–27. [CrossRef]
26. Willemin, S.; Arrachart, G.; Lecren, L.; Larionova, J.; Coradin, T.; Clérac, R.; Mallah, T.; Guérin, C.; Sanchez, C. Immobilisation of single molecule magnets in mesoporous silica hosts. *New J. Chem.* **2003**, *27*, 1533–1539. [CrossRef]
27. Clemente-León, M.; Coronado, E.; Forment-Aliaga, A.; Amorós, P.; Ramírez-Castellanos, J.; González-Calbet, J.M. Incorporation of Mn_{12} single molecule magnets into mesoporous silica. *J. Mater. Chem.* **2003**, *13*, 3089–3095. [CrossRef]
28. Aulakh, D.; Bilan, H.K.; Wriedt, M. Porous substrates as platforms for the nanostructuring of molecular magnets. *CrystEngComm* **2018**, *20*, 1011–1030. [CrossRef]
29. Coradin, T.; Larionova, J.; Smith, A.A.; Rogez, G.; Clérac, R.; Guérin, C.; Blondin, G.; Winpenny, R.E.; Sanchez, C.; Mallah, T. Magnetic nanocomposites built by controlled incorporation of magnetic clusters into mesoporous silicates. *Adv. Mater.* **2002**, *14*, 896–898. [CrossRef]
30. Friedman, J.R.; Sarachik, M.; Tejada, J.; Ziolo, R. Macroscopic measurement of resonant magnetization tunneling in high-spin molecules. *Phys. Rev. Lett.* **1996**, *76*, 3830. [CrossRef] [PubMed]

magnetochemistry

Review

Review of Magnetic Shape Memory Polymers and Magnetic Soft Materials

Sanne J. M. van Vilsteren, Hooman Yarmand * and Sepideh Ghodrat *

Department of Sustainable Design Engineering, Faculty of Industrial Design Engineering, Delft University of Technology, 2628 CE Delft, The Netherlands; S.J.M.vanVilsteren@student.tudelft.nl
* Correspondence: h.yarmand@tudelft.nl (H.Y.); s.ghodrat@tudelft.nl (S.G.)

Abstract: Magnetic soft materials (MSMs) and magnetic shape memory polymers (MSMPs) have been some of the most intensely investigated newly developed material types in the last decade, thanks to the great and versatile potential of their innovative characteristic behaviors such as remote and nearly heatless shape transformation in the case of MSMs. With regard to a number of properties such as shape recovery ratio, manufacturability, cost or programming potential, MSMs and MSMPs may exceed conventional shape memory materials such as shape memory alloys or shape memory polymers. Nevertheless, MSMs and MSMPs have not yet fully touched their scientific-industrial potential, basically due to the lack of detailed knowledge on various aspects of their constitutive response. Therefore, MSMs and MSMPs have been developed slowly but their importance will undoubtedly increase in the near future. This review emphasizes the development of MSMs and MSMPs with a specific focus on the role of the magnetic particles which affect the shape memory recovery and programming behavior of these materials. In addition, the synthesis and application of these materials are addressed.

Keywords: magnetic shape memory polymers (MSMP); magnetic soft materials (MSMs); shape recovery

Citation: van Vilsteren, S.J.M.; Yarmand, H.; Ghodrat, S. Review of Magnetic Shape Memory Polymers and Magnetic Soft Materials. *Magnetochemistry* **2021**, *7*, 123. https://doi.org/10.3390/magnetochemistry7090123

Academic Editors: Cătălin-Daniel Constantinescu and Sabina Lesz

Received: 28 May 2021
Accepted: 3 September 2021
Published: 9 September 2021

1. Introduction

Magnetic shape memory polymers (MSMPs) belong to the group of shape memory materials, a group that can change their shape back to their "remembered" shape when they are exposed to a stimulus [1]. Most commonly, this stimulus is a temperature change [2,3]. However, there are also shape memory materials that are activated by, for example, light [4,5], pH [6–9], or electricity [10–12]. There is an interest in magnetically activated shape memory materials in the biomedical domain because they require no heat and can be remotely controlled, which makes them safer and less drastic to bring inside a body [13]. Internal heating is also favorable for other applications, such as to speed up the curing process of thermoset adhesives by the use of induction-heated magnetic particles by an AC field [14].

Even though there are multiple shape memory materials that are activated by magnetic fields (e.g., gels, fluids, etc., [1]), this review will focus on MSMPs as well as MSMs. These two materials both exist out of a polymer or elastomer with incorporated magnetic particles that will activate the shape memory effect of the material when being placed in an external magnetic field. The difference between the two, however, is how the magnetic particles activate the shape change of the material.

MSMPs are essentially shape memory polymers whose shape memory effect is stimulated by heat. In the case of the MSMPs, magnetic particles are incorporated in the shape memory polymer. When the material is placed in an external alternating magnetic field, the magnetic particles heat up due to induction heating. The heated particles heat the shape memory polymer from the inside and when the activation temperature is reached, the shape memory effect is activated [15–17].

MSMs exist out of an elastomer with incorporated magnetic particles. The magnetic fields of these magnetic particles are set in specific magnetization patterns inside the elastomer during the fabrication process. When the magnetic soft material is placed inside an external static magnetic field, the magnetic fields of the magnetic particles align with the external magnetic field. This creates micro torques in the elastomer and pulls the elastomer matrix in a programmed shape [15,16].

MSMPs and MSMs in literature have many different names. This is especially the case for MSMs. Here we chose the names of MSMPs and MSMs through the study by Ma, Wu, Ze, Kuang, Zhang, Qi and Zhao [16], because this study worked with both of these materials. MSMs are not the same as soft magnetic materials. In MSM, the word 'soft' refers to the material body in which the magnetic particles are incorporated, like elastomers and silicones. In the term soft magnetic material, the word 'soft' refers to the magnet and its magnetic properties (it implies that the materials have a very coercive field, i.e., it has a very narrow hysteresis loop).

This review focuses on MSMPs and MSMs and the research done on their composition, manufacturing techniques, design considerations, computer models to predict and design shape transformations, and finally their applications.

2. MSMP Filled with Magnetic Fillers

In this section, different MSMPs will be addressed. Their shape change is caused by induction heating of the incorporated magnetic particles by AC magnetic fields. The section itself will focus on research on the performance of the material and the influence of the magnetic particles. The section is further divided into subsections addressing different types of magnetic particles.

2.1. MSMPs Filled with Fe_3O_4

Schmidt [17] tried to make remotely controllable shape memory polymers. She made a shape memory polymer with incorporated magnetic particles (Fe_3O_4). These magnetic particles are used to heat the shape memory polymer by induction heating the particles by an AC field. The Massart and Cabuil method was employed for the experimental part [18]. Magnetite nanoparticles between 2 and 12 wt% were used for magnetic heating. She succeeded in remotely activating the shape memory effect with the magnetic particles (see Figure 1), while also maintaining basic thermal and mechanical properties of the polymer matrix.

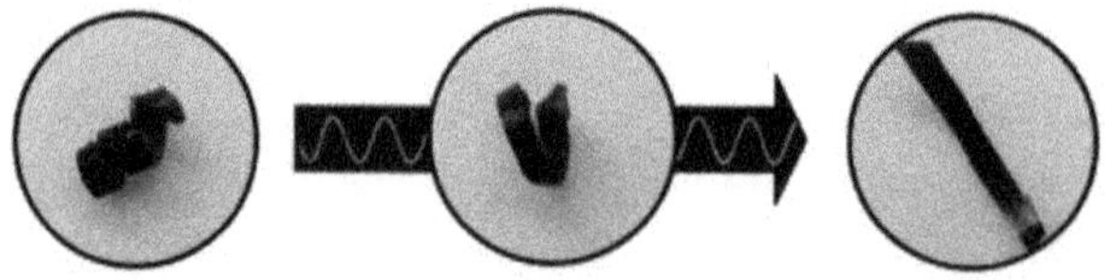

Figure 1. Deformed magnetic shape memory polymer, induction heated magnetic shape memory polymer in the process of recovering its memorized shape, recovered memorized shape [17].

Mohr, et al. [19] also performed research to remotely activate the shape memory effect of a shape memory polymer. Shape changing from a temporary shape (corkscrew like spiral) to a permanent shape (plane stripe) happened within 22 s (see Figure 2). In this study, they made two types of MSMPs. One had a base of polyetheruerthane (TFX) and the other a biodegradable multi block copolymer (PDC). For the magnetic micro particles, Fe_3O_4 was used. They found that the 88 °C maximum temperatures reached by induction heating were dependent on the added magnetic particles and the shape made with the MSMP. All experiments were done with a constant cooling and heating rate of 10 K/min. Moreover, the shape recovery capabilities of induction-heated MSMPs with an AC magnetic field seemed very similar to that for conventional shape memory polymers activated by an environmental heat change, which makes them a promising candidate for smart implants and controlled medical instruments.

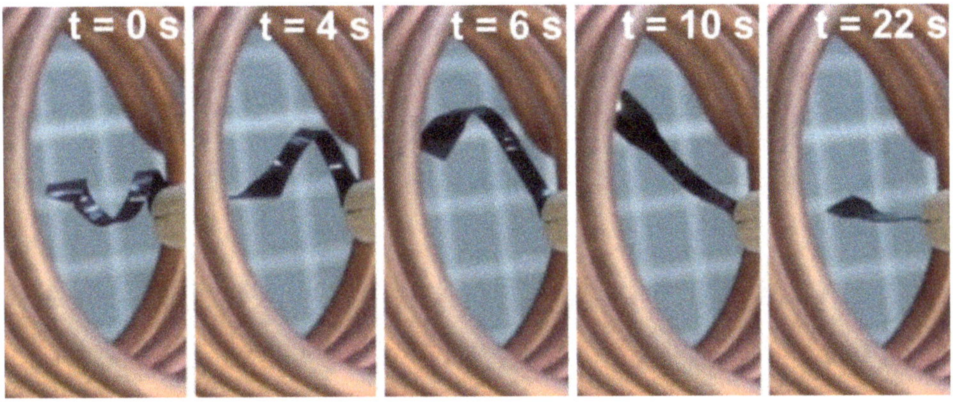

Figure 2. Shape memory effect through induction heating [19].

Instead of using randomly distributed magnetic micron particles, Leng, et al. [20] fabricated a MSMP with the magnetic powder laying in chains by applying a 0.03 T field by an AC magnetic field. They confirmed the formation of clear chain structure in chained samples. These chains increase the electrical conductivity in the shape memory polymer. Due to this, a lower electric voltage is needed to heat the MSMP and activate the shape change. Furthermore, they reported that alignment of magnetic particle chains significantly improves the mechanical properties of the SMP.

In addition, Yakacki, et al. [21] did further research on the influence of the percentage of magnetic particles (Fe_3O_4) incorporated in the polymer matrix. There is a direct influence of the percentage of the added particles on the induction heating rate of the material by an AC magnetic field. The polymer did not influence the heating of the particles. Higher percentages of the magnetic powder also lowered the glass temperature of the material. Materials with a lower percentage of magnetic particles had more crosslinks and therefore had higher glass temperature. Finally, when a large percentage of magnetic particles was added to the polymer matrix, the material had some strains that could not be repaired and the material was more brittle.

Razzaq et al. [22] also investigated the thermal and electrical performance of a MSMP with various wt% of Fe_3O_4 particles. They found that the thermal conductivity of the polymer matrix increased with the increase of magnetic particle fraction. The thermal conductivity improved from 0.19 to 0.60 W/mK with the increase of magnetite volume fraction but the polymer specific electrical resistivity showed a reversed trend. The changes of resistivity could be due to a conversion of the morphology of the polymer matrix which leads to a faster separation of magnetic particles.

Yu, et al. [23] developed a MSMP made from poly(ε-caprolactone) (c-PCL) with incorporated Fe_3O_4 nanoparticles (5–25 wt%). One part of their study consisted of researching the results of crosslinking the polymer. These crosslinks increased the shape recovery of the polymer up to 70%. The MSMP was also activated by both hot water and by an alternating magnetic field. Figure 3 gives an intuitive contrast of the shape-recovery process in both an AC magnetic field and hot water. They observed that both magnetic nanoparticles and hot water stimulated the shape recovery of the helicoidal specimens. However, shape recovery time was significantly different. Thus, this revealed that the reactivity of c-PCL/Fe_3O_4 was better in hot water compared to an AC magnetic field. They believed that the faster recovery time was due to higher thermal transfer and lower heat loss in hot water.

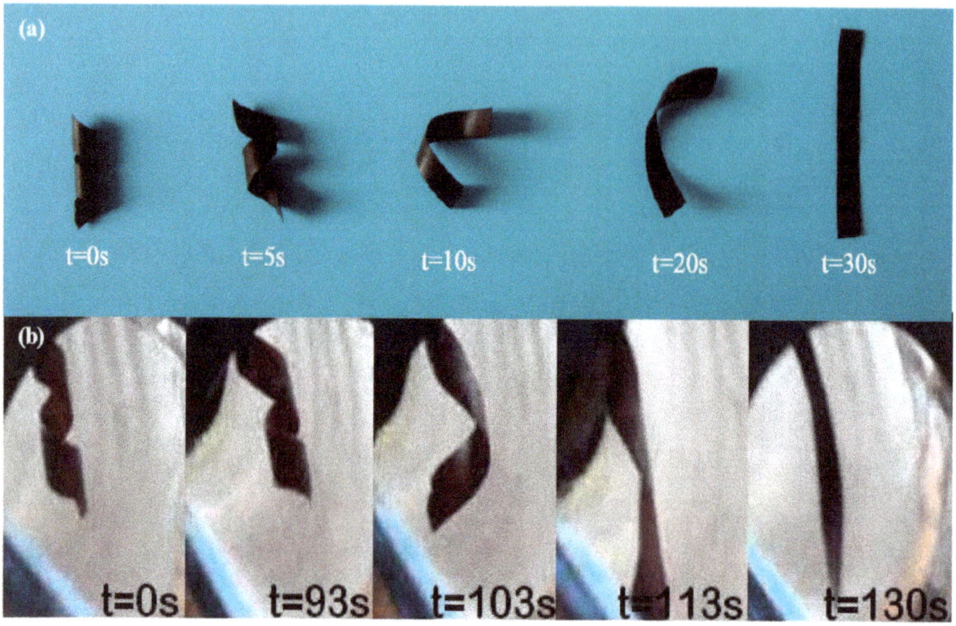

Figure 3. (above, (**a**)) The shape memory effect activated in a heated environment (hot water), (below (**b**)) the shape memory effect remotely activated by an alternating magnetic fields [23].

Weigel, et al. [24] studied the effect of different parameters on the shape memory effect of MSMPs initiated by induction heating through an AC magnetic field. Parameters that were studied were the properties of the magnetic particles, what they were made of, how they were distributed in the polymer matrix, the heat transport conditions, and the surface to volume ratio. The investigated materials were polyetherurethane and a biodegradable multiblock copolymer (PDC) with integrated Fe_3O_4 particles. They found that with the addition of 10 wt% of magnetic particles in the polymer, the shape memory improved and the mechanical properties of the polymer were not changed.

Instead of making a one-way shape recovery MSMP, Kumar, et al. [25] made a MSMP that was triple-shape and could memorize two shapes; the shape memory effect was activated by an AC magnetic field. The magnetic triple-shape polymer was made of poly(3-caprolactone) (PCL) and poly(cyclohexyl methacrylate) (PCHMA) with incorporated Fe_3O_4 particles. The best triple-shape memory effect was achieved by a composite with 40 wt% PCL.

Yang, et al. [26] made a MSMP out of a polynorbornene copolymer and magnetite (Fe_3O_4) with a size of 55 nm. The shape memory effect was activated at a temperature of 51 °C, by the induction heating of magnetite by an AC magnetic field. Through the addition of the magnetic particles, the shape recovery of the material declined. The recovery time for the polymer was 186 s. The added magnetic particles also increased the thermal conductivity of the polymer.

Cai, et al. [27] also made a MSMP out of poly(ε-caprolactone)-polyurethane with incorporated Fe_3O_4. The nanocomposites were made by in situ polymerization at a molar ratio of PCL:MDI:BDO (1:4:3) with various content of Fe_3O_4 nanoparticles. The magnetic particles were homogeneously distributed in the polymer matrix. The shape memory effect was tested both by the activation of hot water and by the induction heating of the magnetic particles by an AC magnetic field (see Figure 4). It was found that the shape recovery started around 40 °C. The best shape recovery was of 97%. Their experiments confirmed that the shape change was faster in hot water compared to the ones heated by magnetic fields. In addition, they reported that shape recovery rate decreased with the more magnetic particles whereas shape recovery time increased.

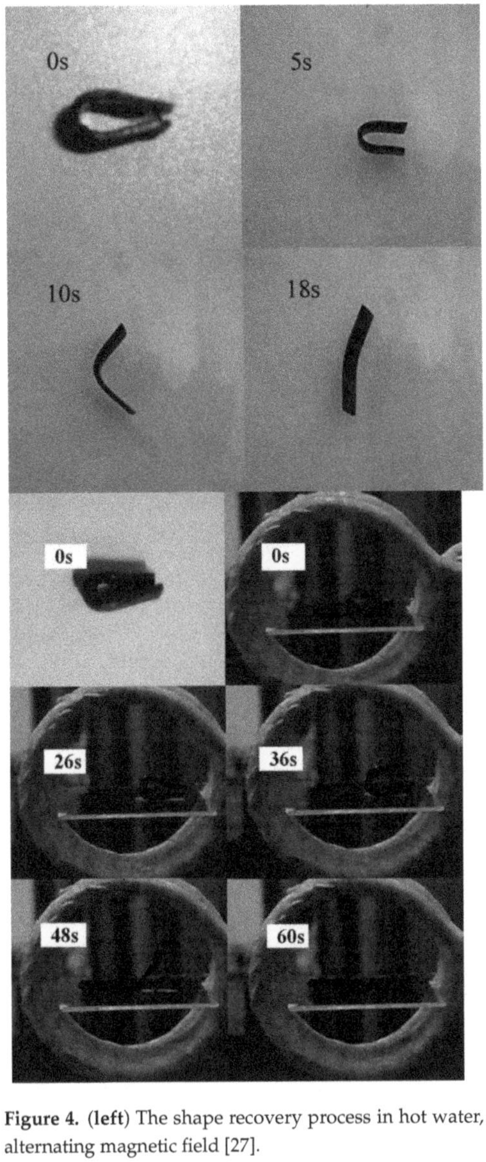

Figure 4. (**left**) The shape recovery process in hot water, (**right**) the shape recovery process in an alternating magnetic field [27].

The development of models to further predict how the magnetic particles influence the shape change of the MSMPs was further studied by Yu, et al. [28]. They used a finite element model that was previously developed to model shape memory polymers. Parameters that they studied were particle size, particle volume fraction, particle heating temperature, and rate to magnetically induce shape recovery behavior. The overall outcome was that an increased volume percentage of the magnetic particles or the smaller the size of the particles increased the heating of the material which stimulated the shape memory effect. They also found that there was a limit to the added volume fraction of particles. When adding particles above this volume percentage, there was no increase in the heating rate of the material.

Liu, et al. [29] developed a dual-response shape memory polymer, which is the same as a MSMP. In this research, the carboxylic styrene butadiene rubber (XSBR)/ferriferrous oxide (Fe_3O_4)/zinc dimethacrylate (ZDMA) was synthesized and tested. The authors investigated how the shape memory effect of the polymer behaved through the two different types of actuation, a heated environment or by induction heating through an AC magnetic field of the magnetic particles (Fe_3O_4) (see Figure 5). The shape recovery ratio of both shape memory activation methods was 100% but with different response times. It is noticeable that the shape recovery rate under magnetic fields was lower at the initial stage and higher in the last phase. This might be due to the fact that heating by an AC magnetic field is a slow process in the early stage and fast in the end phase.

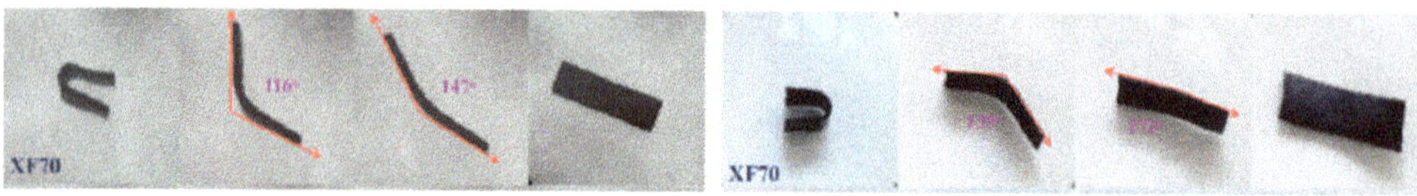

Figure 5. (**Left**) shape memory activated by a heated environment, (**right**) shape memory activated by an AC magnetic field [29].

Zhang, Wang, Zheng, Liu and Leng [13] made a MSMP out of PLA with incorporated Fe_3O_4 particles, that is activated by an AC magnetic field. For this material, the shape memory behaviors were studied through a lens for biomedical applications. It was found that the developed bone support structure (see Figure 6) could very quickly return to its remembered shape, in a few seconds, and the surface temperature of the induction heated material was uniform and around 40 °C, which makes the material safe to use inside the human body. They reported that the shape recovery rate was about 96%. The bones could recover their initial shape within 100 s under a magnetic field. However, they confirmed that the shape recovery time in a hot water was much faster (about 10 s).

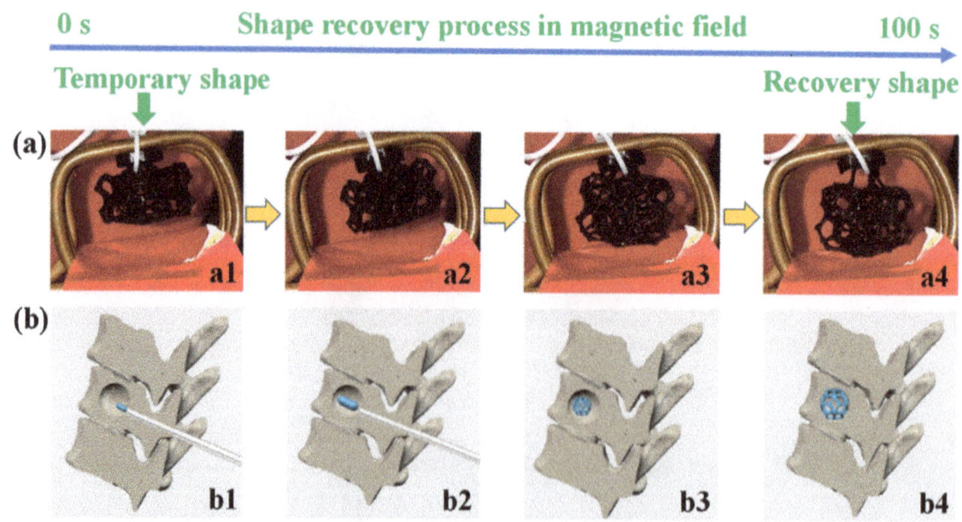

Figure 6. The shape recovery of a printed structure (above (**a1**–**a4**)), the simulation of the shape recovery of a bone support structure in a bone (below, (**b1**–**b4**)) [13].

Zhao, et al. [30] also studied bone tissue scaffolds made out of PLA with incorporated Fe_3O_4 particles that were activated by an AC magnetic field. They mixed 20 wt% Fe_3O_4 with PLA. One of the noticed advantages of MSMPs is that in the context of biomedical design, these structures can be implemented in the body with minimally invasive surgery. Furthermore, this study found that the mechanical structure of the scaffolds was stable and could improve the attachment.

2.2. MSMP Filled with NdFeB

Golbang and Kokabi [31] fabricated a MSMP made out of polyethylene with incorporated NdFeB particles (5, 15, 40 wt%) and 2 wt% organoclay (cloisite 15A) that was activated by an AC magnetic field. The effect of different weight percentages of added NdFeB particles was studied. It was found that the highest generated heat by inductive heating was reliant on the magnetic particles. The sample that had a full shape recovery had 15 wt% of NdFeB particles.

2.3. MSMP Filled with Carbonyl Iron

In the study by Hassan, et al. [32], a MSMP was made with the use of high projection stereolithography with embedded iron carbonyl particles (CIPs). They used three-dimensional printing technology for the fabrication of novel functionally graded MSMPs. They mixed benzyl methacrylate as the monomer with a cross linker, and used poly(ethylene glycol) dimethacrylate (PEGDMA) and phenyl-bis(2,4,6-trimethylbenzoyl)phosphine oxide as the crosslinker and photoiniator, respectively. Through the addition of the magnetic particles, they measured an increased thermal conductivity. Other tests they performed were thermo-mechanical and microstructural surface characterization. The outcome of these tests showed that MSMPs are good for remote controllability, an improvement in the thermal conductivity after embedding the CIPs and close interaction between the particles and the matrix. TGA and DMA results confirmed that the MSMPs have acceptable thermal stability and high mechanical strength.

2.4. MSMP Filled with Nickel Zinc Ferrite

With an increasing interest in making medical devices with shape memory polymers, Buckley, et al. [33] made a shape memory polymer with incorporated nickel zinc ferrite ferromagnetic particles that could be activated though induction heating by an alternating magnetic field. This interest in shape memory polymers for medical devices is because of the remote activation. The challenge for this type of device is to activate the shape memory effect safely and effectively in the body. They developed two in-house prototype devices to elaborate the impact of the added magnetic particles on SMP recovery. Their experiments highlighted that addition of 10 vol% of the magnetic particles made the material heat up uniformly and fast. This volume percentage did not compromise the shape recovery of the shape memory polymer.

In addition, Zhang, et al. [34] fabricated a MSMP with embedded nickel powder. The maximum volume percent of 20% was added to the base polymer. They observed that the homogeneous dispersion of powders had a noticeable effect on the thermal and mechanical abilities of the composite. The more nickel powder was added, the more the tensile strength of the polymer decreased. Two types of materials were made. One only had the added nickel particles. The other had the added nickel particles but it was treated before by a silane coupling agent. The samples with the silane coupling agent showed a better distribution of the particles throughout the material.

2.5. MSMP Filled with Ni-Mn-Ga

Aaltio, et al. [35] also made shape memory polymers with incorporated magnetic particles (Ni-Mn-Ga powder) laying in chains, like Leng, Lan, Liu, Du, Huang, Liu, Phee and Yuan [20]. They studied the damping capabilities of the material. Their developed material was tested on mechanical, structural, and magnetic properties and showed high passive damping for low frequencies. They also compared the damping of their developed material with the same polymeric matrix without magnetic particles and concluded that there was a significant difference between the two materials. They found an increment in the average grain size of the powder particles by the heat treatment. Their experiments confirmed that the magnetization improved more significantly parallel to the chained MSM particles rather than perpendicular to the chains.

2.6. MSMP Filled with Magnetite

Razzaq, et al. [36] studied the mechanical properties of shape memory polymers (polyurethane with 10–40 vol% magnetite). They mentioned the possibility of using AC induction heating. It was found that, with the inclusion of the magnetic particles, the shape memory polymer became more thermally stable but the glass temperature of the material decreased when more magnetite was added. Their report also concluded that the storage modulus of the material became higher when the amount of magnetite increased.

In a study by Puig, et al. [37], a MSMP was made out of magnetite particles that were coated with oleic acid and were embedded in an epoxy system inspired by diglycidylether of bisphenol A modified with oleic acid. The magnetic particles had a homogeneous distribution throughout the material. During the shape activation by an alternating magnetic field of the material, the surface became 25 °C warmer than when the material was not activated. This temperature was enough to activate the shape memory effect. They found the two-step reaction method to be essential to make a well dispersion of magnetics particles in the reactive solvent.

2.7. MSMP Filled with Iron

How the thermal and thermo-mechanical properties of the shape memory polymer were influenced by the addition of magnetic particles was also investigated by Cuevas, et al. [38]. They measured the shape recovery at 90%. Unlike other research, they found that the largest-sized sponged microparticles had a high capacity for the induction heating by an AC magnetic field of the material. Other outcomes of this research were that the stiffness of the shape memory polymer increased when adding more particles and that the material then also became more brittle.

2.8. MSMP Filled with Magnetite or Iron Oxide

Magnetite and iron oxide embedded in silica were compared as magnetic particles in a shape memory foam (Thermoset epoxy DP5.1) [39]. The shape memory effect was activated by the induction heating of the iron oxide by an AC magnetic field. Of these two particles, the iron oxide showed a better performance. It was found that adding a 10 wt% of magnetic particles to the foam did not influence the viscoelastic and thermo-mechanical properties of the material. In the performed experiments, the shapes were recovered within 10 to 20 s. Their results also confirmed that heat transfer between the filler nanoparticles and the bulk foam had a major role in increasing heating performance.

3. MSM

This section will focus on MSMs whose shape change is activated by the alignment of the incorporated magnetic particles in a soft substrate with an external magnetic field. This section will go into the composition of the material and the research carried out on it. In the history of these materials, silicones with embedded whole hard magnets are also taken into account because their working principle is the same as that of MSMs.

A predecessor of the magnetic soft material is presented in a study by Lagcore, et al. [40]. In that study, commercial epoxy resin bars were fabricated with embedded disc formed magnets (4 mm diameter and 90 μm thickness) (strontium ferrite powder). An 80% volume fraction of the hard ceramic ferrite powder was embedded in a commercial epoxy resin. The external magnetic field that was used to change the shape came from a planar coil. A finite element analysis and an analytical model were carried out, which showed good match results compared to the experimental tests.

Another example in which whole permanent magnets were added to a soft component can be seen in the study by Khoo and Liu [41], in which a micropump was developed from an elastomer with an incorporated magnet. By applying a magnetic field, the magnet moves and takes the elastomer with it creating a volume difference on either side of the elastomer. More than 80 mm displacement in the presence of the magnetics particles was reported. During the experiments, the external magnetic field was induced by a permanent disk-shaped NdFeB magnet that was moved near and away from the micro pump to vary the activation flux density.

Another example of hard magnets attached to each other with their magnetic fields pointed towards different directions can be seen in a study performed by [42]. They made different magnetic blocks, of which the magnetic axes (or directions) were orientated in different ways, by curing the materials under a static magnetic field of a permanent magnet. The external magnetic field in the experiment was homogeneous. By moving the permanent magnet, the external field changed its direction. Here, they have not yet made the transition to soft materials or attaching the two squares together by their sides instead of the wired connection.

Erb, et al. [43] also developed a type of magnetic self-shaped soft materials. In the study, they made a material that had a magnetization pattern existing out of two regions. They used a static permanent magnet during the curing process. They proposed a robust and universal technique to replicate this unusual shape-changing mechanism of original objects in an artificial bioinspired composite. They made it by first placing the uncured material in one mold and curing it under a permanent magnetic field. After that, more resin was added and cured under a permanent magnetic field in a different direction. Then, with prepared materials, they made bio-inspired structures. Hydrogel substrate containing aluminum oxide layers coated in Fe_3O_4 particles was added to a Teflon mold above which a permanent magnet was swiveled. The mold was cured with the magnetic field. To make a multi-layer structure, a second hydrogel substrate was merged to the mold and another orientation of magnetic field was imposed. The second layer of the solution was cured at a different orientation, after which the bilayer was extracted from the mold, Figure 7.

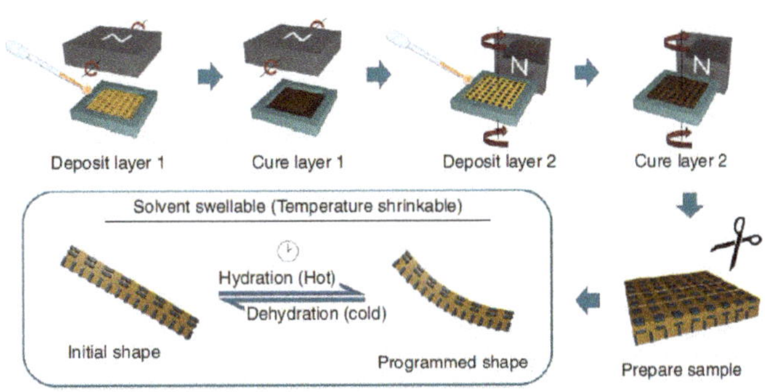

Figure 7. The production process of the magnetic soft material [43].

Lum, et al. [44] were one of the first to develop a computer model to predict the shape memory behavior of MSMs. However, the model is not able to calculate all the shapes of the material over time and is restricted to only being able to define the shape with a 1D Fourier series. In addition, a fabrication process for the magnetic soft material was also presented, as shown in Figure 8. First, a beam of an elastomer (Ecoflex 10-00) with aluminum (non-magnetic; it had the same size as the NdFeB particles so that the elastomer had the same mechanical properties) particles was cast (Figure 8A,B). Depending on the preferred magnitude of shape change, a pater was laser cut out of this strip and filled with

Ecofelx 10-00 with incorporated NdFeB (magnetic) particles (Figure 8C,D). After it was cured, the beam of magnetic soft material was folded in a jig (Figure 8E) and magnetized with a strong B field (~1 T) to get the right magnetization profile (Figure 8(Fi,Fii)). The external magnetic field was induced by an electromagnet.

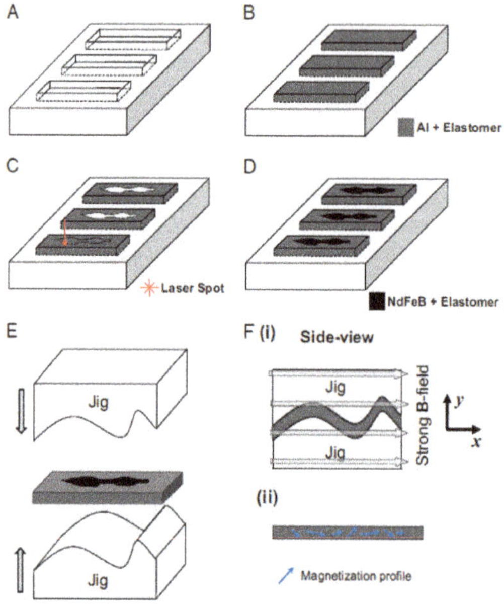

Figure 8. The manufacturing process of magnetic soft materials in the study by Lum, Ye, Dong, Marvi, Erin, Hu and Sitti [44]. (**A**) a negative mold; (**B**) the passive component; (**C**) laser cutting; (**D**) NdFeB and Elastomer poured and cured in the mold (**E**) the beam bent into the jig; (**Fi**) the beam is magnetized with a strong B field (~1 T); (**Fii**) the beam is removed from the jig.

Kim et al. [15] reported one of the first 3D printing magnetic responsive materials by incorporating magnetic microparticles (NdFeB) in a soft compound and printing non-uniform magnetization patterns. This was previously done by embedding whole magnets into a soft compound. Their printing method was based on direct ink writing. During the printing procedure, a static magnetic field was applied at the nozzle to reorient the particles along that field (see Figure 9a) generated by either a permanent magnet or an electromagnet. The magnetic shield between the electromagnet and the printed material weakened the magnetic field from the electromagnet, so that the printed magnetic patterns were not disturbed by it. This ink with its created magnetic field directions could then be used to print complex structures, Figure 9b, which enabled structures that were not possible to make before. For the shape change activation, a permanent magnet was used. The magnetic responsive soft material could react fast, untethered, and in a revered way. They found that with the increase of the applied field at the nozzle tip, the magnetic moment density of produced samples through nozzles improved. Furthermore, the effect of nozzles' diameters on the magnetic moment density was investigated with their custom-made set-up.

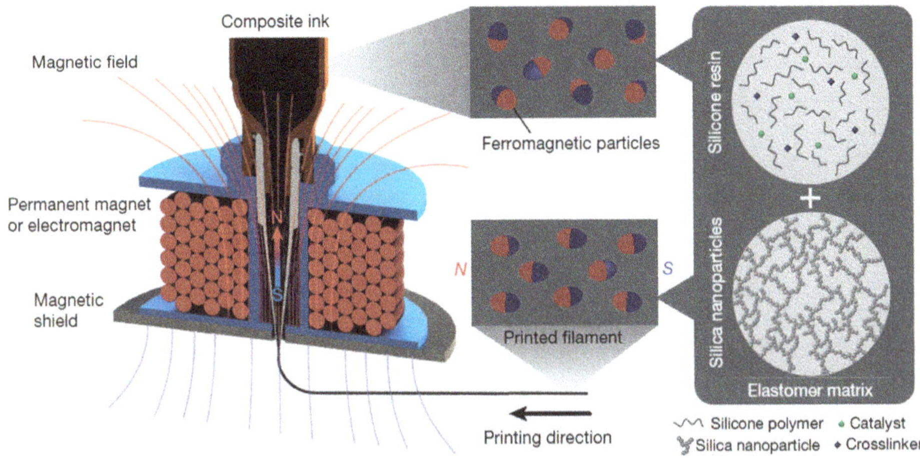

(**a**)

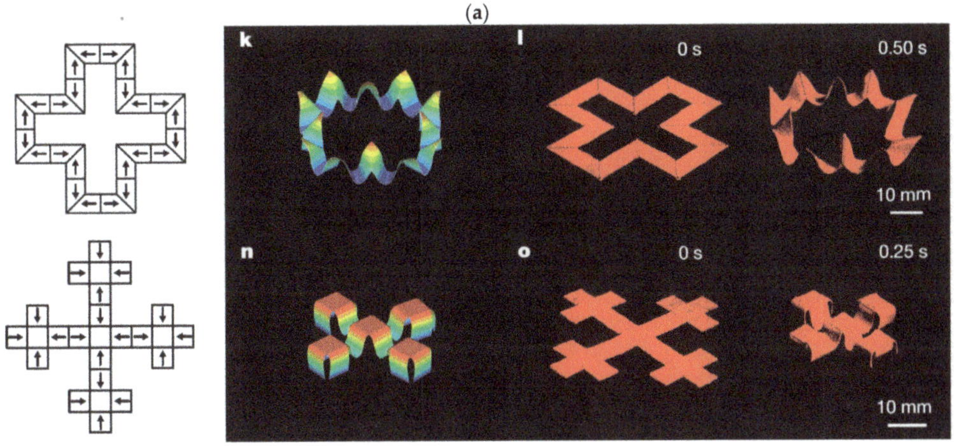

(**b**)

Figure 9. (**a**): The manufacturing process of magnetic soft material developed by Kim, Yuk, Zhao, Chester and Zhao [15] by which the magnetic particles oriented with the magnetic field that was applied at the nozzle. (**b**): Complex structures printed out of magnetic soft material [15].

Instead of using direct ink writing with a 3D printer to create non-uniform magnetization patters, Xu, et al. [45] developed another system based on ultraviolet lithography. Like the 3D printer set up by Kim, Yuk, Zhao, Chester and Zhao [15], this ultraviolet lithography setup is also able to make more complex non-uniform magnetization patterns. The permanent magnet under the plate on which the UV resin with incorporated magnetic particles (NdFeB) lays rotates and changes the orientation of the particles in the uncured resin. When the hall sensors detect that the particles are oriented in the right position, the part of the UV resin that should have this direction for the particles is cured by the exposure to precise target UV light. The design concept for a microgripper magnetically actuated with six-degree-of-freedom can be seen in Figure 10. The six degrees of freedom actuator refers to movement in three dimensions added to grasping, pitch, and yaw motion. The microgripper includes a base segment and multiple arms to encapsulate the cargo. Once the arms are wrinkled, the gripper could be considered as a rigid body and could locomote using the magnetic gradient pulling force or rolling motion (Figure 10A). The shifting steps of a green triangular prism-shaped with the designed microgripper is shown in Figure 10B.

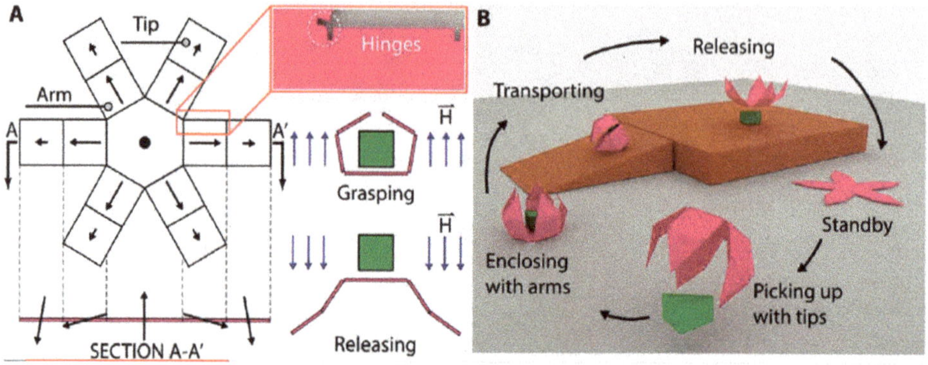

Figure 10. (**A**) Geometry and working mechanism of microgripper; (**B**) the cargo transportation task [45].

Roh, et al. [46] also made structures with MSMs. The 3D printed objects could float on water. In one of the important experiments in this study, structures were printed with varying stiffness. The relation between deformations and material stiffness was reported by Roh et al. This varying stiffness showed different deformations in the activated material through a static magnetic field from an electromagnet. In places where the material was softer, the deformation was bigger and larger, while in the stiffer regions, there was less deformation.

Another study that used varying stiffness in the elastomers was performed by Lantean, et al. [47]. They employed digital light processing to print programmable magnet polymeric materials with acceptable magnetic and mechanical properties. To test their material, a permanent NdFeB magnet was used and moved manually. They varied of the mechanical properties from soft to hard by combination of urethane-acrylate resins and butyl acrylate as a diluent medium. This made it possible to create straight surfaces in the cube. They also changed the magnetization strength of parts of the created objects by changing the weight percentage of the added magnetic particles Fe_3O_4 between 0% and 6%.

Moreover, soft magnetic objects were fabricated to mimic animal movements by Venkiteswaran et al. [48]. Isotropic powder was used, made from praseodymium-iron-boron (PrFeB) magnets and a silicone rubber polymer (1:1 mass ratio). The fabricated soft magnetic strips were made in different formations to show different types of motions that are possible to create. To create the magnetic soft material strips, a magnetic field of 1T of an electromagnet was used. The motion of the robots was activated by static magnetic fields induced by six electromagnets with a magnetic field strength from 10 to 35 mT. By switching different pairs of coils on and off, a rotating external magnetic field was created. The robots with a length of 4 mm had a speed between 0.15 and 0.37 mm/s. Their interesting outcomes showed that bio-mimetic characteristic patterns were followed by all the specimens. Reliable movement of the specimens, from one end of the workspace to the other, is other exciting achievement [48].

Wu, et al. [49] developed different NdFeB particles types of joints for MSM. These joints are symmetric joints and asymmetric joints. The difference between them is a gap between the same two magnetic poles that lay next to each other. The difference of the gap makes the shape transformations of the material, round or angular. Combinations of these types of joints can enable more different types of shape transformations (see Figure 11). The shape transformations were activated by a homogeneous magnetic field induced by electromagnetic coils. The field strengths and directions were changed to obtain the intended shape change during the study. The S–S–S combination (either a "W" or a "M" configuration) was achieved with folding under the changing magnetic fields with amplitude B of 4 mT (Figure 11a,d,e). For the S–A–S combination (Figure 11b,f,g), an "M" mode with B = 6 mT and an arc mode with B = −23 mT was obtained. These required differences in magnetic fields were due to the higher bending stiffness as compared to the folding stiffness. The same trend can be seen in the A-S-A combination (Figure 11c,h,i).

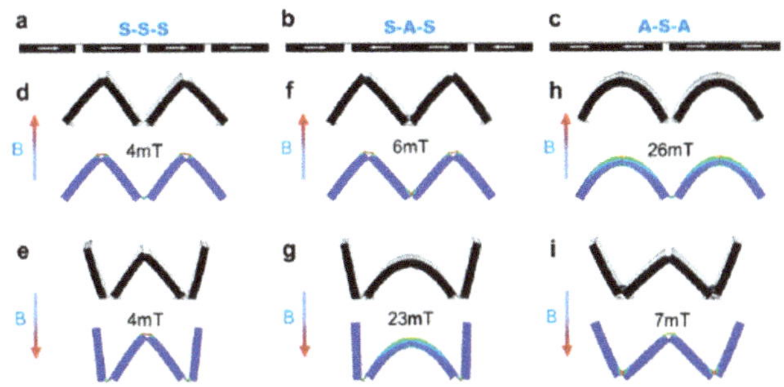

Figure 11. A combination of symmetric (S) and asymmetric (A) joints [49]. (**a**) S−S−S (**b**) S−A−S (**c**) A−S−A. (**d–i**) Simulated deformations for the joint types in (**a–c**) with positive (**d**,**f**,**h**) and negative (**e**,**g**,**i**) magnetic field.

Ma [16] developed a magnetic multi-material printing technique to combine the MSMP and the MSM into one object. The printing was performed with two printing heads under a static magnetic field induced by a permanent magnet for different materials. MSMPs and MSMs could be structurally integrated into the design by adjusting the on-off switching among the syringes and the printing directions. The magnetization of the printed substrate and the direction of the printing ink can be seen in Figure 12a. The created multi-material was able to transform in multiple shapes through the different activation methods of the different materials. They used a one-dimensional MSM/MSMP stripe to investigate the mechanical behavior and four possible deformation modes (Figure 12b,c). They found that only the MSM deformed under actuation magnetic fields at room temperature and the MSMP stayed without deformation due to the higher stiffness (mode 1-U shape). At the elevated temperature (above room temperature), both MSM and MSMP were deformed (Mode 2-W shape). Subsequently, by keeping the magnetic field and cooling the stripe to below room temperature, the MSMP part was locked into the deformed shape whereas the MSM part changed to its original shape once the magnetic field was eliminated (Mode 3 with a peak in the center). In the end, they applied a reversed magnetic field and obtained a wavy configuration (mode 4). The shape memory effect of the MSMP was activated by the induction heating of the incorporated magnetic particles (Fe_3O_4) by an AC magnetic field. The MSMs, on the other hand, transformed to their programmed shape through the alignment of NdFeB magnetic particles with a specific external magnetic field.

A magnetic shape memory and stiffening composite elastomer combined material was made by Testa, et al. [50]. The elastomer base incorporated carbonyl iron particles. When the material was placed in a uniform magnetic field from a permanent magnet, the particles aligned with the magnetic field and stiffened the material up to almost 30-fold so that the magnetic shape memory composite elastomer could carry a weight under the influence of a magnetic field but was not capable of doing this without the magnetic field.

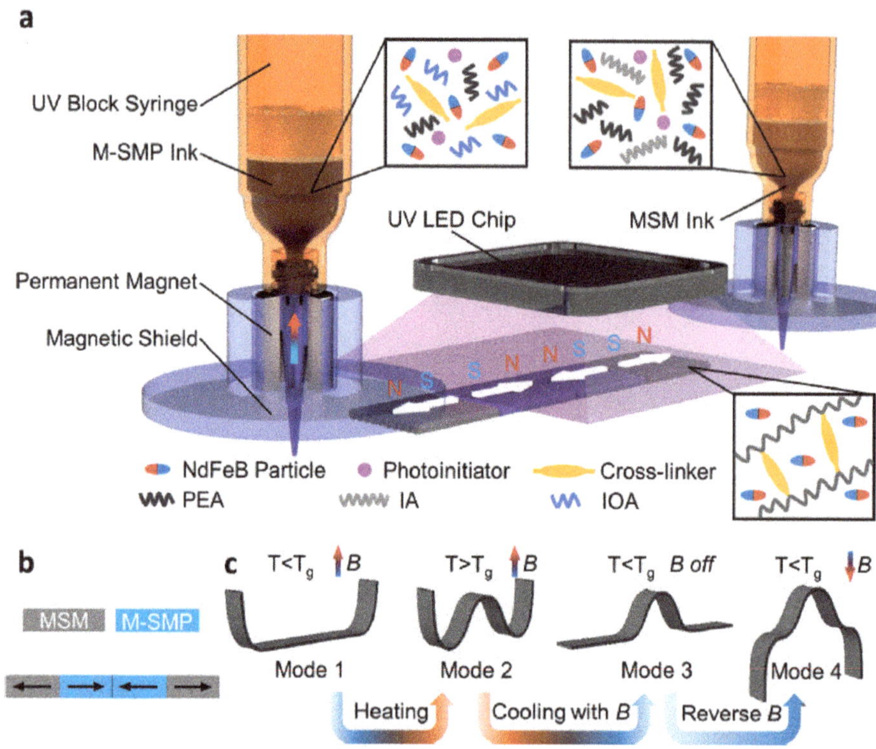

Figure 12. (**a**) The multi-material printing setup with two printer heads for the two different materials. The permanent magnet around the nozzle orients the magnetic particles while printing. (**b**) The pattern printed with the different materials and the directions of the magnetic fields [16].

Another interesting study on composite MSMs was done by Scheerbaum, et al. [51]. They made a polymer (polyester) with incorporated Ni-Mn-Ga particles. The magnetic particles were aligned by being cured under the influence of a magnetic field from a permanent magnet. They found that the perfect orientation of the MSMs was obtained once the magnetic field was applied during the curing of the substrate. They also found that the compressed MSMs magnetized more easily once the magnetic particle direction was parallel in comparison to the perpendicular direction. One of the applications for this material could be magnetic field-controlled dampers. It was also suggested to make this composite with a softer matrix to make actuators of the material.

A good example of hard magnets that have the same working principles as MSMs can be seen in a study by Alharbi, et al. [52]. The square-shaped magnets had their magnetic poles mirrored to the magnet next to them. This formed a zig-zag shape when a DC magnetic field generated by a neodymium permanent magnet was applied over the magnets. In their experimental study, they embedded NdFeB in a soft silicone base and coated e-thread antennas on it. They achieved a stable magnetic property as well as good programmable polarity. They performed mechanical and thermal testing to confirm their results. Various volume percentages of NdFeB particles (5 to 30 vol%) were examined, the results showing a tiny reduction in flexibility due to increasing magnetic particles volume fraction.

In addition, Qi, et al. [53] presented a method to design biomimetic applications from magnetic responsive soft material. This method was developed to design the preferred programmed shape fast. They used inspiration from nature and started designing from the preferred programmed shape and the direction of the magnetic field. From there on, they determined the direction of the magnetic fields that were incorporated in the silicone. The results showed that the soft materials 3D printed filament had reversible, fast, and acceptable deformation properties. They employed the Fused Deposition Modeling (FDM) technique to print the oriented magnetic structural elements and encapsulated mold. To activate the shape change, both a permanent magnet controlled by a linear motor and an electromagnet were used. By the movement of the permanent magnet and the current change in the electromagnet, the magnetic flux density was adjusted to make the material move.

4. MSMP Composite

In this section, some studies are mentioned that show a combination of MSMPs and MSMs by means of their activation principles. These materials are also very interesting but could not be placed in the two previous sections. These combined structures could not be described in the above-mentioned sections due to their specific characterizations.

Ze, et al. [54] presented a material that can have multiple shape manipulations, which can also be locked in place. To do this, they used an amorphous shape memory matrix with two types of incorporated magnetic microparticles (see Figure 13). For the shape change, the matrix was first softened by the inductive heating of Fe_3O_4 particles by an AC field. When the material was soft, the shape could be manipulated by applying a DC magnetic field generated by an electromagnet on which the NdFeB particles moved to align their magnetic fields. These fields had a nonuniform magnetization pattern through which the shape was activated. The stripe could not deform under an actuation magnetic field (Ba) at room temperature because of stiffness of the ink (Figure 13a). Once a high AC magnetic field (B_h) was added, the induction heating in a magnetic particle made the MSMP composite hot (above room temperature) and dropped the modulus of the MSMP noticeably. Subsequently, a tiny B_a could simply stop the stripe (Figure 13b). When the inductive heating was removed, the polymer cooled down and locked its shape in place (Figure 13c). To allow a different actuation, the magnetization pattern of the MSMP can be reconfigured (Figure 13d). The unique advantages of this material are the shape locking abilities without the constant need for a magnetic field.

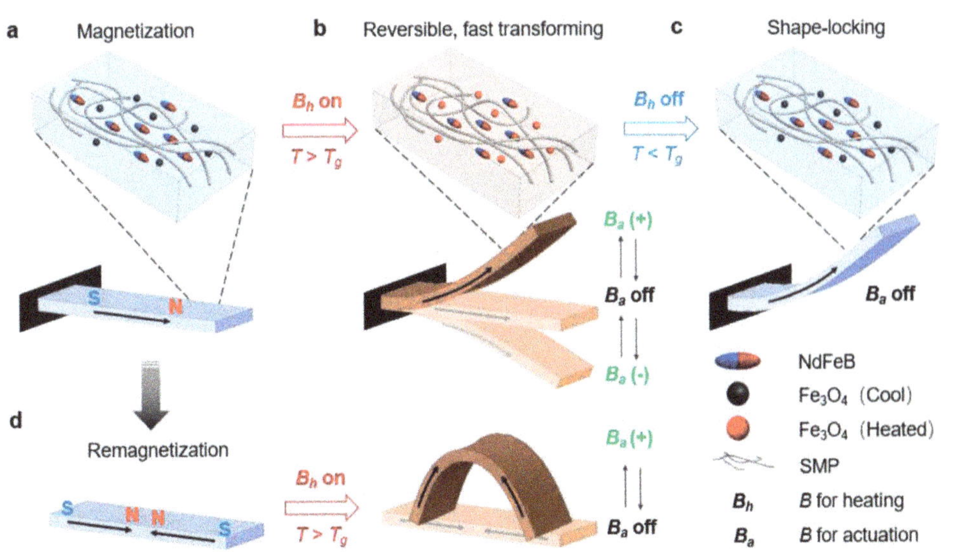

Figure 13. Magnetic shape memory polymer composite. (**a**) Before actuation. (**b**) After induction heating. (**c**) Shape locking of the MSMP composite [54].

Zhang, et al. [55] developed a combination between shape memory materials and MSMs by incorporating magnetic particles laying in magnetization fields in a liquid crystal elastomer. First, the elastomer had the shape of half a circle which could walk with the help of an alternating magnetic field. When the sample reached the heated liquid (around 70 °C), the shape of the elastomer transformed in a helix and the material could use this shape to swim while also being exposed to a magnetic field. The combination of these two materials could allow for even more complex movements and shape transformations. The magnetic field for the movement was a rotating field made from a Halbach array mounted on a stepper motor.

A performance metric of the magnetic particles and variety of them for both MSMP and MSM application has been summarized in Tables 1 and 2. The most-used particle for MSMP application is Fe_3O_4 and for MSM is NdFeB, based on most research reports.

Table 1. Summarized performance metrics of the magnetic particles.

Parameter Magnetic Particles	MSMPs	References	MSM	References
Kind of particle that is used the most often	Fe_3O_4	[13,16,17,21,22,24–31]	NdFeB	[19,45,46,50,53,56–58]
Size	Smaller size and larger volume fraction for faster shape recovery. The largest-sized sponged microparticles are more effective for induction heating. Mostly 10 vol% or wt% particles in SMP without influencing mechanical properties.	[29] [39] [25,34] [40]	Mostly used ≈5–10 μm	[19,45,46,49,58]
Hard or soft magnetic particles			Hard magnetic particles for complex shape change and remembering complex transformation shapes. Soft magnetic particles but only for elongation or compression.	[46,49,58,59] [46,58]
Curie temperature	Fe_3O_4 500–600 °C		PrFeB 345 °C	[49]
Coercivity	10 wt% nanomagnetite 55.5 A cm^{-1}	[40]	80 wt% NdFeB particles (10 μm) 7.9 (10^3 Oe) PrFeB 520-600 kA/m	[58] [49]
Saturation magnetization	MagSilica (50 and 50-H8) 22–32 A m^2 kg^{-1} MagSilica (50–85) 44–56 A m^2 kg^{-1} 10 wt% nanomagnetite 3.7 mT	[25] [25] [40]	80 wt% NdFeB particles (10 μm) 87.3 (emu/g) PrFeB 1600 kA/m	[58] [49]
Nature of field	AC	[13,16,17,27,28]	DC	[53,60]

Table 2. Variety of Magnetic Particles.

Magnetic Particles in MSMPs	References	Magnetic Particles in MSMs	References
Fe_3O_4	[13,17,19–21,23–30]	NdFeB	[15,44,45,49,52,56]
NdFeB	[31]	Fe_3O_4	[46,47]
Carbonyl iron	[32]	Carbonyl iron	[50,53]
Nickel zinc ferrite	[33,34]	Strontium ferrite powder	[57]
Ni-Mn-Ga	[35]	Permalloy	[41]
Magnetite	[36,37]	Superparamagnetic colloidal nanocrystalclusters	[42]
Iron	[38]	Al_2O_3 platelets coated in iron oxide	[43]
Iron oxide	[57]	PrFeB	[48]
		Ni-Mn-Ga	[51]

5. Modeling

The available modeling studies of MSMPs and MSMs are very limited. This section will go further into studies that developed computer models to predict the shape memory effect. It is divided into two sub-sections. One focuses on the computer models developed for MSMP, the other is about the computer models developed for MSMs.

5.1. Modeling of MSMPs

Further development of MSMPs was done by creating models to predict real-life behavior [22]. The material was composed of thermoplastic polyurethane with different volume fractions of Fe_3O_4 particles (0–40 vol% magnetite particles) and fabricated through injection molding. In this research, they compared the measured values of the samples with multiple theoretical models, such as Rayleigh, Lewis–Nielsen, Hashin–Shtrikman, and Agri-Uno correlations for thermal conductivity, to see if there was a correlation. They found good agreement between experimental and theoretical data. However, they did not develop a new computational model and used available theoretical correlation to evaluate their results.

Conti, et al. [58] also reported on modeling and simulating the behavior of MSMP composites. They studied how the magnetic field changed the shape of the composites. Their results were based on macroscopic material properties. In the conclusion, it was pointed out that the orientation of the magnetic particles was more important than the alignment. The shape of the particles was also important and elongated particles had better properties than the round-shaped particles. Finally, a softer polymer gave better results than the incorporated magnetic particles. They developed a numerical method based on the boundary-element method to solve the model for two dimensions and with a linear geometrical matrix.

Moreover, in the study by Yu, Westbrook, Kao, Leng and Qi [28], a model was developed to further predict how the round magnetic particles influence the shape change effect of MSMPs. A finite element model that was previously used to predict the shape change effect of shape memory polymers was altered to accommodate this. They investigated several parameters including particle size, the volume of the magnetic particle fraction, the heating of the particles, and the speed of the shape recovery process. They found that smaller particle sizes or more volume fractions could bring faster shape recovery rate. In addition, they pointed out that there were critical points for each parameter.

5.2. Modeling of MSM

In the study by Crivaro, et al. [59], a finite element analysis model was developed to predict the required magnetic field strength to activate magnetic soft material. The finite element analysis model was also developed to determine if the made mechanism was bistable. They used magneto active elastomer (MAE) patches bonded to a polydimethylsiloxane (PDMS) substrate that was activated by a DC magnetic field generated by an electromagnet. The magnetic soft material that they used was made out of a 70 vol% Dow Corning HS II RTV silicone rubber compound. The catalyst was added in a 20:1 weight ratio. The other 30% of the volume consisted of the magnetic particles (325 mesh M-type barium hexaferrite). Their experimental outcome validated the proposed model, although they observed some issues with the fabrication procedure to enhance the design of the mechanisms involved in making an accurate prediction field to actuate the device.

In the study by Lum, Ye, Dong, Marvi, Erin, Hu and Sitti [44], a computer model was developed to predict the behavior of MSMs. The formulae they used were based on the deflection of a beam. Restrictions of their model that were pointed out were that the model does not show the shape transformation of the beam of magnetic soft material over time, but only its end state of the shape recovery. The shape transformation was activated by an electromagnet.

In the research by Kim, Yuk, Zhao, Chester and Zhao [15], aside from developing the 3D printer system for MSMs, they also developed a computer model to predict the shape changes of complex structures. The model was based on the torques created in the material by the embedded magnetic particles when the material was placed in an external magnetic field and how those torques created stresses in the elastomer surrounding them. The shape change of the material was activated by the use of a manually moved permanent magnet.

Zhao, et al. [60] continued the studies of Kim, Yuk, Zhao, Chester and Zhao [15]. They developed a constitutive model in a finite-element framework to design and predict the behavior of MSMs. They used permanent magnets and a pair of electromagnets for shape activation. With this model, complex structures can be designed out of MSMs which have almost the same shape transformations as their real-life objects (see Figure 14). They adopted a nonlinear field theory to explain the finite deformation coupled with magnetic fields. They developed an eight-node continuum brick element to explore the induced deformation magnetically in MSMs. They investigated three different scenarios: (i) extreme bending up to 180°, (ii) large bending up to 90°, and (iii) small-deflection beam bending. They remarked that the good effective media for a magnetically shape response material could also be a "non-classical" media such as polar media.

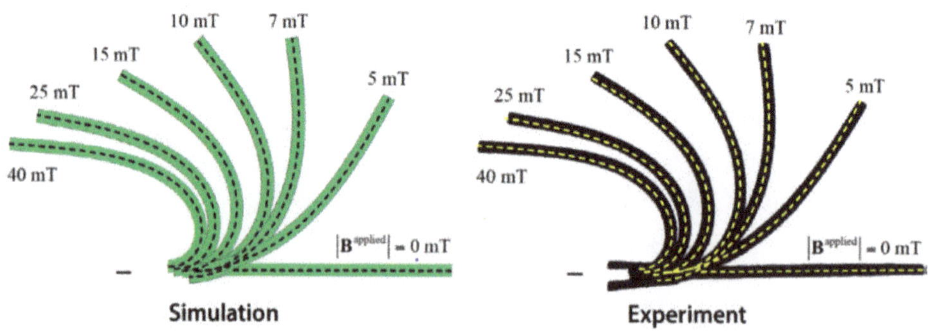

Figure 14. Validation of the developed finite element analysis model [60].

Chen, et al. [61] also developed a computational model that can be used for the development of transformation and movements of MSMs that were activated by cylindrical permanent NdFeB magnets. This was based on the torque equilibrium and iteration in a computational model. The model was tested by developing physical prototypes that could be compared with the prediction of the shape transformation. A model was developed for static shapes as well as the locomotion of magnetic soft material objects.

The summaries of modeling of MSMPs and MSMs are given in Table 3.

Table 3. Summarized models of MSMP and MSM.

Polymer Materials	Magnetics Particle	Method	Remark/Type	References
Polyurethane	Fe_3O_4	Theoretical models	Thermal conductivity/good agreement experiment/theory	[22]
NA	NA	Boundary-element	Solve geometrical matrix	[58]
NA	NA	Finite element model	Effect of particle size, magnetic particle fraction	[28]
Polydimethylsiloxane/ silicone rubber	Barium hexaferrite	Boundary element method	Energy descent algorithm developed/behavior of the macroscopic material	[59]
Ecoflex 00-10	NdFeB	Theoretical model/2D Fourier series	Magnetization profile/time-varying shapes	[44]
SE 1700 and Ecoflex Part B	NdFeB	Finite-element analysis	Validate experimental results	[15]
Polydimethylsiloxane (PDMS)	NdFeB	Nonlinear field theory/developed a constitutive model	Macroscopic behavior of materials	[60]
Polydimethylsiloxane (PDMS)	NdFeB	Iterative computational model	Precise prediction of magnetics soft martial transformations	[61]

6. Application

This section emphasizes various applications of MSMPs and MSMs and is divided into two parts. One of the most important advantages of these materials is that they can be remotely activated and have no or almost no surface heat. Differences between the two are that MSMPs are probably stiffer than magnetic soft material, do not change their shape as fast, and transform only one way. MSMs, on the other hand, are less stiff, can transform fast, and are reversible. These differences can also be seen in their applications.

In MSMPs, the active memory material is the polymer, which has a typical response time of tens of seconds [23]. In MSMs, the memory effect is induced by magnetic particles, which have a much faster response time (<1 s). In the research by Kim et al. [15], it could be seen that the shape change was completed within a fraction of a second (0.50 s and 0.25 s).

Additionally, in MSMPs, the magnetic particles need some time to increase the temperature of the shape memory polymer. The time it takes to reach this temperature depends on the type of shape memory polymer [1] and multiple particle parameters such as their size and volume fraction in the material [28]. That is why there is some delay in the activation of the shape memory effect. The research by Mohr et al. [19] (Figure 2) showed that the shape change of the MSMP under an alternating magnetic field took 22 s. In Figure 3, the shape changes took 130 s [23].

6.1. MSMP Applications

Applications for MSMPs are mostly thought of in the biomedical domain. Advantages of the MSMPs over the shape memory polymers in this domain are that the shape change can be remotely activated by an AC field and that the environment does not have to heat up. Zhao, Huang, Liu, Wang, Leng and Liu [30] also mentioned that shape memory polymers could be used for minimally invasive surgeries, because of being able to compress the shape before implementing or using it inside a body. Even though the environment does not have to heat up for the activation of MSMPs, the material itself should. Zhang, Wang, Zheng, Liu and Leng [13] developed a MSMP that had a uniform surface temperature of 40 °C during the activation of the shape change. With this temperature that is close to the body temperature, it would not harm the body in which the material is used. They developed a bone support structure that would be brought into the body in a compressed state. When it was placed at the right location in the body, a magnetic field was applied to heat the particles and make the bone support structure expand. Bone support structures were also developed in the study by Zhao, Huang, Liu, Wang, Leng and Liu [30]. In the study by Buckley, McKinley, Wilson, Small, Bennett, Bearinger, McElfresh and Maitland [33], other biomedical devices were made that were activated by an alternating magnetic field, such as a flower-shaped endovascular thrombectomy device and an expandable MSMP foam device for aneurysm embolization.

6.2. MSM Applications

As described earlier, in the study by Khoo and Liu [41], they developed micropumps out of an elastomer with an incorporated magnet. The advantages of the magnetic soft material that were used are the fast and reversible shape transformation. These are unique characteristics of the magnetic soft material. Another unique thing about this application is that the pumps could be made very small. This is partially due to the reduction of components needed to make a pump. This reduction in product components is also one of the advantages of using MSMs. In the experiment, the pumps were activated by a manually controlled permanent magnet.

Grippers and other star-shaped objects are also a recurring application of MSMs. One of the grippers was made by Xu, Zhang, Salehizadeh, Onaizah and Diller [45] and rotated by a rotating field generated by an electromagnet. Another gripper that was made to catch a ball was developed by Kim, Yuk, Zhao, Chester and Zhao [15] and activated by a manually controlled permanent magnet. In the study by Ze, Kuang, Wu, Wong, Montgomery, Zhang, Kovitz, Yang, Qi and Zhao [54], they used two types of magnetic particles. First, Fe_3O_4 particles were used to inductively heat and soften up the polymer matrix, and second, the particles (NdFeB) were used to change the shape through the alignment with the particles with the external magnetic field generated by an electromagnet. They found out that without locking the shape (by removing the magnetic field that inductively heats the material), the force of the gripper was not enough to carry the ball. While locking the shape, the same material was able to carry it.

Ren, et al. [62] developed an untethered jellyfish-inspired soft milli swimmer. It was made from a magnetic soft material existing out of Ecoflex 10-00 and NdFeB particles with a size of 5 μm. The swimmer was activated by a magnetic field induced by a custom-made electromagnet coil system that, through a difference in magnitude, frequency, and direction changes, changed the swimming behavior. In this research, it was highlighted how the swimmer can perform different tasks, such as drug delivery and moving through small tubes.

Another gripper-like robot made of magnetic soft material was developed by Goudu, et al. [63]. This gripper was made out of a biodegradable matrix for the use of the gripper in a human body with incorporated iron oxide particles. The robot was also activated by the direction and frequencies of an external magnetic field as was the case by Ren, Wang, Hu and Sitti [62]. The magnetic field was generated by a custom-made electromagnetic coil setup. By enclosing the magnetic soft material structure around an object, rolling it can deliver the object to a specific place.

A frequent application for MSMs is a strip that is able to swim for the biomedical field [64–66]. These strips often have sinusoidal magnetic field patterns. The type of magnetic particles in all three studies was NdFeB, except for one which used NdPrFeb [64]. Even though the three studies had similar magnetization patterns, only the study by Hu, Lum, Mastrangeli and Sitti [65] also showed other ways of locomotion (e.g., walking, crawling, jumping). All the robots were activated by frequently changing and rotating magnetic fields generated by different types of DC coil setups. The type of locomotion was determined by the orientation of the external magnetic field and its strength.

The summarized application for MSMPs and MSMs are given in Table 4.

Table 4. Summarized applications for MSMPs and MSMs.

Polymer Materials	Magnetics Particle	Application/Function	Remark/Type	Refrences
PLA/Fe_3O_4	Fe_3O_4	Bio medical/bone support	MSMP	[13]
PLA	Fe_3O_4	Medical surgery	MSMP	[30]
Polyurethane	Nickel zinc ferrite	Medical device	MSMP	[33]
Polydimethylsiloxane (PDMS)	Ni80Fe20	Micropumps	MSM	[41]
UV resin (flexible type, GC3D-EBE)	NdFeB	Grippers	MSM	[45]
SE 1700 and Ecoflex Part b	NdFeB	Grippers	MSM	[15]
Acrylate mixed resin	Fe_3O_4 and NdFeB	Grippers	MSM	[54]
Ecoflex 10-00	NdFeB	Jellyfish-inspired milliswimmer	MSM	[62]
Gelatin methacryloyl/lithium phenyl	Fe_3O_4	Milli-grippers	MSM	[63]
Ecoflex 00-50	NdPrFeB	Biomedical-swimming robot	MSM	[64]
Ecoflex 00-10	NdFeB	Biomedical-swimming robot	MSM	[65]
Ecoflex 00-50	NdFeB	Biomedical-swimming robot	MSM	[66]

7. Summary and Conclusions

In the last twenty years, MSMPs and MSMs have been developed to serve the purposes of a number of specific design applications. This review reports on the research of these two materials. Parameters that have been studied are their material composition, shape recovery, and shape recovery simulation, with specific emphasis on manufacturing methods and applications. The magnetic shape memory effect activation makes it able to activate shape change remotely and with (almost) no surface heat. This makes shape memory materials useful for more applications than the heat-activated shape memory polymers by conduction or convection. MSMPs often have only one remembered shape and no reversible shape change activated by the magnetic field. It can also take some time to activate the shape change. However, these drawbacks can be avoided by MSMs which offer fast and reversible shape changes. In comparison to MSMPs, MSMs have not yet been investigated to the same level of detail. As MSMs are of a more complicated nature (both in production and application), they pose a major challenge for the future of these materials. In addition, 3D printing is a promising manufacturing technique for both MSMPs and MSMs, although there is not yet sufficient knowledge available about the optimum printing parameters.

In addition, the effect of the magnetic properties of the magnetic components (Curie temperature, coercivity, remanence) on the shape memory function of these materials is as yet largely unknown and needs to be the subject of future research.

Although several successful applications have already been reported (predominantly in the biomedical field), it can be concluded that magnetic shape memory alloys (both MSMs and MSMPs) are merely at the beginning of their developing stage and still need to find their way to large-scale market applications.

Author Contributions: Conceptualization, S.J.M.v.V., H.Y. and S.G.; writing—original draft preparation, S.J.M.v.V. and H.Y.; writing—review and editing, S.G.; supervision S.G. All authors have read and agreed to the published version of the manuscript.

Funding: This research was supported by an internal funding from Delft University of Technology (TU Delft).

Institutional Review Board Statement: Not applicable.

Informed Consent Statement: Not applicable.

Data Availability Statement: Not applicable.

Conflicts of Interest: The authors declare no conflict of interest.

References

1. Bengisu, M.; Ferrara, M. *Materials that Move Smart Materials, Intelligent Design*; Springer: Cham, Switzerland, 2018.
2. Lendleind, A.; Kelch, S. Shape-Memory Polymers. *Angew. Chem. Int. Ed.* **2002**, *41*, 2034–2057. [CrossRef]
3. Pandey, A.; Singh, C.; Singh, S.; Jha, K.; Prakash, C. 3D printed biodegradable functional temperature-stimuli shape memory polymer for customized scaffoldings. *J. Mech. Behav. Biomed. Mater.* **2020**, *108*, 103781. [CrossRef]
4. Lendlein, A.; Jiang, H.; Jünger, O.; Langer, R. Light-induced shape-memory polymers. *Nature* **2005**, *434*, 879–882. [CrossRef] [PubMed]
5. Yu, Y.; Ikeda, T. Photodeformable Polymers: A New kind of Promising Smart Material for Micro and Nano-Applications. *Macromol. Chem. Phys.* **2005**, *206*, 1705–1708. [CrossRef]
6. Li, Y.; Chen, H.; Liu, D.; Wang, W.; Liu, Y.; Zhou, S. pH-Responsive Shape Memory Poly(ethylene glycol)−Poly(εcaprolactone)-based Polyurethane/Cellulose Nanocrystals Nanocomposite. *Appl. Mater. Interfaces* **2015**, *7*, 12988–12999. [CrossRef] [PubMed]
7. Han, X.-J.; Dong, X.-Q.; Fan, M.-M.; Liu, Y.; Li, J.-H.; Wang, Y.-F.; Yuan, X.-J.; Li, B.-J.; Zhang, S. pH-Induced Shape-Memory Polymers. *Macromol. Rapid Commun.* **2012**, *33*, 1055–1060. [CrossRef] [PubMed]
8. Chen, H.; Li, Y.; Liu, Y.; Gong, T.; Wang, L.; Zhou, S. Highly pH-sensitive polyurethane exhibiting shape memory and drug release. *Polym. Chem.* **2014**, *5*, 5168–5174. [CrossRef]
9. Song, Q.; Chen, H.; Zhou, S.; Zhao, K.; Wang, B.; Hu, P. Thermo- and pH-sensitive shape memory polyurethane containing carboxyl groups. *Polym. Chem.* **2016**, *7*, 1739–1746. [CrossRef]
10. Courty, S.; Mine, J.; Tajbakhsh, A.R.; Terentjev, E.M. Nematic elastomers with aligned carbon nanotubes: New electromechanical actuators. *Europhys. Lett.* **2003**, *65*, 654. [CrossRef]
11. Cho, J.W.; Kim, J.W.; Jung, Y.C.; Goo, N.S. Electroactive Shape-Memory Polyurethane Composites Incorporating Carbon Nanotubes. *Macromol. Rapid Commun.* **2005**, *26*, 412–416. [CrossRef]
12. Lehmann, W.; Skupin, H.; Tolksdorf, C.; Gebhard, E.; Zentel, R.; Krüger, P.; Lösche, M.; Kremer, F. Giant lateral electrostriction in ferroelectric liquid-crystalline elastomers. *Nature* **2001**, *410*, 447–450. [CrossRef] [PubMed]
13. Zhang, F.; Wang, L.; Zheng, Z.; Liu, Y.; Leng, J. Magnetic programming of 4D printed shape memory composite structures. *Compos. Part A Appl. Sci. Manuf.* **2019**, *125*. [CrossRef]
14. Chaudhary, R.; Chaudhary, V.; Ramanujan, R.V.; Steele, T.W.J. Magnetocuring of temperature failsafe epoxy adhesives. *Appl. Mater. Today* **2020**, *21*. [CrossRef]
15. Kim, Y.; Yuk, H.; Zhao, R.; Chester, S.A.; Zhao, X. Printing ferromagnetic domains for untethered fast-transforming soft materials. *Nature* **2018**, *558*, 274–279. [CrossRef]
16. Ma, C.; Wu, S.; Ze, Q.; Kuang, X.; Zhang, R.; Qi, H.J.; Zhao, R. Magnetic Multimaterial Printing for Multimodal Shape Transformation with Tunable Properties and Shiftable Mechanical Behaviors. *Appl. Mater. Interfaces* **2020**, *13*, 12639–12648. [CrossRef] [PubMed]
17. Schmidt, A.M. Electromagnetic Activation of Shape Memory Polymer Networks Containing Magnetic Nanoparticles. *Macromol. Rapid Commun.* **2006**, *27*, 1168–1172. [CrossRef]
18. Massart, R.; Cabuil, V. Synthèse en milieu alcalin de magnétite colloïdale: Contrôle du rendement et de la taille des particules. *J. De Chim. Phys.* **1987**, *84*, 967–973. [CrossRef]
19. Mohr, R.; Kratz, K.; Weigel, T.; Lucka-Gabor, M.; Moneke, M.; Lendlein, A. Initiation of shape-memory effect by inductive heating of magnetic nanoparticles in thermoplastic polymers. *Proc. Natl. Acad. Sci. USA* **2006**, *103*, 3540–3545. [CrossRef]
20. Leng, J.S.; Lan, X.; Liu, Y.J.; Du, S.Y.; Huang, W.M.; Liu, N.; Phee, S.J.; Yuan, Q. Electrical conductivity of thermoresponsive shape-memory polymer with embedded micron sized Ni powder chains. *Appl. Phys. Lett.* **2008**, *92*, 014104. [CrossRef]
21. Yakacki, C.M.; Satakar, N.S.; Call, K.; Likos, R.; Hilt, J.Z. Shape-Memory Polymer Networks with Fe_3O_4 Nanoparticles for Remote Activation. *J. Appl. Polym. Sci.* **2009**, *112*, 3166–3176. [CrossRef]
22. Razzaq, M.Y.; Anhalt, M.; Frormann, L.; Weidenfeller, B. Thermal, electrical and magnetic studies of magnetite filled polyurethane shape memory polymers. *Mater. Sci. Eng.* **2007**, *444*, 227–235. [CrossRef]
23. Yu, X.; Zhou, S.; Zheng, X.; Guo, T.; Xiao, Y.; Song, B. A biodegradable shape-memory nanocomposite with excellent magnetism sensitivity. *Nanotechnology* **2009**, *20*, 235702. [CrossRef] [PubMed]
24. Weigel, T.; Mohr, R.; Lendlein, A. Investigation of parameters to achieve temperatures required to initiate the shape-memory effect of magnetic nanocomposites by inductive heating. *Smart Mater. Struct.* **2009**, *18*, 025011. [CrossRef]
25. Kumar, U.N.; Kratz, K.; Wagermaier, W.; Behl, M.; Lendlein, A. Non-contact actuation of triple-shape effect in multiphase polymer network nanocomposites in alternating magnetic field. *J. Mater. Chem.* **2010**, *20*, 3404–3415. [CrossRef]
26. Yang, D.; Huang, W.; He, X.; Xie, M. Electromagnetic activation of a shape memory copolymer matrix incorporating ferromagnetic nanoparticles. *Polym. Int.* **2012**, *61*, 38–42. [CrossRef]
27. Cai, Y.; Jian, J.-S.; Zheng, B.; Xie, M.-R. Synthesis and properties of magnetic sensitive shape memory Fe_3O_4/poly(ε-caprolactone)-polyurethane nanocomposites. *J. Appl. Polym. Sci.* **2012**, *12*, 49–56. [CrossRef]
28. Yu, K.; Westbrook, K.K.; Kao, P.H.; Leng, J.; Qi, H.J. Design considerations for shape memory polymer composites with magnetic particles. *J. Compos. Mater.* **2012**, *47*, 51–63. [CrossRef]
29. Liu, C.; Huang, J.; Yuan, D.; Chen, Y. Design of a High-Strength XSBR/Fe_3O_4/ZDMA Shape-Memory Composite with Dual Response. *Ind. Eng. Chem. Res.* **2018**, *57*, 14527–14534. [CrossRef]

30. Zhao, W.; Huang, Z.; Liu, L.; Wang, W.; Leng, J.; Liu, Y. Porous bone tissue scaffold concepts based on shape memory PLA/Fe_3O_4. *Compos. Sci. Technol.* **2021**, *203*, 108563. [CrossRef]
31. Golbang, A.; Kokabi, M. Magnetic Field Actuation of Shape Memory Nanocomposites. *Adv. Mater. Res.* **2010**, *123–125*, 999–1002. [CrossRef]
32. Hassan, R.U.; Jo, S.; Seok, J. Fabrication of a functionally graded and magnetically responsive shape memory polymer using a 3D printing technique and its characterization. *J. Appl. Polym. Sci.* **2018**, *135*, 45997. [CrossRef]
33. Buckley, P.R.; McKinley, G.H.; Wilson, T.S.; Small, W.; Bennett, W.J.; Bearinger, J.P.; McElfresh, M.W.; Maitland, D.J. Inductively Heated Shape Memory Polymer for the Magnetic Actuation of Medical Devices. *IEEE Trans. Biomed. Eng.* **2006**, *53*, 2075–2083. [CrossRef]
34. Zhang, D.; Liu, Y.; Leng, J. Magnetic field activation of SMP networks containing micro nickel (Ni) powder. In Proceedings of the Second International Conference on Smart Materials and Nanotechnology in Engineering, Weihai, China, 8–11 July 2009.
35. Aaltio, I.; Nilsén, F.; Lehtonen, J.; Ge, Y.; Spoljaric, S.; Seppälä, J.; Hannula, S.-P. Magnetic Shape Memory—Polymer Hybrids. *Mater. Sci. Forum* **2016**, *879*, 133–138. [CrossRef]
36. Razzaq, M.Y.; Anhalt, M.; Frormann, L.; Weidenfeller, B. Mechanical spectroscopy of magnetite filled polyurethane shape memory polymers. *Mater. Sci. Eng.* **2007**, *471*, 57–62. [CrossRef]
37. Puig, J.; Hoppe, C.E.; Fasce, L.A.; Pérez, C.J.; Piñeiro-Rendondo, Y.; Bañobre-López, M.; López-Quintela, M.A.; Rivas, J.; Williams, R.J.J. Superparamagnetic Nanocomposites Based on the Dispersion of Oleic Acid-Stabilized Magnetite Nanoparticles in a Diglycidylether of Bisphenol A-Based Epoxy Matrix: Magnetic Hyperthermia and Shape Memory. *J. Phys. Chem. C* **2012**, *116*, 13421–13428. [CrossRef]
38. Cuevas, J.M.; Alonso, J.; German, L.; Iturrondobeitia, M.; Laza, J.M.; Vilas, J.L.; León, L.M. Magneto-active shape memory composites by incorporating ferromagnetic microparticles in a thermo-responsive polyalkenamer. *Smart Mater. Struct.* **2009**, *18*, 075003. [CrossRef]
39. Vialle, G.; Prima, M.D.; Hocking, E.; Gall, K.; Garmestani, H.; Sanderson, T.; Arzberger, S.C. Remote activation of nanomagnetite reinforced shape memory polymer foam. *Smart Mater. Struct.* **2009**, *18*, 115014. [CrossRef]
40. Lagcore, L.K.; Brand, O.; Allen, M.G. Magnetic microactuators based on polymer magnets. *J. Microelectromechanical Syst.* **1999**, *8*, 2–9. [CrossRef]
41. Khoo, M.; Liu, C. Micro magnetic silicone elastomer membrane actuator. *Sens. Actuators A* **2001**, *89*, 259–266. [CrossRef]
42. Kim, J.; Chung, S.E.; Choi, S.-E.; Lee, H.; Kim, J.; Kwon, S. Programming magnetic anisotropy in polymeric microactuators. *Nat. Mater.* **2011**, *10*, 747–752. [CrossRef]
43. Erb, R.M.; Sander, J.S.; Grisch, R.; Studart, A.R. Self-shaping composites with programmable bioinspired microstructures. *Nat. Commun.* **2013**, *4*, 1712. [CrossRef] [PubMed]
44. Lum, G.Z.; Ye, Z.; Dong, X.; Marvi, H.; Erin, O.; Hu, W.; Sitti, M. Shape-programmable magnetic soft matter. *Proc. Natl. Acad. Sci. USA* **2016**, *113*, E6007–E6015. [CrossRef] [PubMed]
45. Xu, T.; Zhang, J.; Salehizadeh, M.; Onaizah, O.; Diller, E. Millimeter-scale flexible robots with programmable three-dimensional magnetization and motions. *Sci. Robot.* **2019**, *4*. [CrossRef] [PubMed]
46. Roh, S.; Okello, L.B.; Golbasi, N.; Hankwitz, J.P.; Liu, J.A.-C.; Tracy, J.B.; Velev, O.D. 3D-Printed Silicone Soft Architectures with Programmed Magneto-Capillary Reconfigurations. *Adv. Mater. Technol.* **2019**, *4*, 1800528. [CrossRef]
47. Lantean, S.; Barrera, C.; Pirri, C.F.; Tiberto, P.; Sangermano, M.; Roppolo, I.; Rizza, G. 3D Printing of Magnetoresponsive Polymeric Materials with Tunable Mechanical and Magnetic Properties by Digital Light Processing. *Adv. Mater. Technol.* **2019**, *4*, 1900505. [CrossRef]
48. Venkiteswaran, V.K.; Samaniego, L.F.P.; Sikorski, J.; Misra, S. Bio-Inspired Terrestrial Motion of Magnetic Soft Millirobots. *IEEE Robot. Autom. Lett.* **2019**, *4*, 1753–1759. [CrossRef]
49. Wu, S.; Ze, Q.; Zhang, R.; Hu, N.; Cheng, Y.; Yang, F.; Zhao, R. Symmetry-Breaking Actuation Mechanism for Soft Robotics and Active Metamaterials. *Appl. Mater. Interfaces* **2019**, *11*, 41649–44658. [CrossRef] [PubMed]
50. Testa, P.; Style, R.W.; Cui, J.; Donnelly, C.; Borisova, E.; Derlet, P.M.; Dufresne, E.R.; Heyderman, L.J. Magnetically Addressable Shape-Memory and Stiffening in a Composite Elastomer. *Adv. Mater.* **2019**, *31*, 6. [CrossRef]
51. Scheerbaum, N.; Hinz, D.; Gutfleisch, O.; Müller, K.-H.; Schultz, L. Textured polymer bonded composites with Ni-Mn-Ga magnetic shape memory particles. *Acta Mater.* **2007**, *55*, 2707–2713. [CrossRef]
52. Alharbi, S.; Ze, Q.; Zhao, R.; Kiourti, A. Magnetoactuated Reconfigurable Antennas on Hard-Magnetic Soft Substrates and E-Threads. *IEEE Trans. Antennas Propag.* **2020**, *68*, 5882–5892. [CrossRef]
53. Qi, S.; Guo, H.; Fu, J.; Xie, Y.; Zhu, M.; Yu, M. 3D printed shape-programmable magneto-active soft matter for biomimetic applications. *Compos. Sci. Technol.* **2020**, *188*, 107973. [CrossRef]
54. Ze, Q.; Kuang, X.; Wu, S.; Wong, J.; Montgomery, S.M.; Zhang, R.; Kovitz, J.M.; Yang, F.; Qi, H.J.; Zhao, R. Magnetic Shape Memory Polymers with Integrated Multifunctional Shape Manipulation. *Adv. Sci. News* **2020**, *32*, 1906657. [CrossRef]
55. Zhang, J.; Guo, Y.; Hu, W.; Soon, R.H.; Davidson, Z.S.; Sitti, M. Liquid Crystal Elastomer-Based Magnetic Composite Films for Reconfigurable Shape-Morphing Soft Miniature Machines. *Adv. Mater.* **2021**, *33*, 2006191. [CrossRef]
56. Zhang, Y.; Wang, Q.; Yi, S.; Lin, Z.; Wang, C.; Chen, Z.; Jiang, L. 4D Printing of Magnetoactive Soft Materials for On-Demand Magnetic Actuation Transformation. *ACS Appl. Mater. Interfaces* **2021**, *13*, 4174–4184. [CrossRef]

57. Lagorce, L.K.; Allen, M.G. Magnetic and mechanical properties of micromachined strontium ferrite/polyimide composites. *J. Microelectromech. Syst.* **1997**, *6*, 307–312. [CrossRef]
58. Conti, S.; Lenz, M.; Rumpf, M. Modelling and simulation of magnetic-shape memory polymer composites. *J. Mech. Phys. Solids* **2007**, *55*, 1462–1486. [CrossRef]
59. Crivaro, A.; Sheridan, R.; Frecker, M.; Simpson, T.W.; Lockette, P.V. Bistable compliant mechanism using magneto active elastomer actuation. *J. Intell. Mater. Syst. Struct.* **2015**, *27*, 2049–2061. [CrossRef]
60. Zhao, R.; Kim, Y.; Chester, S.A.; Sharma, P.; Zhao, X. Mechanics of hard-magnetic soft materials. *J. Mech. Phys. Solids* **2019**, *124*, 244–263. [CrossRef]
61. Chen, Z.; Lin, Y.; Zheng, G.; Yang, Y.; Zhag, Y.; Zheng, S.; Li, J.; Li, J.; Ren, L.; Jiang, L. Programmable Transformation and Controllable Locomotion of Magnetoactive Soft Materials with 3D-Patterned Magnetization. *ACS Appl. Mater. Interfaces* **2020**, *12*, 58179–58190. [CrossRef]
62. Ren, Z.; Wang, T.; Hu, W.; Sitti, M. A Magnetically-Actuated Untethered Jellyfish-Inspired Soft Milliswimmer. In Proceedings of the Robotics: Science and Systems, Freiburg im Breisgau, Germany, 22–26 June 2019.
63. Goudu, S.R.; Yasa, I.C.; Hu, X.; Ceylan, H.; Hu, W.; Sitti, M. Biodegradable Untethered Magnetic Hydrogel Milli-Grippers. *Adv. Funct. Mater.* **2020**, *30*, 2004975. [CrossRef]
64. Zhang, J.; Jain, P.; Diller, E. Independent Control of Two Millimeter-Scale Soft-Bodied Magnetic Robotic Swimmers. In Proceedings of the IEEE Conference on Robotics and Automation (ICRA), Stockholm, Sweden, 16–21 May 2016.
65. Hu, W.; Lum, Q.Z.; Mastrangeli, M.; Sitti, M. Small-scale soft-bodied robot with multimodal locomotion. *Nature* **2018**, *554*, 81–85. [CrossRef]
66. Diller, E.; Zhuang, J.; Lum, G.Z.; Edwards, M.R.; Sitti, M. Continuously distributed magnetization profile for millimeter-scale elastomeric undulatory swimming. *Appl. Phys. Lett.* **2014**, *104*, 174101. [CrossRef]

MDPI
St. Alban-Anlage 66
4052 Basel
Switzerland
Tel. +41 61 683 77 34
Fax +41 61 302 89 18
www.mdpi.com

Magnetochemistry Editorial Office
E-mail: magnetochemistry@mdpi.com
www.mdpi.com/journal/magnetochemistry

www.ingramcontent.com/pod-product-compliance
Lightning Source LLC
Chambersburg PA
CBHW041947260926
51850CB00018B/1889
* 9 7 8 3 0 3 6 5 8 3 6 7 9 *